M

BIOCHEMICAL APPROACHES TO AGING

BIOCHEMICAL APPROACHES TO AGING

Morton Rothstein

Division of Cell and Molecular Biology
State University of New York at Buffalo
Buffalo, New York

1982

ACADEMIC PRESS

A Subsidiary of Harcourt Brace Jovanovich, Publishers

New York   London
Paris San Diego San Francisco São Paulo Sydney Tokyo Toronto

ACADEMIC PRESS, INC.
111 Fifth Avenue, New York, New York 10003

United Kingdom Edition published by
ACADEMIC PRESS, INC. (LONDON) LTD.
24/28 Oval Road, London NW1 · 7DX

Library of Congress Cataloging in Publication Data

Rothstein, Morton, Date
    Biochemical approaches to aging.

    Includes index.
    1. Aging. 2. Biological chemistry. I. Title.
[DNLM: 1. Aging. WT 104 R847b]
QP86.R65   574.3'72   82-6751
ISBN 0-12-598780-3   AACR2

PRINTED IN THE UNITED STATES OF AMERICA

82 83 84 85   9 8 7 6 5 4 3 2 1

To the memory of my father, Sam Rothstein, and to my mother, Etta. To my wife, Jean, and to my children, Zel, David, and Marilyn, who have made my life a happy one

Contents

Preface

As a biochemist engaged in research on aging, I have long felt that a serious need exists for a book that brings the biochemical literature together in a comprehensive manner. I also believe that such a book should be prepared by a single author. In my biased view, multi-authored volumes dealing with aging, though they can better describe specialized biochemical areas, lack broad coverage and an appropriate sense of interconnection. This criticism also applies to published symposia. Generalized books on aging are even less satisfactory. They tend to be theoretical, and they certainly do not provide a detailed analysis of the literature.

Based on the perceived need, this book presents the results of a thorough survey of the existing literature of the biochemistry of aging through 1980. I have tried to organize the mass of material into digestible pieces, which nonetheless maintain some of the original detail. To obtain an independent viewpoint, I have derived the information mainly from original research papers, but reviews and books have been consulted. In interpreting these papers, I have tried as much as possible to make use of original data and to take into account the author's own interpretation.

In considering the material in this book, one does not have to read too far to discover that a great deal of the published literature on the biochemistry of aging is contradictory. I have had little choice but to report all sides of controversial findings, unless, of course, there are obvious technical deficiencies that militate against a particular viewpoint. Unfortunately, such faults are not usually apparent in published research. Therefore, where there is an unresolved contradiction, I have attempted to remain neutral. Otherwise I would be obliged to take sides on the basis of a personal belief in the superiority of one group of investigators over another. I have tried to avoid this contretemps, though some degree of favoritism may have subconsciously crept into the writing. On the other hand, where I felt it was justified by my evaluation of the data, I have chosen one side or the other.

As to references, I have attempted to be complete but not exhaustive. Papers that do not, in my opinion, add to the information available have been purposely omitted. Others, without a doubt, have been omitted by oversight. Still others have been omitted because the old animals were too young to be considered senescent. In fact, many otherwise good papers have not been quoted because the "old" animals were 12 or 15 months of age. For the oversights, I apologize.

In writing a reasonably comprehensive book, one finds that some areas of research are represented by few publications. It is an unfortunate consequence that such papers receive greater individual emphasis than those in areas that are well documented. Thus, the length of discussion associated with various papers is not necessarily a mark of quality or even of importance. It simply may encompass all we know of some facet of aging research.

It is my opinion that, by and large, papers published in "aging" journals are frequently of questionable quality. I feel that many would not pass rigorous review in the more narrow periodicals devoted to biochemistry or cell biology. Introductions are sometimes sketchy and inaccurate; discussion often goes beyond the range of the data; conclusions are often based on meager data; reports often are fragmentary rather than profound. Though I might have personal reservations about the significance of some of the reported results, there is rarely enough evidence to the contrary to make reasonable, let alone unequivocal, disclaimers. To a degree, we have no consumer protection. It is a case of "let the buyer beware." The researcher's credo should be "read with care; accept with caution." To take the available papers and forge a clear and comprehensive picture of aging at the molecular level has, to my regret, been impossible. The best I could do was to organize the material in a context where it could be seen as part of a larger, if not coherent, picture. I hope I have at least been able to provide a background, *pro* and *con*, of current work from which future investigators can proceed to more clear-cut findings. I also hope that I have made it clear that there is controversy in many areas in which it is believed that agreement exists.

Undoubtedly, I will be called to task by experts in each field for misrepresenting their areas. The cell biologists, the lipid chemists, and the DNA experts will most certainly find my description of their respective specialty lacking in some degree of sophistication or, at worst, woefully inadequate. I can only suggest that they read the other parts of the book and skip their areas of expertise.

This is not a book of theory or abstract ideas, but a detailed account of the status of major areas in which significant biochemical research on aging has been carried out. As such, it is intended to serve gerontologists in a wide range of biological disciplines. It should provide both background information and a picture of the current status of research for active investigators, for those interested but not currently working in aging, for those who are in areas peripheral to biochemistry such as physiology or pharmacology, or for those who might teach a course on aging or wish to enrich a biochemistry course with material pertinent to aging.

Having made it clear that this book should be of value to almost everybody, let me hasten to add that I will be satisfied if it serves a useful function for my colleagues—the biochemists and cell biologists who are earnestly engaged in aging research.

To those friends in the field who were pleased to know I was working on this book, I am grateful for their implied confidence that I would do a good job. I am also grateful to those friends who answered my questions and sent helpful preprints of recent articles. Thanks in particular are due to the State University of New York at Buffalo for granting me a sabbatical leave, which provided uninterrupted time to get the book started; to Dr. Eric Barnard for arranging my appointment as Visiting Professor in the Biochemistry Department at Imperial College in London and acquiring for me a private office in the Lyon-Playfair Library; to Dr. Penny Barnard for her general kindness and patience during my frequent use of her office and telephone; to Dr. L. M. Franks of the Imperial Cancer Research Fund for his thoughtfulness in arranging for my periodic use of that Institution's library and for several pleasant and most helpful discussions. Last, I would like to thank the National Institute on Aging for providing for the last nine years financial support for my research, thereby giving me a permanent interest in aging.

I would be remiss if I did not add a special note of gratitude to Mrs. Joyce Hough for her patience in making multitudes of corrections during typing of the manuscript and especially for her "eagle-eye" in locating misplaced, misdated, and misspelled references.

Morton Rothstein

Chapter 1

Introduction

The period from birth to adolescence encompasses a bewildering variety of biological changes. A small, helpless, toothless, often hairless, bundle of uncoordinated humanity develops into a large individual provided with an astonishing endowment of neurophysiological, emotional, and intellectual abilities. In a general sense, the rate at which these developmental changes occur slows with increasing age. Following adolescence, further changes take place very slowly, but continue over a large number of years. In fact, the outward manifestations proceed with such smoothness, with such little day-to-day change, that they pass quite unnoticed. It is only if one looks at a photograph taken perhaps 5 or 10 years earlier, or one sees an acquaintance for the first time in several years that one perceives, in a flash of recognition, that definite changes in appearance have occurred. Particularly during the early mature years, aging changes may be slight. Physiological and neurological changes take place so subtly that only when the utmost is demanded from the organism, can deficiencies be noted. The athlete reaching the 30s, with a few outstanding exceptions, loses the ultimate edge in timing and reflex action. Fortunately, the physical capabilities humans possess are so far above what is needed to sustain a reasonable level of activity that one can lose a substantial degree of function without threat to life. One might draw the analogy of two cars in good condition, one capable of 100 mph and one 60 mph. As long as it is only necessary to travel at 30 or 40, there is literally no difference. It is only under conditions of great demand that age becomes a serious factor.

Though there is often the perception that aging accelerates late in life, it must be recognized that there is great individual variation. Some people appear to age quickly, others appear "ageless"—for a while. Eventually, all will die.

By some definitions, including the author's, the changes from maturity through senescence constitute the "aging" process. As such, it has puzzled and fascinated man from the earliest written expression and undoubtedly before that—almost certainly since the ability arose for abstract thought. The fact that each newborn babe has the absolute certainty of following the typical human design for growth, development, and senescence has been grist for the philosopher's mill through the ages. The unswerving repetition of the pattern is described in the oft-quoted lines from Shakespeare's "As You Like It," written in 1599–1600.

All the world's a stage,
And all the men and women merely players:
They have their exits and their entrances;
And one man in his time plays many parts,
His acts being seven ages. At first the infant,
Mewling and puking in the nurse's arms.
And then the whining school-boy, with his satchel
And shining morning face, creeping like snail
Unwillingly to school. And then the lover,
Sighing like furnace, with a woeful ballad
Made to his mistress' eyebrow. Then a soldier,
Full of strange oaths and bearded like the pard,
Jealous in honour, sudden and quick in quarrel,
Seeking the bubble reputation
Even in the cannon's mouth. And then the justice,
In fair round belly with good capon lined,
With eyes severe and beard of formal cut,
Full of wise saws and modern instances;
And so he plays his part. The sixth age shifts
Into the lean and slipper'd pantaloon,
With spectacles on nose and pouch on side,
His youthful hose, well saved, a world too wide
For his shrunk shank; and his big manly voice,
Turning again toward childish treble, pipes
And whistles in his sound. Last scene of all,
That ends this strange eventful history,
Is second childishness and mere oblivion,
Sans teeth, sans eyes, sans taste, sans everything.

In previous times, though relatively few individuals lived to become old, the infirmities of senescence were well noted. As a consequence, many of the adjectives in Roget's "Thesaurus" which describe aging have an unpleasant connotation: declining, senile, run to seed, doddering, decrepit, stricken in years, wrinkled, having one foot in the grave, doting.

Aging was, until recent times, probably accepted more with a sense of the inevitable. However, as is undoubtedly the case today—or else why would we try and improve our life-spans?—people in the Renaissance period were not so passive that they did not dream of extended, if not eternal youth. They placed their hopes in elixirs and magic potions. The Fountain of Youth could only have seemed a reality because of an intense desire not to grow old. In this modern day, the promise of a Fountain of Youth lies not in a magical bath, but in the ability of bioscientists to uncover the cause of aging and to successfully intervene in the process. How is this challenge being met? The physiologist measures the decline of function, the statistician tells us how long, the epidemiologist, where, the physician treats (as well as possible), but it remains for the biochemist to determine the changes that occur at the molecular level whose sum total is what others are observing.

To slightly rephrase the question—How are the biochemists meeting the challenge of aging? To answer truthfully, not very well! Other than providing several unsupported theories, little meaningful work on aging was performed until around the early 1960s. At that time, measurements of biochemical processes, such as changes in the rate of RNA synthesis, possible changes in DNA structure, and differences in enzyme levels, began to be made. The work, in general, suffered from lack of technical finesse. It tended to be scattered and narrow in scope and was not followed up so as to create any sort of "critical mass" upon which other work could feed. Perhaps part of the blame lay in the minimal amount of financial support that was available for research in aging. In the 1960s, there was only an "Aging Branch" at The National Institute for Child Health and Human Development. Whatever the reason, the 1960s were more an hors d'oeuvre than an entree. Therefore, the chapters in this book tend to deal little with research carried out before 1970.

In spite of the sparsity of meaningful experimental results in the 1960s, at least three proposals arose during that period which have had a continuing influence on aging research. One is the Free Radical Theory of aging proposed by Harmon in 1956. The theory proposes that peroxidation of unsaturated fatty acids results in formation of free radicals and that there is consequent cross-linking of biomolecules, which results in an accumulation of damage. The idea of damage by derivatives of oxygen, which include hydrogen peroxide and superoxide and hydroxyl radicals, continues to generate interest in aging research. Indeed, a relationship has been shown between lipid peroxidation and formation of age pigments (lipofuscin). There is little question that biological systems can generate free radicals, and there has been continued attention to free radical generation *in situ* and its possible effects on aging. Occasional papers continue to be published on the effects of feeding antioxidants, such as vitamin E, on age pigments (they are reduced in amount) and life-span (no effect in mammals).

The second event of substance that emerged in the 1960s was Orgel's Error Catastrophe Hypothesis. Modified in 1970, the theory suggested that an error in the protein-synthesizing system would produce proteins with errors in their sequence. If some of these proteins, in turn, were involved in protein synthesis, then a new generation of erroneous proteins would be formed. The errors would thus be amplified until a "catastrophe" occurred. The theory served to stimulate considerable research on aging at the molecular level. As a result of work through the 1970s, the idea that a substantial level of errors is generated in proteins during aging is no longer feasible. The altered proteins, which have been isolated from aged animals, appear to result from postsynthetic modification—actually a change in conformation without the formation of new covalent bonds. Unfortunately, in spite of the contrary evidence, many papers on aging continue to refer to "errors." Errors that are a result of the aging process have yet to be shown to exist.

The third event from the 1960s that has continued to have consequences in aging research is the proof by Hayflick and Moorhead that diploid cells in culture have a limit to the number of times they can divide. This conclusion was borne out by other investigators, and as a result, the old dogma that cells in culture are immortal was shattered. Hayflick noted the slowing and eventual cessation of cell division after a certain number of population doublings and referred to the phenomenon as *cell senescence*. As a result, a considerable amount of work has evolved using cells in culture as a model system for the study of aging. The system continues to be studied in considerable detail by a number of capable investigators. Though there is no doubt as to its value as a biological system worthy of study, there is as yet no conclusive evidence, pro or con, that the decreasing ability of cells to divide with increasing passage number is analogous to aging *in vivo*.

By the end of the 1970s, some of the biochemical work on aging had become more sophisticated and certain features, though still unclear, began to emerge. It is this time period, the last 10 years, which this volume deals with in detail. Although our understanding of the aging process has progressed little beyond that in the previous decade, a base of information is being created that can serve as a jumping-off point for more profound investigations. Thus, it is known that mitochondrial metabolism is changed subtly in senescence, that certain proteins become altered and others do not, and that hormone receptors are often reduced in number. Moreover, model systems that are useful for aging studies have been developed more fully. On the other hand, there is still no agreement on such basic processes as changes in gene function or chromatin structure. Perhaps equally important, in a negative sense, is the realization that dramatic changes are not to be expected—one is looking for fine adjustments in an extremely complex system.

Since aging studies borrow heavily from techniques developed in other biochemical areas, we are, to a degree, waiting on more general advances in other disciplines. Research in aging will advance in lockstep with progress in molecular biology, cancer, biophysics, etc. For all that, research in aging has begun to expand noticeably in degree and quality. One may expect that the next 10 years will show a multiplication of the progress made in the last 10 years. Whether or not our new knowledge will permit successful intervention remains to be seen. A healthy old age for all (rather than an extended life span) with a consequent elimination of the emotional and economic stress involved in the case of the infirm aged is indeed a goal worth striving for.

Chapter 2

Model Systems for the Study of Aging

I. OVERVIEW

Ideally, experimental animals for the study of aging should possess characteristics that can be related to the phenomenon as it appears in humans. They should avoid, as far as possible, problems which though seemingly peripheral, may nonetheless pose hidden difficulties for the researcher busily investigating his favorite aging phenomenon. An obvious example of such a difficulty is the prevalence of tumors in old rats, but dietary preferences, obesity, or perhaps even lack of exercise in old animals could result in altered endocrine states which would in turn reflect an altered metabolic condition not truly a function of aging. Favorable characteristics for experimental animals would include (1) a reasonably short life-span so that one would not have to wait for long periods to obtain old animals, nor would the expense of their maintenance be too great; (2) genetic homogeneity; (3) large enough size for adequate amounts of tissue; (4) absence of diseases; (5) controlled environmental situation, especially diet; (6) convenient and inexpensive care; and (7) adequate background of biological information. Undoubtedly, one could make a longer list of desirable features. However, even if all of these preferences were to be met, one would still have to consider the fact that old animals are survivors and may therefore not be good examples of the biological changes which occurred in those animals which died. For example, if a rat is fated to die at 22 months of age and by chance is used for an experiment at 21 months, it might be quite different from another 21-month-old rat destined to live to 36 months. The latter, might appear quite "young" in its biochemical aspects. In the end, no matter which animals are used for aging studies, one must make a number of compromises and attempt to cancel out the uncertainties by using statistically significant numbers of samples (often difficult) or to find that a biochemical feature can be related to more than one tissue. Even better would be to find the same age-related change in more than one animal species. The latter reasoning implies that certain basic aging phenomena occurring at the molecular level are common to many if not all species, though they may be manifested differently. If such were not the case, studies of aging would be impossible, as we would for example, not be able to apply to humans results obtained from rats.

It is obvious that nonhuman model systems for aging are necessary if the biochemical aspects of aging are to be studied in depth. Human beings meet few of the above criteria of suitability for animal models. They are inexpedient not only because of large genetic variations, but environmental factors including diet, behavior, place of work, etc., interpose huge differences among the samples. Not the least of the inconvenient factors is that humans live too long and individual members of the species are reluctant to part with appropriate organs for study, except postmortem.

The obvious model for study would appear to be laboratory rodents which are readily available, easily grown under standardized conditions, have a relatively short life-span, are highly inbred, and about which much is known. Indeed, it is safe to say that rodents are the most utilized model system for aging studies. Nonetheless, as will be seen below, their use is not as straightforward as would appear on the surface. The problems involved in use of aged rodents have led investigators to seek other model systems, which include insects, free-living nematodes, protozoans, and cells in tissue culture. These systems are not only simpler, but some of them may exaggerate certain facets of the aging process and thus provide an advantageous system for study. Of these nonrodent model systems, cells in culture are the most studied. The heavy emphasis has derived from the fact that when cells are made to divide repeatedly, they exhaust their potential for further growth. This phenomenon has been termed *cellular senescence*. Much effort has been put into characterizing cells in the course of *in vitro* aging and recently, into comparing senescent cells with aging *in vivo*.

One advantage of studying age-related changes in more than one type of animal is that any changes common to aged specimens can confidently be ascribed to "aging" and not to some characteristic of the type of animal being studied. In fact, it is not a bad idea for biochemists to insist that changes ascribed to aging be documented in more than one species.

II. MAMMALIAN SPECIES

Though occasional studies involving aging have been performed on dogs, guinea pigs, rabbits, and tissues from cows versus calves, the bulk of current biochemical research is carried out using rats and mice. Fox (1980) has recently provided background material and a review of aging studies performed with rabbits. Vincent *et al.* (1980) provide a background for the use of the Mongolian gerbil.

Since rats and mice bear the major weight of research in aging, the more data available regarding the mortality characteristics of each strain and the effect of nutrition and environmental conditions, the more useful the animals for aging research. The pathology of aging animals is, of course, an important consideration. One must know if they have tumors, if there is near-renal failure, or if there are other damaging conditions which might reflect on measurements of biochemical parameters, such as levels of various metabolites, hormone response, or protein synthesis and degradation. On the other hand, the average biochemist cannot afford either the time or expense involved in having each aged animal from which tissue is taken examined by a trained pathologist. One must depend greatly on outward appearance. Thus, study of liver mitochondria, purification of an enzyme, isolation of chromatin, or synthesis of RNA is assumed to be normal for animals of any given age unless there are overt symptoms of disease. Fortunately, the strains of rats and mice used most often show good survival characteristics so that it is not difficult to obtain healthy animals in the 28- to 30-month range.

Long term considerations such as diet or environmental conditions certainly affect the life-span of laboratory animals. In this respect, a number of papers have documented the effects of altered intake of protein or carbohydrate. Caloric restriction (Chapter 11) has a marked influence in extending life-span. Barrier-reared (pathogen-free) animals tend to be longer-lived than conventionally reared animals, thus implicating infectious agents in shortening life-spans. Cohen (1978) suggested that barrier-reared animals should not be held for more than 30 days under conventional conditions. However, Hollander (1976), based on work with two strains of rats maintained at the Institute for Experimental Gerontology in The Netherlands, proposed that, for convenience and to reduce expense, rats be born and reared under strict specific pathogen-free conditions and then maintained under "clean conventional" conditions.

In undertaking research in aging, it should be borne in mind that two rats of the same age are not necessarily the same physiologically. Moreover, it should be remembered that old animals are survivors and therefore they will have minimized those aging changes which have been fatal to their littermates. The best way around the problem is to use a statistically significant number of individual animals for each experiment, although this will prove costly. Strain differences are also a problem. There is no guarantee that all strains emphasize the same aging characteristic to the same degree. For example, the age-related delay of induction of glucokinase (Chapter 10, Section XIII) in male Sprague-Dawley rats (Adelman, 1972) is reported to be shortened in male Wistars and nonexistent in female Wistars (David and Pfeifer, 1973). Especially, strains of animals that have relatively short life-spans may not have time to develop those aging characteristics which show up late in longer-lived species. That is, the former may have an acute metabolic flaw which kills them early, leaving perfectly normal those consequences of aging which would have developed later. The investigator looking for these late-developing traits in a short-lived strain might see no age-related differences.

A. Rats

A colony of aged Fischer 344 rats is maintained at Charles River Breeding Laboratories under the aegis of the National Institute on Aging. They are available with permission, to grantees at a basic cost (as of 1980) of $3.16/month of age. These animals are barrier-reared, providing assurance that they are specific pathogen-free. Coleman *et al.* (1977) provided a detailed report of age-related pathological changes in these animals, including neoplasms, respiratory, cardiac, urinary system, splenic, adrenal gland, pancreatic, adenohypophyseal, thyroid and parathyroid, hepatic, digestive system, testicular, muscle, central nervous system, and ocular lesions. The animals studied had a 50% mean survival of 29 months and maximal survival of 35 months. Mortality was negligible until about 20 months. Goodman *et al.* (1979) listed neoplastic and nonneoplastic lesions for Fischer 344 rats maintained at various testing laboratories. As to other strains, neoplasms in aging Sprague-Dawley rats are described by Sims (1967) and Berg (1967); aging Wistars are described by Boorman and Hollander (1973); Long-Evans rats by Durand *et al.* (1964); Osborn-Mendel rats by Goodman *et al.* (1980). One should be aware of specific pathologies. For example, by 900 days 70% of Sprague-Dawley female rats have mammary tumors. Similarly, by 30–33 months, 66% of male Fischer 344 rats have testicular tumors. Practically all Fischer rats have some degree of renal pathology.

There is considerable variation in life-span from laboratory to laboratory, even when the same strain of rats is grown on the same diet.

Hoffman (1978) provides survival data on Fischer 344, Wistar rats (conventional and barrier-reared), Sprague-Dawley, and Long-Evans rats. Data for male and female animals are provided. Masoro (1980a) provides a table of life-spans of various colonies of these strains and includes information on environmental conditions, weight, diet, and the median life-spans of animals in colonies maintained in different laboratories. There is often considerable variation in the life-spans. For example, for Fischer rats the extremes reported for median life-span are 22 and 29 months. Figure 2.1 shows a typical set of survival curves for specific pathogen-free Fischer 344 rats (Hoffman, 1978). In general, barrier-reared Fischer 344 rats have a median life-span of 29 months. Nonbarrier-reared animals are somewhat shorter-lived, showing median life-spans of 26–27 months. Figure 2.2 illustrates the difference in life-span between conventionally raised and specific pathogen-free Wistar rats (Hoffman, 1978, using data from Paget and Lemon, 1965). Reports on the life-span of Sprague-Dawley rats vary from around 24 months to a reported 30 months for barrier-reared animals. Male Sprague-Dawley rats have been reported to have a shorter life-span than females (Hoffman, 1978), although this conclusion is not firmly established.

As noted above, there are differences between rat strains which may be of significance to the investigator. Wistar rats are usually "outbred." That is, they are mated randomly with other Wistars. An "inbred" strain

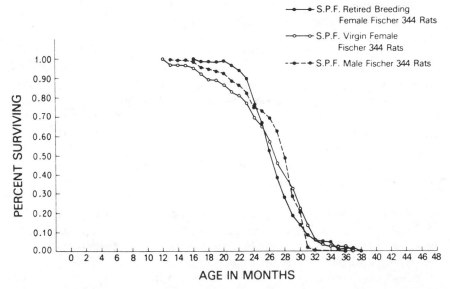

Fig. 2.1. Survival curves for specific pathogen-free Fischer rats. From Hoffman (1978).

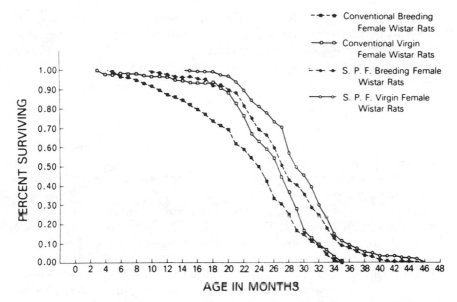

Fig. 2.2. Life-span of conventionally raised and specific pathogen-free Wistar rats. From Hoffman (1978).

is one that is maintained by brother–sister mating, e.g., Fischer 344 rats. The former type can vary considerably from one source to another.

Male rats tend to be heavier than females. Weight in male Sprague-Dawleys has been reported to increase through 2 years of age to a level of about 750–880 g and females to 600 g. Masoro (1980a), however, found that male Sprague-Dawleys in his facility only attained about 450 g and females, 290 g. Males housed three to four per cage were a bit larger than singly housed animals. Fischer 344 rats tend to be smaller than Sprague-Dawleys. In the former, increases in weight are very slow after 16 months and minimal after 20 months. The maximal mean weight for the barrier-reared Fischers (male) at Charles River was 454 g. Nutritional effects again show their importance because these same rats, transferred when young to San Antonio, achieved a maximal mean body weight of 569 g, but median life-span was 23½ instead of 29 months (Masoro, 1980a). Conditions in the animal rooms, other than diet and number of rats per cage, were reportedly identical.

Lean body mass also differs with age in different species. Sprague-Dawleys tend to become obese in old age, but Fischer 344 rats remain relatively lean. Obviously, research on age versus lipid metabolism must take such observations into account. However, there is no significant relationship between weight attained and life-span (Masoro, 1980b).

In general, unless the investigator is maintaining his own colonies, it is best to purchase aged animals grown under specific pathogen-free conditions just previous to their projected use. In that way, the animals will reflect known backgrounds without the complications of subsequent changes in environment.

B. Mice

The use of mice in aging research involves the same considerations applicable to rats. For the biochemist who must purchase animals, there are two interrelated disadvantages; there is less tissue available but costs per animal are the same.

Longevity in mice appears to be strain-specific and, therefore, depends heavily on genetic background. Russell (1979) provided a chart of the genetic origins of inbred strains of mice. Information on physiological and endocrine effects, neoplasms, and other pathology are all described in several chapters in "Development of the Rodent as a Model System of Aging," published by the National Institutes of Health. A description of the pathology of a colony of aging C57BL mice has been reported by Rowlatt *et al.* (1976) and for CD-1 Ham/1CR mice by Homburger *et al.* (1975). Cohen (1968) pointed out a number of factors that may influence life-span, some of which may be surprising to the biochemist unfamiliar with aging colonies. The factors include disease, exercise, nutrition, cage type, cage size, number of animals per cage, cage area per animal, isolation, handling, fighting, routine, temperature, lighting, and noise.

In general, the survival curves for mice are similar to those for rats, although more strains are available, including short-lived strains. Mean life-spans for 16 strains are listed by Russell (1966) and vary from 256 (AKR/J) to 706 days (LP/J) for virgin females and 272 (AKR/J) to 539 days (C57BL/6J) for males. For several species males appeared to be shorter-lived, but this is not always the case. Storer (1966) obtained generally greater mean life-spans for a large number of strains of virgin female mice, some being considerably longer than those reported by Russell (1966). In fact, the mean life-span of male C57BL/6J mice was reported to be 743 and 792 days in 1970 and 1971, respectively. In general, male mice were not significantly shorter-lived than females. More recently, Kunstyr and Levenberger (1975) provided longevity statistics for conventionally reared C57BL/6J mice. They obtained quite long survivals, the average life-span being 878 days for males and 794 days for females. They also observed that over the years there has been an increase in the longevity of the male versus the female of this strain. Rowlatt *et al.* (1976)

provided a detailed report on age changes (organ, weight, tumor incidence, pathology) of a strain of C57BL mice which remained healthy into old age. A typical survival curve is shown from this paper (Fig. 2.3). Mean survival age was 809 days for males and 763 days for females. These animals, too, were raised under conventional conditions. Other life-span curves for A/J, BALB/cJ, C57BL/6J, and DBA/2J inbred strains and the hybrid combinations are given by Goodrick (1975). Sex differences in longevity were not found except for the BALB/cJ strain in which the females were longer-lived. A comparison of the longevity of C57BL/6J mice and mutations that differed in body weight was also made by Goodrick (1977). Mutant mice with high body weight did not show differences in longevity, although obese mice had shorter life-spans. All the mutant strains were shorter-lived than the control groups. Slowing the rate of growth by a low-protein diet increased the life-span significantly (Goodrick, 1978).

C. Comment

That there are measurable effects on the longevity of rodents which are brought about by various factors which include diet, disease, or environmental conditions should be apparent from the information given above. What the effects are with respect to biochemical parameters is not known. Thus, there is always a chance of anomalous findings, particularly when few animals are used for experiments. The unidentified factors affecting longevity may cause particular problems in that they probably would affect the more volatile metabolic reactions—

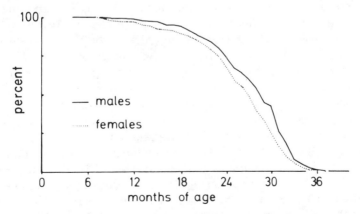

Fig. 2.3. Survival curves of a strain of C57BL virgin mice. Percentage alive were plotted at the end of each month. From Rowlatt *et al.* (1976).

those which can fluctuate widely even under nonaging conditions. Thus, changes in enzyme levels may have less to do with age than with adaptation of the animal to environmental conditions. Perhaps a slight sense of insecurity is the best approach to using rodents for aging studies, along with the use of adequate sampling and an experimental plan which takes several age groups into account.

III. CELLS IN CULTURE

As a model system for the study of aging, the use of cells in culture has generated much research. On the surface, the system offers considerable convenience and flexibility in that some cell types may be stored for long periods in liquid nitrogen. Moreover, cells can be derived from almost any animal species, from many types of tissue, and from young and old donors. The main weakness of the system is the unproved relationship between cells in culture and aging *in vivo*.

For the researcher, there is a cell repository maintained on contract with the National Institute on Aging so that investigators may readily obtain various types of cells, including those from subjects with age-related diseases. The repository is maintained at the Institute for Medical Research, Copewood Street, Camden, New Jersey 08103.

A. Background

The use of tissue culture as a model for aging studies has its genesis in the discovery of Hayflick and Moorhead (1961) that vertebrate fibroblasts have a limited potential for continued division. Previously, based on the early work of Carrel (1912) with chick fibroblasts, it was assumed that cells could be cultivated indefinitely. Failure of others to maintain cells in culture was thought to be a result of deficiencies in the medium or other environmental failures. Swim and Parker (1957) drew attention to the fact that in all 51 cultures they studied the cells first grew well, but eventually ceased to divide. The authors pointed out that this characteristic of cells was more general than usually recognized. Hayflick and Moorhead (1961) and Hayflick (1965) established with certainty that human cells have a finite life-span. Hayflick and Moorhead (1961) observed that cells from a primary culture divided rapidly (Phase I) and continued to divide unabated for many cell divisions. This period is termed Phase II. Eventually, after about 45 doublings for WI-38 (embryonic human lung) cells (using a 2:1 split ratio), the cells will not reach confluence, their number decreasing with each passage. This condition is termed Phase III. The pattern is shown in Fig. 2.4.

The investigator should keep in mind that comparisons of "young" and "old" cells may sometimes refer to cultures in different levels of Phase II, or as is often the case, to cells in Phase II versus cells in Phase III. The criterion for the latter is inability of the cells to reach confluency after an extended time period.

The number of population doublings depends upon the size of the inoculum or "split." A 2:1 split means that one-half of the cells harvested at a given passage are placed back into culture. When a monolayer covers the vessel (confluency), the number of cells will have doubled. Unfortunately, it turns out that many cells do not divide at all and others divide repeatedly to make up the deficit. Therefore, though a 2:1 split results in a doubling of the population, it does not mean that each cell has divided once. Of course, a different split, e.g., 4:1, would represent the equivalent of two doublings. The notation for the process is cell population doubling, population doubling level, or simply, passage number.

Although it is now accepted as normal that vertebrate cells, with a few possible exceptions, have a finite life-span, the change in view from

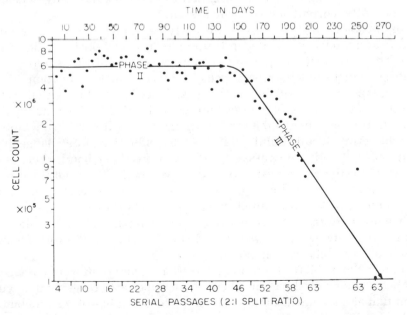

Fig. 2.4. Cell counts of WI-44 cells during serial passage. (Hayflick, 1965, by permission of Academic Press.)

"immortality" to "limited growth potential" was rather a dramatic one. Those cells that do proliferate indefinitely are assumed to be cancerous, precancerous, or to have chromosomal aberrations. The best explanation for the reported immortality of Carrel's chick fibroblasts (now known to have a limited life-span) seems to be the one suggested by Hayflick (1965) that the periodic addition of fresh chick serum during subcultivation inadvertently provided innocula of fresh cells to the cultures.

Hayflick and Moorhead (1961) and Hayflick (1965) expressed the idea that cessation of cell division represents "cellular senescence." Since that time, considerable "aging" research has been performed using diploid fibroblasts, particularly WI-38, a fibroblast from fetal human lung. In addition to a multitude of research papers, there exist many review articles, chapters, and books that deal with the use of cells in culture as a model system for studying aging. Useful examples are recent comprehensive review articles by Martin (1977), Hayflick (1980), and Cristofalo (1980). "Aging" in less commonly studied cell types has recently appeared in a volume edited by Nichols and Murphy (1979).

Given the large volume of published material and the inclusion of biochemically oriented research in the appropriate chapters of this volume, the following material is designed to provide a modestly detailed outline of the current status of the system.

For the biochemist interested in studying aging, the use of tissue culture as a model system depends upon its true relationship to aging *in vivo*. There are many arguments, pro and con, about the propriety of relating changes in late-passage cells to intact animals or of comparing late-passage cells to cells in old subjects. It seems fair to say that most investigators using the *in vitro* system recognize that "cellular aging" in their cultures may not be the same as aging *in vivo*. However, one notes that the terms "young" and "old" cells are often used interchangeably with "early-" and "late-passage" cells. On the other hand, there is a tendency for the biochemist to consider work with tissue culture as not dealing truly with "aging." To avoid such arguments, each chapter in this volume, where warranted, contains a section dealing specifically with work performed on early- versus late-passage cells. One should distinguish between this system and one in which cells from young versus old subjects are compared.

Though the inability of cells to divide after a given number of passages has been unequivocally established, there appears to be general agreement that aging animals do not run out of cells because of their inability to divide. The observation that, on the average, cells from old as compared to young individuals undergo fewer population doublings is not impressive because the deficit can only be seen by the application of statistical methods. The loss of doubling potential with age of the donor was first suggested by Hayflick (1965), confirmed by Martin *et al.* (1970) and subsequently by others. Though the loss of dividing potential is real, the scatter in the data is large and the degree of loss, particularly between young adults and aged people, is modest. Most of the reduction in replication capacity appears at early ages. For example, roughly calculated from the data of Martin *et al.* (1970), the number of population doublings is reduced from 40 to 28 using tissue samples (skin) from 10 to 20 versus 70 to 80-year-old age groups, respectively; LeGully *et al.* (1973) found a weak but significant correlation between population doubling potential and donor age of cells derived from liver; Schneider and Mitsui (1976) reported 45 and 34 population doublings for skin fibroblasts from donors of 21–36 and 63–92 years of age, respectively; Goldstein *et al.* (1978) reported a reduction from 55 to 40 doublings at 20 versus 80 years of age, respectively. These differences are even less substantial when subjects suffering from clinical disorders (Schneider and Chase, 1976), especially diabetes (Gleason and Goldstein, 1978), are removed from the statistics. One should also bear in mind that the cell complement in skin biopsies from young and old subjects may differ (Harper and Grove, 1979). If in 80-year-old subjects, the cells can still undergo even 70% of their original potential for doubling, there does not appear to be a serious danger of running out of cells because of age. Moreover, there is a justifiable argument that it is the postmitotic cells that must survive for the entire life-span of the animal. It can be claimed that replicating cells simply need to produce new and effective duplicates of themselves.

A feature which makes it hard to relate late-passage cells to aging, is found in the report that, at any passage number, cells can be rescued and transformed into an "immortal" line of tumor cells by use of SV-40 virus, "fixing" some of the characteristics at a particular passage level (Kaji and Matsuo, 1980). The ability of this virus to transform even late-passage cells is well documented. It seems possible, however, that the virus may simply infect some of the small proportion of viable (dividing) cells remaining in the "old" cultures.

B. The Nature of Cells in Culture

The use of tissue culture for the study of aging is complicated by a number of questions about the nature of the cultures. First of all, there is some question as to the origin of the cells. Franks and Cooper (1972), Franks and Wilson (1977), and Franks (1982) pointed out that as cultures divide and redivide, a selection takes place in which specialized cells will tend to die out. In this regard, Franks and Cooper (1972) examined 15 human embryo lung cultures at 11 different generations and observed that two predominant cell types were seen in all cultures. Differentiated epithelial cells died out after a short period. The authors concluded that the cultures are derived from endothelial cells and pericytes of small blood vessels. They point out that the structure of cells in culture from other tissues and even other animals (mouse, rat, hamster) appear to be similar in structure. Though biochemical variations in different cell types are substantial, they do not appear to be related to tissues of origin. Part of the problem is that cells in culture are forced to adapt to the conditions imposed by the medium and may do so with a standard response to yield undifferentiated patterns, regardless of origin. Franks (1982) demonstrated that cultures from prostate, salivary gland, colon, skin, and bladder lose differentiated cells and are repopulated by stem cells. Another problem in carrying out comparative studies is that there may be significant variation from biopsy to biopsy, even from the same human subject (Martin, 1977; Harper and Grove, 1979). In terms of biochemistry, it must be considered that one may be studying responses of cells that are, in a sense, being stressed by highly unnatural conditions of growth. In fact, cultures of cells exist in a number of metabolic states, the two most obvious being dividing and nondividing. The proportion of the latter increases with passage number until not enough cells can be produced to reach confluence (Phase III) and eventually no cells divide. Besides the more usual cells (embryonic human lung, skin, and foreskin) used for study, similar behavior has been noted in cultures of chick embryo cells (Macieira-Coelho and Lima, 1973), human glial cells (Blomquist *et al.*, 1980), mouse embryo cells (Meek *et al.*, 1977), and rat embryo cells (Meek *et al.*, 1980).

It has been shown repeatedly that the life-span of cells in culture is determined by the cumulative number of population doublings and not by calendar time in culture. Cells from different sources have different potentials for division, the value being 50 ± 10 population doublings for WI-38 cells. Phase III cells can survive for long periods of time as shown by Matsumura *et al.* (1979b), who maintained WI-38 cells in this condition for more than 1 year. The cells did not divide or become transformed, although morphological changes took place, particularly an increase in multinucleated cells. Macieira-Coelho (1976) pointed out that cells from different sources have different tendencies to become transformed either spontaneously or by chemical agents. In order of increasing probability of transformation are cells derived from chicks (no tendency), humans, cows, rabbits, hamsters, rats, and mice.

Clearly, the idea that all cells divide at equal rates until they are "senescent" and then slow down until they finally stop does not represent the real picture. On the contrary, cultures consist of heterogeneous populations of cells at different stages of the cell cycle (Macieira-Coelho, 1974; Macieira-Coelho and Loria, 1974; Absher and Absher, 1976) and probably contain cells that exist at different passage numbers. Thus, a cell culture typically consists of a mixture of cells, some of which are dividing rapidly, some slowly or transiently undergoing division (Macieira-Coelho and Loria, 1974; Burmer and Norwood, 1980), and some not dividing. The inability to divide probably reflects a true cell characteristic and is not a function of a deleterious environment. No one has yet succeeded in getting such nondividing cells to divide normally. Even cultures starting from cloning experiments are heterogeneous (Bell *et al.*, 1978; Merz and Ross, 1969; Absher *et al.*, 1975). Moreover, Bell *et al.* (1978) found interdivision times for human diploid fibroblasts to be heterogeneous for cells at all levels of population doubling. Individual cells in a population of fibroblasts vary considerably in their doubling potential (Smith and Hayflick, 1974; Martin *et al.*, 1974; Smith and Whitney, 1980). The last of these papers elegantly explores the division potential of cloned cells. Cells were taken at random at 16, 26, and 36 population doublings from the parent clone which had been developed from a single cell. The subclones possessed doubling potentials which varied dramatically. Two cells arising from a single metabolic event also yielded cells with greatly different reproductive abilities. It was found that two such cells could vary by as many as eight population doublings. In this respect, Harley and Goldstein (1978) observed that in circular outgrowths of human skin fibroblasts, the cells at the circumference divided more frequently than cells located further toward the center. Centrally located cells were found to exist at a much lower generation level. The authors noted that these results can help explain the differences in proliferative capacity found in subcultured cells.

From the above considerations, it can be seen that the "ages" at which cells are compared reflect average values of the cells in various metabolic states. One would expect and one indeed finds that the greatest difference exists between Phases II and III because the later period is particularly marked by reductions in the number of cells undergoing rapid division and in the number of cells that retain a high division potential. In addition, there is an increase in cell cycle time. These changes are presumably going on to some degree at all passage levels, becoming predominant in what is known as Phase III.

As to culture conditions, growth of cells may be modulated by extracellular components such as growth factors, hydrocortisone, volume of culture medium, and the amount of serum added. With regard to serum, which is usually fetal calf serum, there may be differences in response from batch to batch. The amount used can greatly affect biochemical parameters such as protein metabolism (Chapter 8, Section IV), and low levels (0.5% in place of the usual 10%) can be used to create "stationary" cultures (Dell'Orco et al., 1973). Ohno (1979) reported that the use of a low level of serum reduced the total number of population doublings of human diploid fibroblasts from 76 to 54. Perhaps the small amount used (0.3 versus the usual 10%) was deleterious. Another consideration is whether or not cells are affected by the trypsin typically used to free them from the walls of culture vessels. Hadley et al. (1979) and Kaji and Matsuo (1980) found that proliferative capacity was unaffected. However, there may be more subtle consequences. As an example, Hughes and Ayad (1980) reported that trypsin affects adenylate cyclase activity, though differently, in Chinese hamster cells, a mouse lymphoma cell, and a hybrid of the two.

The reasons for the decreasing proportion of dividing cells in culture at late passage and the reason for the finite life-span of diploid fibroblasts have drawn much attention and several explanatory theories have been proposed. In brief, some sort of genetic program seems to be the most favored idea. Thus, Cristofalo (1972), Martin et al. (1974, 1975), Bell et al. (1978), and Kontermann and Bayreuther (1979) propose that the aging of cells in culture is a differentiation process. That is, cells differentiate to a terminal type. The recent findings of Smith and Whitney (1980) (see above) are also compatible with this idea. Holliday et al. (1977) developed a commitment theory in which uncommitted (potentially immortal) cells are diluted by their committed (terminally differentiated) offspring and are lost in the serial subculturing process. This idea has been refuted by Smith and Whitney (1980) and Harley and Goldstein (1980). Other ideas include some mutation (Nichols, 1975), error catastrophe (Chapter 9, Section III), virus infection, loss of gene repression (Smith and Lumpkin, 1980), and a stochastic model (Good and Smith, 1974; Smith and Whitney, 1980). Reis et al. (1980) summed up the situation and agreed that there is at least a stochastic element in the loss of division potential. In truth, the reason for cessation of cell division is not known and there are not as yet any clear lines of approach to the mechanism involved.

C.   Measuring Cell "Age"

Cristofalo and Sharf (1973) described a useful procedure for judging the "age" of cells in culture by measuring the proportion that became labeled with [^{3}H]thymidine (Thymidine Labeling Index). Those cells not undergoing division (no DNA synthesis) did not become labeled. The percentage of unlabeled WI-38 cells yielded a linear relationship to the proportion of the life-span, which had been completed under the given culture conditions. Cristofalo (1976) subsequently reviewed the method and its limits. Vincent and Huang (1976) obtained somewhat different results with WI-38 cells, observing no fall-off in doubling until after 40 population doublings.

A number of other relationships have been noted, which can be used as criteria of cell aging. Recently, Smith *et al.* (1980) proposed a system for determining the replicative age of cell cultures. The authors found a linear relationship between the population doublings remaining and the percentage of cells able to form colonies of 16 cells. For chick embryo cells, the relationship held for cells able to form colonies of 64 cells. The relationships can be expected to apply only to cells cultured under the conditions reported, paying careful attention to matching the calf serum used in each set of cultures. Matsumura *et al.* (1979b) described a method for determining the doubling potential of cultures based upon the number of cells that are *not* labeled by [³H]thymidine. They thus claim to avoid the problems of radiation noted by Cristofalo (1976).

D.   Extension of Cellular Life Span by Hydrocortisone

It was first reported that cortisone (Macieira-Coelho, 1966) and hydrocortisone (Cristofalo, 1970) extended the life-span of human cells in culture. Subsequently, Cristofalo and co-workers made several studies of the phenomenon. Early-passage cells were shown to be most responsive to the hormone, yielding a life extension of 30–40%, even at very low concentrations (14 μM) (Cristofalo, 1970; Cristofalo and Kabakjian, 1975). There was no effect on senescent cultures. In fact, if hydrocortisone is withdrawn from treated cells that have gone byond their normal (unstimulated) life-span, they die out very quickly (Macieira-Coelho and Loria, 1974). Though hydrocortisone caused increased proliferation in human fetal lung and fetal foreskin cells, it was inhibitory for other cell types, for example, African green monkey (kidney), Syrian hamster (embryo), C3H mouse (arcolar), frog, and minnow (Rosner and Cristofalo, 1979). Interestingly, adult lung, transformed fetal lung, and fetal but not adult skin cells were inhibited. The response appears to be due to the presence of high-affinity binding sites which are specific for the steroid. These sites decreased about 40% with cell age, presumably explaining the reduced effect of the hormone on senescent cultures. Many compounds with related chemical structures are inactive except for dexamethasone, triamcinolone, and prednisolone.

The enhanced growth appears to be cell cycle dependent (Cristofalo *et al.*, 1979), and there is an influence exerted by medium in which cells have already been grown. That is, there is a given time (0–6 hours) after refeeding quiescent cells during which they respond to hydrocortisone. As perhaps might be expected, transcriptional activity increases in this period (3–5 h) after hormone addition to confluent (arrested) cells (Ryan and Cristofalo, 1979). The effects of hydrocortisone on WI-38 cells in culture have been summed up by Cristofalo and Rosner (1979).

E. Are Late-Passage Cells Analogous to Tissues Aging *in Vivo*?

In the years since Hayflick and Moorhead (1961) established that diploid cells in culture had a finite life-span, many studies have been carried out that deal with the characteristics of the *in vitro* system. The relationship between aging in animals and "aging" of cells in culture has had less attention, although it is now receiving growing emphasis. After all, the degree of similarity between "aging" *in vitro* and aging *in vivo* is the critical aspect of using cells in culture as a model system. The initial designation of late-passage cells as "senescent" (Hayflick and Moorhead, 1961) appeared to receive general acceptance, although biochemists have been inclined to look askance at the system. In fact, there have tended to be distinct "tissue culture" and "intact animal" sessions at meetings that deal with aging.

Starting with Schneider and Mitsui (1976), several investigators compared biochemical differences between early- and late-passage cells with findings in cells (usually skin fibroblasts) obtained from young and old donors. They also compared various properties of late-passage cells with those of cells from old donors. In interpreting these studies, one should bear in mind the possibility that biopsy material may provide a different population of skin fibroblasts from young and old donors (Harper and Grove, 1979).

How similar are the findings *in vitro* and *in vivo*? There are areas both of agreement and of disagreement. Schneider and Mitsui (1976) and Schneider (1979) confirmed that such parameters as average rate of fibroblast migration, *in vitro* life-span, and replication rate were decreased in cultures of skin fibroblasts developed from an old donor group (63–92 years) compared to a young group (21–36 years). Whereas these changes are reminiscent of early Phase III cells, the results are considerably less dramatic in scope and in general show substantial overlap in values between cells from young and old donors. As the authors point out, the variation may well be a result of different physiologic age among the donors. Smith *et al.* (1978) obtained a sharper distinction, observing that only 2% of cells from old donors could produce colonies of 256 cells or more, whereas the majority of cells from young donors could do so. Goldstein and Harley (1979) in a review of work on cultured cells from subjects with age-associated diseases (diabetes, Werner's Syndrome, progeria) noted that both late-passage cells and cells from old donors had a reduced response to nonsuppressible insulinlike activity as measured by stimulation of DNA synthesis. Schneider *et al.* (1979a) also found agreement between *in vivo* and *in vitro* aspects of sister chromatin exchanges (SCE) in which pieces of chromosomes exchange positions. Using differential staining techniques to analyze DNA damage (SCE) after incorporation of bromodeoxyuridine into metaphase chromosomes, the authors noted that the baseline values for SCE do not vary significantly either with passage number of human fetal fibroblasts or with donor age for skin fibroblasts. The ability of mitomycin C and several other DNA-damaging reagents to induce SCE is decreased with age in both systems. Interestingly, in the *in vitro* system, the decreased SCE appeared in middle passage before the onset of Phase III. SCE are also reduced in bone marrow cells and spleen cells from old rats.

Schneider and Smith (1981) recently summed up the findings from *in vitro* versus *in vivo* studies. The information in Table 2.I, taken from this article, shows some of the characteristics of skin fibroblasts derived from young and old donors. These differences are reminiscent of those in early- versus late-passage cells.

Table 2.I. Properties of Skin Fibroblast Cultures Derived from Young and Old Human Subjects[a]

	Subjects	
Replication parameter	20–35 years	65+ years
Onset of senescent phase (PD)[b]	35.2 ± 2.1 (23)	22.5 ± 1.7 (21)
In vitro life-span (PD)	44.6 ± 2.5 (23)	33.6 ± 2.1 (21)
Cell population replication rate (hours)	20.8 ± 0.8 (18)	24.3 ± 0.9 (18)
Percentage replicating cells	87.1 ± 1.6 (7)	79.6 ± 2.5 (7)
Cell number at confluency ($\times 10^4$ cells/cm^2)	7.31 ± 0.42 (18)	5.06 ± 0.52 (18)
Percentage of cells able to form colony of 16 cells[c]	69.0 ± 3.3 (9)	48.0 ± 4.4 (8)
Sister chromatid exchanges/cell	67.9 ± 1.6 (7)	56.1 ± 1.4 (6)

[a] Numbers within parentheses indicate number of cell cultures examined; values are mean ± standard error of the mean. See Schneider and Smith, 1981. Reprinted by permission.

[b] PD, Population doublings.

[c] Two weeks after plating at low cell densities.

Recently, it has been shown that human diploid fibroblasts derived from fetal tissues (lung, heart, liver, skin, and muscle) increased their absorption of concanavalin-treated red blood cells with increased passage number (Aizawa *et al.*, 1980). A similar increase was noted in skin fibroblasts from human donors of increasing age, from 0 to 89 years old.

In contrast to the above results, several changes, which occur in late-passage cells, are not observed in cells derived from old donors. Robbins *et al.* (1970) early reported that increased residual bodies in late-passage cells (embryonic lung or skin) are not observed in skin sections of aging subjects. Moreover, the increase in protein content, RNA content and size of late-passage cells observed by several investigators have no counterpart *in vivo*. Lin and Chang (1979) observed that the specific activity of tRNA methylase declines in late-passage cells (fetal lung and skin fibroblasts), whereas, in skin fibroblasts from donors ranging from 3 months to 94 years, there was no change in the enzyme activity after an initial drop between fetal and newborn samples.

From the data available, the evidence that "old" cells *in vitro* are or are not analogous to cells from old donors is not overwhelming. Some features agree, some do not. Little clarification can be obtained from examination of the biochemical characteristics of early- versus late-passage cells found in the various chapters of this volume as equivalent experiments in old animals are not generally available for comparison. It must be said that biochemical data is still relatively sparse for the tissue culture system so that comparisons are sketchy in any case. Moreover, though the point is not generally stressed, it would seem wise for comparative purposes to emphasize those changes that occur between early and late Phase II. Phase III cells presumably do not exist in aged animals and, therefore, comparison of Phase II with Phase III cells has no anology *in vivo*.

All in all, it is not surprising that there are proponents and opponents of the *in vitro* system as a model for aging studies. Schneider *et al.* (1979b) summarized some of the factors involved in studying cellular replication and its relation to aging. Evans (1979) marshaled evidence to support the idea that organisms age because of problems in the non-dividing cells, suggesting that aging is a result of the absence of mitosis, rather than a cause of that condition. Hayflick (1979, 1980) has continued to express no doubt that cells in culture manifest true symptoms of aging.

It is premature to take a firm stance, one way or the other, on the propriety of the *in vitro* system as a model for aging *in vivo*. Problems include the fact that when cells from young and old donors are prepared, they are generally passaged a few times before use. They might, therefore, tend to adapt to the *in vitro* situation and show characteristics similar to those of cultured cells. One must also be careful about comparing human embryonic lung cells with skin fibroblasts. Alternatively, the fact that there exist substantial biochemical differences between the *in vitro* and *in vivo* systems means that although cells in culture may turn out to be a valid model for studying the underlying causes of aging, they are not an accurate marker for more superficial metabolic characteristics.

F. Biochemical Changes in Late-Passage Cells

Those biochemical changes in late-passage cells, which lie in the mainstream of current work, are dealt with in the appropriate chapters (e.g., nucleic acids, proteins, membranes, etc.). Current reports, however, continue to cover a wide range of topics, some of which lie outside of these areas. As a few examples, Hatcher *et al.* (1976) observed a cell surface bound protease activity, which, when tested in a number of cell types, decreased following mitosis. Sun *et al.* (1979) observed that 5'-nucleotidase activity increased with increased population doubling. Vorbrodt *et al.* (1979) noted that the level of this plasma membrane bound enzyme is greatest in confluent "old" cells. Honda and Matsuo (1980) noted that growth of late-passage WI-38 cells (45–52 doublings) is more sharply inhibited by hyperbaric oxygen than those early-passage cells (32–44 doublings). Balin *et al.* (1978) had earlier concluded that oxygen toxicity does not play a role in limiting the life-span of WI-38 cells. Bhargava *et al.* (1979) reported that late-passage human uterine smooth muscle cells in culture possess an increased number of epidermal growth factor receptors. Murota *et al.* (1979) observed a decline in 6-ketoprostaglandin F_1 production in aging human fibroblasts. Hayflick (1980) summarized exhaustively the biochemical parameters which increase, decrease, or remain the same in late-passage cells, although no distinction is made as to whether the changes occur in Phase III or late Phase II.

G. Relationship of Population Doubling Potential to Life-Span

A number of investigators have explored the possibility that the potential for cell doubling in a species may be related to its life-span. Hayflick (1977) assembled data which included the doubling potential for cells from the Galapagos tortoise, man, mink, chicken, and mouse. The number of doublings obtained from the cells of mammals was in the order of their life-span, but the chicken was out of place. Stanley *et al.* (1975) found no relationship between cell doubling potential and life-span in a number of mammals ranging from human to Australian marsupials. However, since the site of the tissue biopsies varied, as did the original age of the donors, the results are not definitive. At this time, there is too little information to either support or negate the relationship. Offhand, it looks like another attractive, but simplistic, idea that has no real rationale and that will not be sustained. It is reminiscent of the proposal of a direct relationship between DNA-repair capacity and life-span, which is still being argued about (Chapter 6, Section IX).

H. Use of Fused Cells

Cell fusion techniques have been utilized in an effort to explore the nature of the finite life span of cells in culture or more specifically, to determine if control of the phenomenon resides in the nucleus or the cytoplasm. The fusion of senescent cells with SV-40 transformed cells (Croce and Koprowski, 1974) or with HeLa cells (Stanbridge, 1976) yielded immortal hybrid lines. Such hybrids (heterokaryons) showed reinitiation of DNA synthesis in the senescent nuclei (Norwood *et al.*, 1975), though heterokaryons composed of "old" and "young" (i.e., nontransformed) cells did not promote similar DNA synthesis (Norwood *et al.*, 1974). The latter results were interpreted to mean that "old" cytoplasm inhibits DNA synthesis in "young" cells. Recently, Norwood *et al.* (1979) fused senescent human diploid cells and a murine cell line and observed incorporation of labeled thymidine in both nuclei. The evidence supports the idea that the labeling in senescent nuclei represents replicative DNA synthesis rather than DNA repair.

A more direct way of studying the influence of young and old nuclei and cytoplasm on cell longevity is to hybridize enucleated cells (cytoplasts) with whole cells to form "heteroplasmons." Cytochalasin B, a product of the mold *Helminthosporium dematiodium*, causes spontaneous enucleation of cells. When used in conjunction with centrifugation, enucleate cells and nuclei with a small amount of cytoplasm (karyoplasts) can be obtained. Wright and Hayflick (1975a,b), who used a somewhat involved chemical treatment to prevent division of unfused cells, obtained suggestive evidence that the ability to maintain continued cell doubling resides in the nucleus. Hybrids composed of "old" cytoplasts fused to "young" cells (in which the cytoplasmic enzymes were inactivated with iodoacetate) showed substantial ability to divide, whereas "young" cytoplasts fused to "old" cells showed little doubling capacity. Hence, only cells with "young" nuclei could continue to divide. The authors noted that the experiments are subject to a degree of ambiguity arising particularly from the severe chemical treatment utilized.

A further simplification of the reconstituted cell system was made by Muggleton-Harris and Hayflick (1976). They fused isolated nuclei (karyoplasts) and enucleated cytoplasms (cytoplasts) of human diploid cells (WI-38) to form reconstituted cells. Various combinations of young and old cytoplasts and karyoplasts were prepared. Though successful fusions were obtained with "old" cytoplasts and "young" karyoplasts, the division of cells was limited. Only "young–young" combinations were able to attain a level of cell division similar to that of the controls. From these results, it appeared that control of continued ability to double resides in both nucleus and cytoplasm. However, there are unresolved questions about the effects of the technique itself, which make such a conclusion equivocal. Nonetheless, Tanaka *et al* (1980) also concluded from fusion experiments that both components of the cell are involved in the retarded DNA synthesis noted in fibroblasts from subjects with Werner's Syndrome.

Muggleton-Harris and Palumbo (1979) developed procedures for introducing individual viable nuclei into cells. When the authors fused nuclei from "young" WI-38 cells, SV-40 transformed cells, and HeLa cells into senescent WI-38 cells, the senescent cells subsequently underwent a limited number (1–7) of doublings. Senescent nuclei fused to senescent cells did not divide. Morphology of the cells retained fibroblastic characteristics. The "immortal" nuclei failed to produce the immortal cells noted by Stanbridge (1976). However, he fused nuclei to whole cells whereas Muggleton-Harris and Palumbo (1979) utilized only the cytoplasm.

Muggleton-Harris and DeSimone (1980) made a further attempt to resolve the question as to whether the control of cellular life-span resides in the nucleus or cytoplasm. Log phase WI-38 cells were fused to SV-40 transformed cells and nuclei from transformed cells were fused to WI-38 cells and to WI-38 cytoplasts, respectively. All of the fusion products showed a doubling potential similar to that of the control WI-38 cells. Clearly, WI-38 cells did not become "immortal" like their transformed "parent." Less than 2% of the WI-38 transformed cell hybrids yielded clones with sustained replication. Controls with homogeneous components behaved like the parental cells. From these results, it seems that whatever governs the limited life-span of cells is a dominant trait. Either such factors lie in the WI-38 cytoplasm, or there are activators of the nucleus in the transformed cytoplasm that are lacking in the nontransformed cells.

There is still much to be learned about the cell fusion system. Recently, Yanishevsky and Stein (1980) found that when "young" and "old" human diploid fibroblasts are fused, the "young" nuclei of the resulting heterodikaryons do make DNA if they are in the S phase at the time of fusion. The data indicate that the senescent cell inhibits a young cell from entering S phase.

Muggleton-Harris (1979) has recently reviewed the area of reassembly of cell components.

IV. FREE-LIVING NEMATODES

Use of free-living nematodes for aging studies has been described in detail by Rothstein (1980). Generally, *Turbatrix aceti* has been used for these studies. However, methods have also been devised for the culture and aging of *Caenorhabditis elegans*. *Turbatrix aceti* grows to a length of approximately 2 mm, is bisexual, and, like other free-living nematodes, is born with all its cells except for the reproductive system. The organism grows to populations of 350,000–400,000/ml in 15–17 days in axenic, defined medium at 30°C (about 400–600 mg wet weight/ml). *Caenorhabditis elegans* is somewhat smaller in size and is hermaphroditic, although a small number of males is produced. This organism grows on *E. coli* or in a semidefined, axenic medium to populations approaching 200,000/ml in 8–9 days at 20°C (Rothstein and Coppens, 1978). The yield is approximately 1 mg dry weight/10^5 organisms. Claims made for the advantages of nematodes as a model system for aging studies include the following: they have short life-spans, 50% survival being 26 days for *T. aceti* in mass axenic culture at 30°C (Hieb and Rothstein, 1975) and, for *C. elegans*, about 16 days at 20°C (Klass, 1977), 13 days (Hosono *et al.*, 1980), or 11.5 days (Bolanowski *et al.*, 1981); essentially all the cells are nonmitotic so that old organisms consist of old cells; there are a few cell types, the bulk consisting of muscle and gut so that the need to use intact organisms is not a great disadvantage; they can be grown in defined (semidefined for *C. elegans*) medium, which can be manipulated with regard to components; atmospheric conditions can be controlled; radioisotopes are readily taken up and incorporated; adequate amounts of tissue can be obtained; synchronization of populations is reasonably simple; organisms can be observed individually or en masse; genetic studies with *C. elegans* are advancing rapidly and offer the potential of being coordinated with aging studies.

A. Synchronization and Aging of Cultures

Since mature organisms produce numerous offspring, young organisms predominate in any culture. Gershon (1970) first reported the synchronization of *T. aceti* in axenic medium containing liver extract. He collected small (young) organisms by passage through a column of glass beads and blocked DNA synthesis by adding fluorouridine deoxyribose (FUdR), aminopterin, or hydroxyurea to the cultures. At 100 μg/ml, FUdR plus 100 μg/ml of uridine inhibited DNA synthesis by 96% and prevented reproduction. Longevity of individual animals was reported to be unchanged. Hieb and Rothstein (1975) further developed the procedure. Young *T. aceti* (estimated at 0–4 days of age) were obtained by filtration of cultures through sterile, stainless steel screens. The small worms, which passed through the screens, were permitted to age in defined medium containing 100 μg/ml of FUdR + 150 μg/ml of uridine. Hydroxyurea was not satisfactory in that it did not consistently prevent reproduction. In the presence of FUdR, the organisms grew in size, but reproduction did not take place. Roux bottles containing 50 ml of medium were found to be most satisfactory for aging the cultures. Populations during aging must be limited to a density of 20,000/ml. The 50% survival of organisms in such cultures is 26 days. The freshly screened organisms are allowed to grow for 24–28 h before addition of FUdR so that growth is not stunted, as reported by Kisiel *et al.* (1972). The fact that the cells (except for reproductive tissues) in nematodes are not dividing presumably protects the organisms from severe damage by the DNA inhibitor. Hieb and Rothstein (1975) also aged *T. aceti* by maintaining young organisms at 37°C to prevent reproduction. This temperature was quite critical: higher temperatures led to early death and lower temperatures permitted reproduction. Even at 37°C, the life-span was considerably shortened, and average growth in size was inhibited.

Rothstein and Sharma (1978) described a repeated screening method of aging *T. aceti*, which did not involve the use of chemical inhibitors. In this procedure, *T. aceti* were screened and the large organisms remaining on top of the screen were placed in fresh medium. After 3 days, they were again screened, thus removing newborn organisms. The screening was repeated every 3 to 4 days, the large, aging organisms being retained on top of the screen. Organisms aged this way and those aged in the presence of FUdR yielded similar results in studies of protein turnover and synthesis (Sharma *et al.*, 1979; Chapter 8). Both methods showed the presence of altered enzymes in old organisms (Rothstein and Sharma, 1978; Chapter 9, Section V). Thus, biochemical characteristics, insofar as they have been studied, do not appear to be altered by use of FUdR in the aging procedure.

Although *C. elegans* has not been utilized extensively for biochemical studies, it has been extensively studied in terms of developmental genetics, and large numbers of behavioral mutants have been "banked" in liquid nitrogen. Methods for obtaining aging populations have been described by Mitchell *et al.* (1979), Gandhi *et al.* (1980), and Hosono (1978). Basically, the methods involve the use of FUdR, but careful studies have been made of the appropriate conditions. The reagent at 0.4 mM prevents egg hatching, but causes no apparent morphological abnormalities (Mitchell *et al.*, 1979). However, Bolanowski *et al.* (1981) did find evidence on photomicrographs of structural changes as well as noting slowing of movement. Much lower concentrations of FUdR (25 µM) can be used if applied just before synchronized cultures reach maturity (Gandhi *et al.*, 1980).

In general, nematodes are easy to grow and handle. They can be obtained in reasonable quantities and can be treated much like bacteria in terms of culture, labeling, and harvesting. They can also be treated and studied as individuals. Hermaphroditic species such as *C. elegans* are useful for genetic and neurobiological studies, particularly as mutants and can be stored for years in liquid nitrogen.

B. Biochemical Studies

The original discovery of an altered enzyme was made when it was found that more antiserum was required to inactivate a given activity of isocitrate lyase in homogenates of old versus young *T. aceti* (Gershon and Gershon, 1970). Subsequently, a number of enzymes (isocitrate lyase, phosphoglycerate kinase, enolase, aldolase) were purified and were found to be altered in old organisms. Triosephosphate isomerase, however, was found to be unaltered. "Young" and "old" enolase were subsequently characterized in considerable detail (Sharma and Rothstein, 1978a,b). The results of these experiments with nematode enzymes support the early contention of Rothstein and co-workers and have culminated in proof that the alterations are conformational in nature, without covalent changes (Sharma and Rothstein, 1980). These experiments have also led to the hypothesis that the changes in enzyme structure result from a slowed protein turnover. Indeed, turnover of enolase and of soluble proteins in *T. aceti* has been shown to slow dramatically with age (Sharma *et al.*, 1979). Thus, the nematodes have been invaluable in invalidating the idea that errors in sequence are responsible for altered enzymes in old organisms. Detailed discussion of altered enzymes, error theory, and protein turnover is presented in Chapters 8 and 9.

Other biochemically oriented work in aging utilizing nematodes includes studies that show that the antioxidants α-tocopherolquinone (Epstein and Gershon, 1972) and centrophenoxine (Kisiel and Zuckerman, 1978) retard the accumulation of age pigments in C. briggsae. The former compound was reported to extend both the mean and maximal life-span by 28 and 31%, respectively. Effects of temperature, food concentration, and parental age on life-span have been described (Klass, 1977). Effects of age on osmotic fragility (increases), movement, and pharyngeal pumping (decreases) have also been reported (Hosono et al., 1980; Croll et al., 1977; Kisiel et al., 1975).

V. INSECTS

The use of insects, although perhaps not a major factor in aging studies, nonetheless represents a steady source of published observations. Recent work mostly involves flies of one type or another, often Drosophila or the housefly (Musca), but occasionally includes other species such as the mosquito (Aedes). Musca has a typical 50% survival of about 19 days at 24°C with a life-span of about 30 days (Sohal, 1976). For Drosophila, the mean life span is 78 days (Miquel et al., 1975), with a maximum of 98 days. These figures vary in different laboratories. As with higher animals, dietary regimen and environmental conditions appear to play a marked role in survival times and thus may affect biochemical characteristics.

Insects can be grown in substantial quantity and analyzed individually, en masse or by removal of specific parts such as head or thorax. Environmental effects on the survival of the organisms, e.g., temperature (Atlan et al., 1976), diet (Driver and Cosopodiotis, 1979; Massie and Williams, 1979), and oxygen level (Miquel et al., 1975) are important. Change of enzyme levels with age have been reported by a number of investigators, (Burcombe, 1972; Beezeley et al., 1974; Mills and Lang, 1976) including changes in enzymes related to oxygen metabolism such as superoxide dismutase (Massie et al., 1980) and peroxidase (Armstrong et al., 1978). The availability of mutants of Drosophila offers a broadened base for such comparisons (Baker, 1978; Ganetzky and Flanagan, 1978).

Adult insects, like nematodes, consist of postmitotic cells. Age pigments are prominent and have been the subject of several studies (Chapter 3, Section III). The demand for rapid energy production has made insect flight muscle the subject of considerable attention over the years. Hence the natural interest in this system for study of aging mitochondria (Chapter 5, Section II). Other studies have dealt with ultrastructural changes and changes with age of various biochemical constituents (Sohal, 1976; Baker, 1976). Work of this nature is discussed in other chapters of this book, where relevant, under the appropriate biochemical topic.

It seems fair to say that research on aging using insects has provided interesting adjuncts to observations made in other animals but that new ground has not been broken. There has really been too little sustained research to permit a coherent picture to develop. Moreover, there appears to be a residue of doubt, perhaps unjustified, that aging in insects may be "different" and thus may be reflected only in part, in higher animals. Even if the biochemical events that cause aging in insects and higher animals are similar in many respects, the resulting manifestations may be so different that the similarities are obscured.

VI. PROTOZOA

There have been a number of studies of "aging" that have attempted to determine whether or not protozoans have finite life-spans or are immortal. For most species, the current status of work is too rudimentary to provide an adequate background for the researcher in aging who wishes to make use of a reasonably well-defined model. A thorough review of the current status of protozoans with respect to age is provided in a chapter by Smith-Sonneborn (1981).

One protozoan, *Paramecium aurelia*, has been studied extensively with regard to aging, mostly by Smith-Sonneborn and co-workers. Sonneborn (1954) originally observed that paramecia age and die unless they undergo either self-fertilization (autogamy) or cross-fertilization (conjugation). Smith-Sonneborn and Reed (1976) showed that the life-span of *P. aurelia* depends upon the number of fissions undergone rather than calendar time. In this respect, the organism is analogous to the doublings of cells in culture, although there are other substantial differences between the systems. It is interesting that the yeast *Saccharomyces cerevisiae* also shows a life-span related to budding rather than to calendar time (Muller *et al.*, 1980). Clones of *P. aurelia* have a life-span of 150–200 fissions. Age is defined as the number of mitotic divisions since cross-fertilization or self-fertilization. These processes set the fission clock back to zero. The organisms show a definite pattern of aging behavior (Sonneborn, 1954): an immature period in which the cell cannot undergo a sexual process; a period of maturity during which sexual processes decline; death. With increasing fission number, there is a gradually decreasing fission rate and an increased frequency of death of the progeny after autogamy. Takagi and Yoshida (1980) recently showed that clones of *P. caudateum* also measure age in terms of fissions and not calendar time. These organisms are longer-lived than *P. aurelia* and have a maximal life-span of 658 fissions.

Rodermel and Smith-Sonneborn (1977) observed a sharp increase in mortality and a reduced viability of progeny at 50–80 fissions for most autogamous clones. The results appear to be due to increased mutation of the germ line, which is carried in the micronucleus. In addition, micronuclear mutations induced by UV irradiation increase with clonal age. At least part of the problem lies in a decreased level of repair with increased "age." In a following publication, Smith-Sonneborn (1979) showed that as the fission age of a clone of *P. tetraurelia* increased, the UV dose required to reduce the life-span decreased. If repair by photoreactivation (which monomerizes induced dimers in the DNA) was permitted, no age-related effects of UV damage were observed. In fact, after photoreactivation, cells treated with UV a second time, particularly old cells, outlived the controls. The first treatment apparently induced resistance to the second treatment. The authors suggested that repair processes induced by UV radiation perhaps correct age-related damage. The fact that UV-induced damage, if unrepaired, reduces life-span suggests that DNA damage is a factor in the life-span of *P. aurelia*.

Biochemical studies are somewhat limited. Klass and Smith-Sonneborn (1976) observed that as fission age increased, macromolecular DNA, RNA synthesis, and DNA template activity decreased as did endocytic capacity (Smith-Sonneborn and Rodermel, 1976). DNA polymerase activity was unchanged with age (Williams and Smith Sonneborn, 1980).

Sundararaman and Cummings (1976a) studied morphological changes in *P. aurelia* with "age" and observed an increase in the number of mitochondria (which also showed structural alterations) and lysosomes. Dense bodies, equated to age pigments, were also present. Macronuclei were also reported to undergo structural changes (Sundararaman and Cummings, 1976b). Ribosome particles too, appear to be altered with age (Sundararaman and Cummings, 1977). Other changes, including morphological characteristics, have been summarized by Smith-Sonneborn (1981).

VII. OTHER

A. Tissue Transplants

There have been attempts at serial transplantation of tissues from young into old animals and vice versa. Krohn (1966) grafted skin by serial transplants in inbred mice. He reported that the longest graft survived nearly 2 times the life-span of the animals. Young *et al.* (1971), using a previously devised mouse mammary transplantation procedure, found that tissues from young and old donors behaved similarly in young hosts. Neither type of transplant grew well in old hosts. Daniel (1972) performed further work on the system. He concluded that the number of cell divisions rather than the passage of time might determine the proliferative capacity of mammary epithelium transplants. In general, there are technical problems with the arrangement such as invasion of the grafts by blood vessels and connective tissues of the host.

A somewhat different approach to transplantation was used by Harrison (1975). He made five sequential, functionally successful transplants of normal marrow cells from young and old mice into anemic mice. There were no significant differences in the proliferative capacity of cells derived from young versus old donors. However, the transplants expired after 3–6 serial transplantations, though they produced erythrocytes for a total of 100 months, 3½ times the mean life-span of the donor mice (Harrison, 1979a). An overview of the system has been provided by Harrison (1979b).

In general, transplantation would seem to have a useful potential for aging studies, but at present it is still in the "how long does it live" stage. Moreover, there are always doubts about the effects of the procedures, as noted by several authors. For the biochemist, the use of serial transplantation does not presently seem to offer a system for general use in aging studies.

B. Red Blood Cells

The mammalian red blood cell, particularly the human erythrocyte, has been considered as a model for aging studies. Since the mature red cell contains no nucleus, it cannot synthesize new components and hence it must survive on its original complement of structural and functional elements. The subsequent deterioration of enzymes and membranes during its life-span of 120–130 days can, therefore, be regarded as a model for deteriorative processes in aging. However, the fact that the cell is designed to last a certain time, the probable lack of repair, and the fact that special mechanisms of degradation are present make this a dubious proposition. As to changes brought about as red cells age, there are numerous reports in the literature ranging from alterations in enzyme content (Landaw, 1976; Seaman et al., 1980) to changes in membrane structure (Kadlubowski and Agutter, 1977).

A second approach is to consider changes that might occur in cells in aging individuals. Such changes would reflect defects in the manufacture of the cells or in their environment. Examples of this approach are the report of Araki and Rifkind (1980) that cells from old subjects have an increased osmotic fragility and that of Hegner *et al.* (1979), who examined erythrocyte ghosts from two groups of human males, one under 30 and the other over 70 years of age. They observed a statistically significant decrease in phospholipid content, membrane fluidity, Na^+,K^+-ATPase and sialic acid in the red cells of the old group. Cholesterol content and fatty acid composition were unchanged. Platt and Norwig (1980) also reported that ATPase activity decreases with increasing donor age.

Survival time of red cells decreased in aged mice (Abraham *et al.*, 1978), and sulfhydryl and reduced glutathione content were also reported to be lower. However, caution should be observed in applying these results broadly, because rat red cells are reported not to be a model for the human red cell with respect to changes in surface properties (Walter *et al.*, 1980).

C.   Hepatocytes

Isolated hepatocytes have been utilized by a few investigators as a model system to study aging. Since the cells are postmitotic, they should reflect the age of the donor. Thus far, studies with this system have concentrated upon protein synthesis (Chapter 8, Section III), and lysosomes (Chapter 5, Section IV). Knook (1980) presented a review of the use of hepatocytes in aging studies.

VIII.   COMMENT

It can be fairly stated that none of the above systems, *in vivo* or *in vitro*, represent completely satisfactory models for studying human aging. For all, to one degree or another, there is the question as to whether the observed age-related changes are reflected in humans. In the case of rodents, there are additional concerns about the true physiological age of individuals, the effects of disease, and the expense of maintenance. Invertebrate or tissue culture systems are less complex and simpler to maintain, but require more validation as being truly representative of mammalian aging *in vivo*. Even so, the problems are more irritants to be overcome than severe blocks to progress. In truth, it is not the difficulties with the experimental animals that account for the lack of rapid advance in uncovering the secrets of aging. It is the extremely subtle nature of the changes that lie hidden below the surface manifestations.

REFERENCES

Abraham, E. C., Taylor, J. F., and Lang, C. A. (1978). *Biochem. J.* **174**, 819–825.
Absher, P. M., and Absher, R. G. (1976). *Exp. Cell Res.* **103**, 247–255.
Absher, P. M., Absher, R. G., and Barnes, W. D. (1975). In "Cell Impairment in Aging and Development" (Cristofalo and Holeckova, eds.) pp. 91–105. Plenum, New York.
Adelman, R. C. (1972). *Adv. Gerontol. Res.* **4**, 1–23.
Aizawa, S., Mitsui, Y., Kurimoto, F., and Matsuoka, K. (1980). *Mech. Ageing Dev.* **13**, 297–306.
Araki, K., and Rifkind, J. M. (1980). *J. Gerontol.* **35**, 499–505.
Armstrong, D., Rinehart, R., Dixon, L., and Reigh, D. (1978). *Age* **1**, 8–12.
Atlan, H., Miquel, J., Helmle, L. C., and Dolkas, C. B. (1976). *Mech. Ageing Dev.* **5**, 371–387.
Baker, G. T. (1976). *Gerontology* **22**, 334–361.
Baker, G. T. III (1978). *Mech. Ageing Dev.* **3**, 367–373.
Balin, A. K., Goodman, D. B. P., Rasmussen, H., and Cristofalo, V. J. (1978). *J. Cell Biol.* **78**, 390–400.
Beezeley, A. E., McCarthy, J. L., and Sohal, R. S. (1974). *Exp. Gerontol.* **9**, 71–74.
Bell, E., Marek, L. F., Levinstone, D. S., Merrill, C., Sher, S., Young, I. T., and Eden, M. (1978). *Science* **202**, 1158–1163.
Berg, B. B. (1967). In "Pathology of Laboratory Rats and Mice" (E. Cotchin and F. J. C. Roe, eds.), pp. 749–786. Davis, Philadelphia, Pennsylvania.
Bhargava, G., Rifas, L., and Makman, M. H. (1979). *J. Cell Physiol.* **100**, 365–374.
Blomquist, E., Westermark, B., and Ponten, J. (1980). *Mech. Ageing Dev.* **12**, 173–182.
Bolanowski, M. A., Russell, R. L., and Jacobson, L. A. (1981). *Mech. Ageing Dev.* **15**, 279–295.
Boorman, G. A., and Hollander, C. F. (1973). *J. Gerontol.* **28**, 152–159.
Burcombe, J. V. (1972). *Mech. Ageing Dev.* **1**, 213–225.
Burmer, G. C., and Norwood, T. H. (1980). *Mech. Ageing Dev.* **12**, 151–159.
Carrel, A. (1912). *J. Exp. Med.* **15**, 516–528.
Cohen, B. J. (1968). In "The Laboratory Animal in Gerontological Research" (T. W. Harris, ed.), Nat. Acad. Sci., Publication 1591, Washington, D.C.
Cohen, B. J. (1978). In "Development of the Rodent as a Model System of Aging" (D. C. Gibson, R. C. Adelman and C. Finch, eds.), Book II, pp. 135–137. U.S. Department of Health, Education and Welfare. Publ No. 79-161.
Coleman, G. L., Barthold, S. W., Osbaldiston, G. W., Foster, S. J., and Jonas, A. M. (1977). *J. Gerontol.* **32**, 258–278.
Cristofalo, V. J. (1970). In "Aging in Cell and Tissue Culture" (E. Holeckova and V. J. Cristofalo, eds.), pp. 83–119. Plenum, New York.
Cristofalo, V. J. (1972). *Adv. Gerontol. Res.* **4**, 45–79.
Cristofalo, V. J. (1976). *Gerontology* **22**, 9–27.
Cristofalo, V. J. (1980). In "Aging, Cancer and Cell Membranes" (C. Borek, C. M. Fenoglio and D. W. King, eds.), pp. 100–114. Thieme, Stuttgart.
Cristofalo, V. J., and Kabakjian, J. R. (1975). *Mech. Ageing Dev.* **4**, 19–28.
Cristofalo, V. J., and Rosner, B. A. (1979). *Fed. Proc. Fed. Am. Soc. Exp. Biol.* **38**, 1851–1856.
Cristofalo, V. J., and Sharf, B. B. (1973). *Exp. Cell Res.* **76**, 419–427.
Cristofalo, V. J., Wallace, J. M., and Rosner, B. A. (1979). *Cold Spring Harbor Conf. Cell Proliferation* **6**, 875–887.
Croce, C. M., and Koprowski, H. (1974). *Science* **184**, 1288–1289.
Croll, N. A., Smith, J. R., and Zuckerman, B. M. (1977). *Exp. Ageing Res.* **3**, 175–189.

Daniel, C. W. (1972). *Adv. Gerontol. Res.* **4,** 167–199.

Davis, L. C., and Pfeifer, W. D. (1973). *Biochem. Bophys. Res. Commun.* **54,** 726–713.

Dell'Orco, R. T., Mertens, J. G., and Kruse, P. F., Jr. (1973). *Exp. Cell Res.* **77,** 356–360.

Driver, C. J. I., and Cosopodiotis, G. (1979). *Exp. Gerontol.* **14,** 95–100.

Durand, A. M., Fisher, M., and Adams, M. (1964). *Arch. Pathol.* **77,** 268–277.

Epstein, J., and Gershon, D. (1972), *Mech. Ageing Dev.* **1,** 257–264.

Evans, C. H. (1979). *Med. Hypotheses* **5,** 53–66.

Fox, R. R. (1980). *Exp. Aging Res.* **3,** 235–248.

Franks, L. M. (1980). *Methods Cell Biol.* **21,** 153–169.

Franks, L. M. (1982). *In* "Techniques in Cellular Physiology", Elsevier, Amsterdam. (*in press.*)

Franks, L. M., and Cooper, T. W. (1972). *Int. J. Cancer* **9,** 19–29.

Franks, L. M., and Wilson, P. D. (1977). *Int. Rev. Cytol.* **48,** 55–139.

Gandhi, S., Santelli, J., Mitchell, D. H., Stiles, J. W., and Sanadi, D. R. (1980). *Mech. Ageing Dev.* **12,** 137–150

Ganietzky, B., and Flanagan, J. R. (1978). *Exp. Gerontol.* **13,** 189–196.

Gershon, D. (1970). *Exp. Gerontol.* **5,** 7–12.

Gershon, H., and Gershon, D. (1970). *Nature (London)* **227,** 1214–1217.

Gleason, R. E., and Goldstein, S. (1978) *Science* **202,** 1217–1218.

Goldstein, S., and Harley, C. B. (1979). *Fed. Proc. Fed. Am. Soc. Exp. Biol.* **38,** 1862–1867.

Goldstein, S., Moerman, E. J., Soeldner, J. S., Gleason, R. E., and Barnett, D. M. (1978). *Science* **199,** 781–782.

Good, P. I., and Smith, J. R. (1974). *Biophys. J.* **14,** 811–822.

Goodman, D. G., Ward, J. M., Squire, R. A., Chu, K. C., and Linhart, M. S. (1979). *Toxicol. Appl. Pharmacol.* **48,** 237–248.

Goodman, D. G., Ward, J. M., Squire, R. A., Paxton, M. B., Reichardt, W. D., Chu, K. C., and Linhart, M. S. (1980). *Toxicol. Appl. Pharmacol.* **55,** 433–447.

Goodrick, C. L. (1975). *J. Gerontol.* **30,** 257–263.

Goodrick, C. L. (1977). *Gerontology* **23,** 405–413.

Goodrick, C. L. (1978). *J. Gerontol.* **33,** 184–190.

Hadley, E. C., Kress, E. D., and Cristofalo, V. J. (1979). *Exp. Gerontol.* **34,** 170–176.

Harley, C. B., and Goldstein, S. (1978). *J. Cell Physiol.* **97,** 509–516.

Harley, C. B., and Goldstein, S. (1980). *Science* **207,** 191–193.

Harper, R. A., and Grove, G. (1979). *Science* **204,** 526–527.

Harrison, D. E. (1975). *J. Gerontol.* **30,** 279–285.

Harrison, D. E. (1979a). *Mech. Ageing Dev.* **9,** 427–433.

Harrison, D. E. (1979b). *Mech. Ageing Dev.* **9,** 409–426.

Hatcher, V. B., Werthein, M. S., Rhee, C. Y., Tsien, G., and Burk, P. B. (1976). *Biochem. Biophys. Acta* **451,** 499–510.

Hayflick, L. (1965). *Exp. Cell Res.* **37,** 614–635.

Hayflick, L. (1977). *In* "The Biology of Aging" (C. E. Finch, L. Hayflick, eds.), pp. 159–186. Van Nostrand Reinhold, Princeton, New Jersey.

Hayflick, L. (1979). *J. Invest. Dermatol.* **73,** 8–14.

Hayflick, L. (1980). *Mech. Ageing Dev.* **14,** 59–79.

Hayflick, L. and Moorhead, P. S. (1961). *Exp. Cell Res.* **25,** 585–621.

Hegner, D., Platt, D., Hacker, H., Schloeder, V., and Breuninger, V. (1979). *Mech. Ageing Dev.* **10,** 117–130.

Hieb, W. F., and Rothstein, M. (1975). *Exp. Gerontol.* **10,** 145–153.

Hoffman, H. J. (1978). *In* "Development of the Rodent as a Model System of Aging" (Gibson, D. D., Adelman, R. C., and Finch, C., eds.), Book II, pp. 19–34. U.S. Department of Health, Education and Welfare. Publ No. 79-161.

Hollander, C. F. (1976). *Lab. Animal Sci.* **26,** 320–328.
Holliday, R., Huschtscha, L. I., Tarrant, and Kirkwood, T. B. L. (1977). *Science* **198,** 366–372.
Homburger, F., Russfield, A. B., Weisburger, J. H., Lim, S., Chak, S. P., and Weisburger, E. K. (1975). *J. Natl. Cancer Inst.* **55,** 37–45.
Honda, S., and Matsuo, M. (1980). *Mech. Ageing Dev.* **12,** 31–37.
Hosono, R. (1978). *Exp. Gerontol.* **13,** 369–374.
Hosono, R., Sato, Y., Aizawa, S. I., and Mitsui, Y. (1980). *Exp. Gerontol.* **15,** 285–289.
Hughes, R. J., and Ayad, S. J. (1980). *Biochem. Biophys. Acta* **623,** 202–209.
Kadlubowski, M., and Agutter, P. S. (1977). *Brit. J. Haematol.* **37,** 111–125.
Kaji, K., and Matsuo, M. (1980). *Mech. Ageing Dev.* **13,** 219–225.
Kisiel, M. J., and Zuckerman, B. M. (1978). *Age* **1,** 17–20.
Kisiel, M., Nelson, B., and Zuckerman, B. M. (1972). *Nematologica* **18,** 373–384.
Kisiel, M. J., Castillo, J. M., Zuckerman, L. S., and Zuckerman, B. M. (1975). *Mech. Ageing Dev.* **4,** 81–88.
Klass, M. R. (1977). *Mech. Ageing Dev.* **6,** 413–429.
Klass, M. R., and Smith-Sonneborn, J. (1976). *Exp. Cell Res.* **98,** 63–72.
Knook, D. L. (1980). *Proc. Soc. Exp. Biol. Med.* **165,** 170–177.
Kontermann, K., and Bayreuther, K. (1979). *Gerontology* **25,** 261–274.
Krohn, P. L. (1966). *In* "Topics of the Biology of Aging" (P. L. Krohn, ed.), p. 125. Wiley, New York.
Kunstyr, I., and Levenberger, H. G. W. (1975). *J. Gerontol.* **30,** 157–162.
Landaw, S. A. (1976). *In* "Special Review of Experimental Aging Research (M. F. Elias, B. E. Eleftheriou and P. K. Elias, eds.), pp. 304–328. EAR, Maine.
LeGully, Y., Simon, M., Venoia, P., and Bourel, M. (1973). *Gerontologia* **19,** 303–313.
Lin, F. K., and Chang, S. H. (1979). *Mech. Ageing Dev.* **11,** 383–392.
Macieira-Coelho, A. (1966). *Experentia* **22,** 390–391.
Macieira-Coelho, A. (1974). *Nature (London)* **248,** 421–422.
Macieira-Coelho, A. (1976). *Gerontology* **22,** 3–8.
Macieira-Coelho, A., and Lima, L. (1973). *Mech. Ageing Dev.* **2,** 13–18.
Macieira-Coelho, A., and Loria, E. (1974). *Nature (London)* **251,** 67–69.
Martin, G. M. (1977). *Am. J. Pathol.* **89,** 484–511.
Martin, G. M., Sprague, C. A., and Epstein, C. J. (1970). *Lab. Invest.* **23,** 86–92.
Martin, G. M., Sprague, C. A., Norwood, T. H., and Pendergrass, W. R. (1974). *Am. J. Pathol.* **74,** 137–154.
Martin, G. M., Sprague, C. A., Norwood, T. H., Pendergrass, W. R., Bornstein, P., Hoehn, H., and Arend, W. P. (1975). *Adv. Exp. Med. Biol.* **53,** 67–90.
Masoro, E. J. (1980a). *Exp. Aging Res.* **6,** 219–233.
Masoro, E. J. (1980b). *Exp. Aging Res.* **6,** 261–270.
Massie, H. R., and Williams, T. R. (1979). *Exp. Gerontol.* **14,** 109–115.
Massie, H. R., Aiello, V. R., and Williams, T. R. (1980). *Mech. Ageing Dev.* **12,** 279–286.
Matsumura, T., Zerrudo, Z., and Hayflick, L. (1979a). *J. Gerontol.* **34,** 328–334.
Matsumura, T., Pfend, E. A., and Hayflick, L. (1979b). *J. Gerontol.* **34,** 323–327.
Meek, R. L., Bowman, P. D., and Daniel, C. W. (1977). *Exp. Cell Res.* **107,** 277–284.
Meek, R. L., Bowman, P. D., and Daniel, C. W. (1980). *Exp. Cell Res.* **127,** 127–132.
Merz, G. A., and Ross, J. D. (1969). *J. Cell Physiol.* **82,** 75–81.
Mills, B. J., and Lang, C. A. (1976). *Biochem. J.* **154,** 481–490.
Miquel, J., Lundgren, P. R., and Bensch, K. G. (1975). *Mech. Ageing Dev.* **4,** 41–57.
Mitchell, D. H., Stiles, J. W., Santelli, J., and Sanadi, D. R. (1979). *J. Gerontol.* **34,** 28–36.
Muggleton-Harris, A. L. (1979). *Int. Rev. Cytol.* Supplement **9,** 279–301.

Muggleton-Harris, A. L., and DeSimone, D. W. (1980). *Somatic Cell Genet.* **6,** 689–698.
Muggleton-Harris, A. L., and Hayflick, L. (1976). *Exp. Cell Res.* **103,** 321–330.
Muggleton-Harris, A. L., and Palumbo, M. (1979). *Somatic Cell Genet.* **5,** 397–407.
Muller, I., Zimmerman, M., Becker, D., and Flomer, M. (1980). *Mech. Ageing Dev.* **12,** 47–52.
Murota, S. I., Mitsui, Y., and Kawamura, M. (1979). *Biochem. Biophys. Acta* **574,** 351–355.
Nichols, W. W. (1975). *Hereditas* **81,** 225–236.
Nichols, W. W., and Murphy, D. G. (1979). *Int. Rev. Cyt.,* Supp. **10.**
Norwood, T. H., Pendergrass, W., and Martin, G. M. (1974). *Proc. Natl. Acad. Sci. U.S.A.* **64,** 551–556.
Norwood, T. H., Pendergrass, W. R., and Martin, G. M. (1975). *J. Cell Biol.* **64,** 551–556.
Norwood, T. H., Rabinovitch, P. S., and Ziegler, C. J. (1979). *Fed. Proc. Fed. Am. Soc. Exp. Biol.* **38,** 1868–1872.
Ohno, T. (1979). *Mech. Ageing Dev.* **11,** 179–183.
Paget, G, F., and Lemon, P. G. (1965). *In* "Pathology of Laboratory Animals" (W. E. Ribelin and J. B. McCoy, eds.), pp. 382–405. Thomas, Springfield, Illinois.
Platt, D., and Norwig, P. (1980). *Mech. Ageing Dev.* **14,** 119–126.
Reis, R. J. S., Goldstein, S., and Harvey, C. B. (1980). *Mech. Ageing Dev.* **13,** 393–395.
Robbins, E., Levine, E. M., and Eagle, H. (1970). *J. Exp. Med* **131, 1211** 1222.
Rodermel, S R., and Smith-Sonneborn, J. (1977). *Genetics* **87,** 259–274.
Rosner, B. A., and Cristofal, V. J. (1979). *Mech. Ageing Dev.* **9,** 485–496.
Rothstein, M. (1980). *In* "Nematodes as Model Biological Systems" (B. M. Zuckerman, ed.), pp. 29–46. Academic Press, New York.
Rothstein, M, and Coppens, M. (1978). *Comp. Biochem. Biophys.* **61B,** 99–104.
Rothstein, M., and Sharma, H. K. (1978). *Mech. Ageing Dev* **8,** 175 180.
Rowlatt, C, Chesterman, F. C., and Sheriff, M. V. (1976). *Lab. Anim.* **10,** 419–442.
Russell, E. S. (1966). *In* "Biology of the Laboratory Mouse" (E. L. Green, ed.), pp. 511–519. McGraw-Hill, New York.
Russell, E. S. (1979). *In* "Development of the Rodent as a Model System of Aging" (D. C Gibson, R. C. Adelman and C. Finch, eds.), pp. 37–44. Dept. Health, Education and Welfare Publ No. 79–161, Washington, D.C.
Ryan, J. M., and Cristofalo, V. J. (1979). *Exp. Cell Res.* **122,** 179–184.
Schneider, E. L. (1979). *Fed. Proc. Fed. Am. Soc. Exp. Biol.* **38,** 1857–1861.
Schneider, E. L., and Chase, G. A. (1976). *Interdiscip. Top. Gerontol.* **10,** 62–69.
Schneider, E. L., and Mitsui, Y. (1976). *Proc. Natl. Acad. Sci. U.S.A.* **73,** 3584–3588.
Schneider, E. L., and Smith, J. R. (1981). *Int. Rev. Cytol.* **69,** 261–270.
Schneider, E. L., Kram, D., Nakanishi, Y., Monticone, R. E., Tice, R. R., Gilman, B. A., and Nieder, M. L. (1979a). *Mech. Ageing Dev.* **9,** 303–311.
Schneider, E. L., Sternberg, H., Tice, R. R., Senula, G. C., Kram, D., Smith, J. R., and Bynum, G. (1979b). *Mech. Ageing Dev.* **9,** 313–324.
Seaman, C., Wyss, S., and Piomelli, S. (1980). *Am. J. Hematol.* **8,** 31–42.
Sharma, H. K., and Rothstein, M. (1978a). *Arch. Biochem. Biophs.* **174,** 324–332.
Sharma, H. K., and Rothstein, M. (1978b). *Biochemistry* **17,** 2869–2876.
Sharma, H. K., and Rothstein, M. (1980). *Proc. Natl. Acad. Sci. U.S.A.* **77,** 5865–5868.
Sharma, H. K., Prasanna, H. R., Lane, R. S., and Rothstein, M. (1979). *Arch. Biochem. Biophys.* **194,** 275–282.
Sims, H. S. (1967). *In* "Pathology of Laboratory Rats and Mice" (E. Cotchin and F. J. C. Roe, eds.), pp. 733–747. Davis, Philadelphia, Pennsylvania.
Smith, J. R., and Hayflick, L. (1974). *J. Cell Biol.* **62,** 48–53.
Smith, J. R., and Lumpkin, C. K. Jr. (1980). *Mech. Ageing Dev.* **13,** 387–392.

Smith, J. R., and Whitney, R. G. (1980). *Science* **207**, 82–84.

Smith, J. R., Pereira-Smith, O. M., and Schneider, E. L. (1978). *Proc. Natl. Acad. Sci. U.S.A.* **75**, 1353–1356.

Smith, J. R., Pereira-Smith, O. M., Braunschweiger, K. I., Roberts, T. W., and Whitney, R. G. (1980). *Mech. Ageing Dev.* **12**, 355–365.

Smith-Sonneborn, J. (1979). *Science* **203**, 1115–1117.

Smith-Sonneborn, J. (1981). *Int. Rev. Cyt.* **73**, 319–354.

Smith-Sonneborn, J., and Reed, J. C. (1976). *J. Gerontol.* **31**, 2–7.

Smith-Sonneborn, J., and Rodermel, S. R. (1976). *J. Cell Biol.* **71**, 575–588.

Sohal, R. S. (1976). *Gerontology* **22**, 317–333.

Sonneborn, T. M. (1954). *J. Protozoal.* **1**, 38–53.

Stanbridge, E. J. (1976). *Nature (London)* **260**, 17–20.

Stanley, J. F., Pye, D., and MacGregor, A. (1975). *Nature (London)* **255**, 158–159.

Storer, J. B. (1966). *J. Gerontol.* **21**, 404–409.

Sun, A. S., Alvarez, L. J., Reinach, P. S., and Rubin, E. (1979). *Lab. Invest.* **41**, 1–4.

Sundararaman, V., and Cummings, D. J. (1976a). *Mech. Ageing Dev.* **5**, 139–154.

Sundararaman, V., and Cummings, D. J. (1976b). *Mech. Ageing Dev.* **5**, 325–338.

Sundararaman, V., and Cummings, D. J. (1977). *Mech. Ageing Dev.* **6**, 393–406.

Swim, H. E., and Parker, R. F. (1957). *Am. J. Hyg.* **66**, 235–243.

Takagi, Y., and Yoshida, M. (1980). *J. Cell Sci.* **41**, 177–191.

Tanaka, K., Nakazawa, T., Okada, Y., and Kumahara, Y. (1980). *Exp. Cell Res.* **127**, 185–190.

Vincent, R. A. Jr., and Huang, P. C. (1976). *Exp. Cell Res.* **102**, 31–42.

Vincent, A. L., Rodrick, G. E., and Sodeman, W. A. (1980). *Exp. Aging Res.* **6**, 249–260.

Vorbrodt, A., Charpentier, R., and Cristofalo, V. J. (1979). *Mech. Ageing Dev.* **11**, 113–125.

Walter, H., Krob, E. J., Tamblyn, C. H., and Seaman, G. V. F. (1980). *Biochem. Biophys. Res. Commun.* **97**, 107–113.

Williams, T. J., and Smith-Sonneborn, J. (1980). *Exp. Gerontol.* **15**, 353–357.

Wright, W. E., and Hayflick, L. (1975a). *Exp. Cell Res.* **96**, 113–121.

Wright, W. E., and Hayflick, L. (1975b). *Fed. Proc. Fed. Am. Soc. Exp. Biol.* **34**, 76–79.

Yanishevsky, R. M., and Stein, G. H. (1980). *Exp. Cell Res.* **126**, 469–472.

Young, L. J. T., Medina, D., DeOme, K. B., and Daniel, C. W. (1971). *Exp. Gerontol.* **6**, 49–56.

Chapter **3**

Free Radicals, Age Pigments, and Lipid Metabolism

I. OVERVIEW

The idea that cellular aging is related to damage of biological structures caused by free radicals holds a number of attractions. In particular, free radical attack can be related to several concepts that have been put forth as causes of aging in their own right. For example, to the extent that it exceeds repair capability, cross-linking resulting from free radical reactions could cause irreversible and cumulative damage to the genetic apparatus, leading to "errors" which might become amplified, or to omissions in information molecules. More direct effects would be the damaging of membranes by peroxidation of phospholipids with a resultant loss of ability to control influx and efflux of metabolites.

The generation of free radicals quite obviously occurs in biological systems and can be brought about by ionizing radiation, UV radiation, or generation by certain enzymatic and nonenzymatic reactions. Biological systems are protected from these effects by a number of radical scavenging molecules (antioxidants such as vitamin E) and by enzyme systems designed to remove superoxide radicals (superoxide dismutase) or peroxides (catalase, peroxidase), which result from free radical reactions. However, these systems do not offer complete protection, as witnessed by the gradual accumulation of age pigment (lipofuscin). These pigments were associated with peroxidation of polyunsaturated fatty acids by the discovery that a product of lipid peroxidation, malonaldehyde, acts as a cross-linking agent with amino groups of proteins to yield a typical fluorescence spectrum.

Research into the effects of free radicals on aging has encompassed a wide range of approaches, from dietary increases in antioxidants (vitamin E is the most popular of these) to proving that free radicals are damaging to biological systems *in vivo.* Antioxidants have been studied both as to their effects on life-span and their ability to reduce the amount of age pigments formed by various tissues. They do little for the former but help with the latter process. A few studies have taken the opposite tack and employed high oxygen concentrations to see if the consequence is a shortened life-span and increased pigment production.

Searches to demonstrate the cross-linking of DNA in old animals have not provided much evidence that such a process is involved in aging. Changes with age of protective systems have been investigated, but outside of a few intriguing hints, these studies have not provided any unequivocal answers. It seems clear that free radicals do form in biological systems, that unsaturated fatty acids are split to form malonaldehyde, which is the cross-linking agent involved in the formation of age pigments, and that large doses of dietary antioxidants can function to reduce the amount of these materials, although extension of life-span is not affected, at least in mammals.

As to metabolism, most work involves the determination of changes in the levels of various lipids, particularly cholesterol, in tissues and in serum. There is a general consensus that these increase with age, although patterns of change vary with species and, it seems, with the experimenter. Investigation of enzymes or regulatory mechanisms has been too sparse to provide a satisfactory explanation for the putative increases in lipid concentration with age.

Taking the above into consideration, it hardly needs stating that much more basic information remains to be obtained before one can begin to put together even a tentative picture as to what effect lipid metabolism has on aging and vice versa.

II. FREE RADICAL FORMATION AND LIPID PEROXIDATION

Harman (1956) proposed that aging might be partly due to the deleterious effects of free radicals. The reactions of such free radicals with respect to polyunsaturation lipids can be outlined as follows.

1. Initiation

(metal or enzyme system)

$$LH + O_2 \quad \rightarrow \quad L\cdot \text{ (free radical)} + HOO\cdot \text{ (perhydroxyl radical)}$$

2. Propagation

$$L\cdot + O_2 \rightarrow LOO\cdot \text{ (lipid peroxy radical)}$$
$$LOO\cdot + LH \rightarrow L\cdot + LOOH \text{ (lipid hydroperoxide)}$$

3. Termination

$$L\cdot + L\cdot \rightarrow L{:}L$$

or

$$LOO\cdot + \text{antioxidant} \rightarrow LOOH + \text{oxidized antioxidant}$$

Oxygen may be converted to superoxide (O_2^-) or hydroxyl (OH·) radicals by the following sequence.

$$O_2 + e \rightarrow O_2^- \text{ (superoxide radical)}$$
$$2H^+ + O_2^- + O_2^- \rightarrow O_2 + H_2O_2 \text{ (hydrogen peroxide)}$$

superoxide

Fe^{3+} dismutase

$$O_2^- + H_2O_2 \rightarrow O_2 + OH^- + OH\cdot$$

(Haber–Weiss reaction)

The lipid radicals may undergo cross-linking reactions with other bio-compounds or decompose to a variety of products, including mal-onaldehyde, a very reactive compound which is involved in the formation of age pigments (Section III). It is cross-linking reactions such as these that have been postulated to bring about age-related damage.

Superoxide radicals are generated by a number of processes, both biological and chemical. The latter include autooxidation of hydro-qiunones, leukoflavins, catecholamines, thiols, reduced dyes, tetrahydropteridines, and ferredoxins. As for enzymes, they include xanthine oxidase, which is much used for experimental radical generation, aldehyde oxidase, dihydroorotic dehydrogenase, NADH and NADPH oxidase, certain flavoprotein dehydrogenases, and various oxygenases and hydroxylases (Fridovich, 1975). Moreover, mitochondria, chloroplasts, microsomes, and granulocytes during phagocytosis are biological sources of free radicals. Radiation, oxygen, ozone, peroxides, and autoxidation of lipids are nonliving sources and a nonenzymatic reaction which forms free radicals is catalyzed by a transition metal ion and ascorbic acid plus oxygen. The postulated and experimental effects of ascorbic acid in peroxidation may be a reflection of this reaction. There is a problem in determining the *in vivo* role of this system and, for that matter, the *in vivo* role of isolated biological systems, since experimental measurements disrupt the cellular structure, releasing catalysts and perhaps changing the original orderly arrangements of the lipids.

Superoxide radical (O_2^-), hydrogen peroxide, and hydroxyl radical (OH·) have all been detected in the various peroxidizing systems. The hydroxyl radical was shown to be formed during oxidation of NADPH by microsomes in the presence of ADP and Fe^{3+}. Pedersen and Aust (1972) demonstrated that the NADPH-dependent lipid peroxidation was catalyzed by NADPH-cytochrome c reductase. The system has been further detailed by Svingen et al., (1979). These authors provide evidence for initiation by an ADP-perferryl ion-catalyzed formation of lipid hydroperoxides. Propagation involves breakdown of the hydroperoxides, catalyzed by any of a number of iron-containing compounds, including cytochrome P-450. Thiobarbituric acid reacting material is formed in this step, along with additional lipid hydroperoxides. Thus, the finding of Baird (1980) that cytochrome P-450 is not necessary for lipid peroxidation by intact microsomal membranes is not contradictory. Baird (1980) also reported the presence of an NADPH-driven lipid peroxidation in rat hepatic nuclei and nuclear membranes.

Pfeifer and McCay (1972) provided early evidence for a lipid peroxidizing system in mitochondria, and Loschen et al. (1971) initially reported the production of O_2^- in mitochondrial membranes. In the latest of several reports, Turrens and Boveris (1980) located the production of this radical at the ubiquinone-cyt b area as well as at the NADH dehydrogenase site. Lippman (1980) has recently designed sensitive, site-specific chemiluminscent probes to measure free radical formation in cell organelles (also see Lippman, 1981, Lippman et al., 1981a and b).

It can be seen from this sampling that there have been many studies dealing with the generation of superoxide and hydroxyl radicals. Their possible effects on biological systems has also been the subject of much study. To mention only a few of the most recent of these, Lesko et al. (1980) pointed out that hydroxyl radical is considered to be responsible for 90% of the damage to DNA caused by ionizing radiation and that H_2O_2 causes strand breakage. The authors found that superoxide radical also caused strand scission. Oberley et al. (1980) discussed the possible relationship of superoxide and superoxide dismutase to cell differentiation, aging, and cancer. DelMaestro et al. (1980) considered the role of free radicals in tissue injury. With respect to aging, Tolmasoff et al. (1980) attempted to correlate the level of superoxide dismutase activity in two rodent and 12 primate species to maximum life-span potentials with negative results.

In spite of the undoubted existence of free radicals in biological systems, there remains considerable controversy over the role played in vivo by the various products. For example, Tien et al. (1981) do not believe that the hydroxyl radical plays a physiological role in lipid perox-idation. Cohen and Cederbaum (1980) feel that this radical is probably involved in the microsomal oxidation of alcohols and that its formation requires electron transport as a necessary part of the system. One may add to this situation reports of cytosolic factors that inhibit or enhance peroxidation (Wright *et al.*, 1981). One thing that remains clear is that peroxidative reactions exist in biological systems. The known buildup of age pigments, which involves malonaldehyde formation (Section III), and the fact that altering antioxidant intake can affect this process cer-tainly strengthens the idea of a continuous but presumably small leak-age of radicals into biosystems. The question becomes not whether tissues can form peroxides or free radicals, but whether or not they do so to an excess over the protective or repair capacity of the various systems designed to remedy the problem. One approach to aging studies is, therefore, to see if these protective systems fail with increasing age.

A. Protective Mechanisms

There are a number of mechanisms involved in protection of tissues against the action of free radicals. Certain enzymes, the presence of antioxidants, and geographical or structural arrangements all play important roles. With respect to aging, the question is whether or not any of these systems become less effective with age, thus permitting increasing damage to cellular components. Alternatively, there may be a steady rate of damage that simply accumulates with time, but does not accelerate with age. In either case, the cells would become burdened, and secondary effects might come into play.

One of the best-studied enzymes concerned with free radicals is su-peroxide dismutase, which was first described by McCord and Fridovich (1969). The enzyme has since been found, in one form or another, to be ubiquitous in tissues of aerobic (but not anaerobic) organisms. It cata-lyzes the following reaction.

$$O_2^- + O_2^- + 2H + \rightarrow H_2O_2 + O_2$$

The reaction goes spontaneously so that steady state concentration of O_2^- is very low. The enzyme is extremely active even at low levels of the radical.

As mentioned above, if excess O_2^- is present, it will lead to formation of hydroxyl radical by the Haber-Weiss reaction.

$$O_2^- + H_2O_2 \rightarrow OH^- + OH\cdot + O_2$$

The hydroxyl radical, which can be detected in various peroxidizing preparations, is an extremely potent agent. To prevent its formation, either O_2^- or H_2O_2 must be kept to minimal amounts. Of course, even if superoxide dismutase is functioning efficiently, hydrogen peroxide will be formed. This product is scavenged by catalases and peroxidases.

$$2H_2O_2 \rightarrow 2H_2O + O_2 \text{ (catalase)}$$
$$H_2O_2 + H_2R \rightarrow 2H_2O + R \text{ (peroxidase)}$$

Catalase is widely distributed in cells, as is glutathione peroxidase. The latter not only works on hydrogen peroxide, but on lipid hydroperoxides, although which of these reactions is more important is, at present, not resolved. Thus, glutathione peroxidase may protect fatty acids before or after they become peroxidized (see below). Selenium, which for years was a mysterious nutritional requirement involved in some way with an antioxidant function related to Vitamin E, turns out to be an essential part of the enzyme. Dietary deficiency of the metal, therefore, has a deleterious role in protection against oxidative damage.

Chio and Tappel (1972) early proposed a role for the level of glutathione (GSH) peroxidase in protection from lipid peroxide formation. They exposed rats to ozone and found increased levels of GSH peroxidase, GSH reductase, and glucose-6-phosphate dehydrogenase. Glutathione peroxidase is believed to reduce lipid peroxides to harmless hydroxy fatty acids (Christophersen, 1969). However, McCay *et al.* (1976) could detect no hydroxy fatty acids in membranes of rat liver microsomes undergoing peroxidation. Moreover, partially purified glutathione peroxidase alone did not prevent lipid peroxidation, as measured by malonaldehyde formation, but required a soluble, heat-labile factor (McCay *et al.*, 1981). The authors believe that glutathione peroxidase defends against free radical attack on the membranes by metabolizing even low levels of H_2O_2. Tappel (1980), referring to current experiments in his laboratory, described the existence of a hydroperoxide phospholipase which might remove peroxidized lipids and thus make them available to the GSH peroxidase system.

Though details of the mechanism of attack and protection are still the subject of speculation, one can conclude that free radicals exist in biological systems and that the enzymes superoxide dismutase, catalase, and GSH peroxidase are present for protective purposes. Further protective roles have been proposed for antioxidants, which include vitamin E, vitamin C, and reduced glutathione. There is considerable experimental work in this respect, particularly dealing with vitamin E (Section III). Figure 3.1 summarizes the arrangement of protective mechanisms for mitochondrially generated O_2^- .

Defense mechanisms	Location
Manganese superoxide dismutase	Mitochondrial matrix
α-Tocopherol, membrane incorporated	Mitochondrial inner membrane
Cupro-zinc superoxide dismutase	Cytoplasm and mitochondrial inner membrane space
Glutathione peroxidase and catalase	Cytoplasm, etc.
Mercaptoamino acids, quinones, glutathione, low concentrations of ascorbic acid and tocopherols in free solution	Serum, tissues, and cytoplasm

Fig. 3.1. Protective mechanisms against superoxide radical. From Lippman (1982) with permission of Raven Press.

In considering free radical generation, one should bear in mind that *in vitro* preparations may be altered in properties and may no longer be acting upon normal substrates or may have lost contact with free radical scavengers. For example, Mead (1976) points out that the cytochrome *P*-450 system which is involved in hydroxylation of many compounds, normally utilizes NADPH, Fe^{2+}, and oxygen. In normal operation, free radicals are neither formed or released by the enzyme complex beause it is a two-electron process. In the case of tissue disruption, the system, not having normal substrates available, may attack other components.

With the above background in mind, what are the effects of aging on the interplay of free radical attack and its prevention? As with many other areas under investigation, the results so far are more exploratory than conclusive. Cross-linking of DNA, an often postulated event, would have dramatic effects on organisms. However, there is no clear evidence that this reaction is a concomitant of aging (Chapter 6, Section VII). As to the effect of aging on radical and peroxide formation, Player *et al.* (1977) reported that the heat-sensitive NADPH-dependent lipid peroxidation system increases substantially with age in rat liver microsomal and mitochondrial fractions. Both mitochondrial and microsomal preparations showed an age-related increase in the production of malonaldehyde at 6 versus 26 months of age (fetal and neonatal preparations were also studied) after shaking at 25°C for 30 minutes. There was, however, no significant difference in the activity of NADPH–cytochrome *c* reductase, an enzyme that has been shown to be involved in generating free radicals. As the authors point out, changes in the lipid composition of the young versus old membranes could have a considerable effect on the production of malonaldehyde. Indeed, Grinna (1977a) reported increases in decosahexenoic acid in the polar lipids of microsomal and mitochondrial fractions from old (24 months) compared to young (6 months) rats. Others have reported a lowered degree of unsaturation with age (Chapter 4). Player *et al.* (1977) also pointed out that the levels of superoxide dismutase and glutathione peroxidase may be affected by use of isolated cell fractions.

Barrett and Horton (1975) found that in freshly prepared rat liver mitochondria from 6- to 24-month-old animals, the amount of thiobarbituric acid reacting material (malonaldehyde) actually decreased with age, the levels being small in all cases (0.191 nmol malonaldehyde/mg mitochondrial protein). On the other hand, the potential for peroxidation increased about 2.3-fold with age, as shown when preparations were shaken in air for 90 minutes. This result may be due to an age-related increase in the double bond content of the old membranes in the mitochondrial lipids. Grinna and Barber (1973) examined lipid peroxidation in aerobically incubated homogenates and microsomal preparations of rat liver and kidney. They measured color formation with thiobarbituric acid and found much lower values in old preparations (24 months) compared to young ones (6 months). Addition of ascorbic acid raised all values. The authors concluded that there was a soluble inhibitor present in the old tissues (not α-tocopherol) and a lowered amount of ascorbic acid. There may be a technical problem in that malonaldehyde can be metabolized by mitochondria. If this activity shows an age-related differential, the results will be in error.

Recently, there have been studies dealing with the effect of age on free radical production in rat heart mitochondria (Chapter 5, Section II). Nohl and Hegner (1978a) obtained evidence for free radical production above the protective enzyme level and found an age-related (3 versus 23 months) increase in superoxide radical production. Nohl (1979) and Nohl and Kramer (1980) studying the inner mitochondrial membrane reported a decrease in the ratio of unsaturated to saturated acids with age, which is presumed to be due to peroxidation losses. Nohl *et al.* (1979) found that in heart mitochondria of 3- versus 24-month-old rats, superoxide dismutase activity was unchanged, but catalase activity increased by nearly 60% and GSH peroxidase by 29% in the older animals. Glutathione was unchanged. Evidence for the presence of these three enzymes in the mitochondrial matrix was reported by Nohl and Hegner (1978b). The increases were attributed to a response in the increased level of H_2O_2 and lipid peroxides observed in old mitochondria (Nohl and Hegner, 1978a). The lack of change in mitochondrial superoxide dismutase activity agrees with the report of Kellogg and Fridovich (1976) that the activity of the rat liver enzyme shows little change between 13 and 26 months. On the other hand, Reiss and Gershon (1976) reported that cytoplasmic superoxide dismutase in rat and mouse liver decreases based on activity per milligram of protein.

Not only the protective enzymes, but the levels of protective antioxidants (see Section IV) are presumed to be involved in preventing peroxidative attack on membranes. Glutathione, especially in conjunction with GSH peroxidase probably plays an important protective role. Stohs *et al.* (1980) reported that there was a statistically significant reduction in the levels of glutathione in blood, kidney, and intestinal mucosa (μg/micrograms per milligram protein) of mice between 18 and 24 months of age. However, the values were essentially unchanged in liver. Hazelton and Lang (1980) reported that in old mice (31 months), GSH in liver, kidney, and heart was 30, 34, and 20% lower, respectively, compared to the levels in the tissues of mature animals (17–23 months). The sharpest decrease appears to occur after about 23 months of age, which might explain why Stohs *et al.* (1980) found relatively small differences. The oxidized form (GSSG) was present at a low level and did not vary with age. Nohl *et al.* (1979) found no change in GSH in rat heart mitochondria at 24 months of age. Abraham *et al.* (1978) had previously shown a decrease in GSH both in old erythrocytes and in erythrocytes from old mice. Hazelton and Lang (1980) proposed that there is a loss of reducing potential with senescence. Support for this idea was provided by Hughes *et al.* (1980), who observed that the total number of SH groups on the outside of rat adipocytes decreased with age.

B. Peroxidative Changes in Lipid Composition

Lipid composition of soluble components or cellular structures may alter with age as a response to altered functional needs or as a result of a deleterious aging process. Obviously, it is not easy to distinguish one process from the other. Changes resulting from lipid peroxidation would certainly seem to be more related to the latter, as they tend to be external events, that is, not enzyme mediated. If the level of lipid peroxidation increases above the protective ability in old animals, one should see a lowering of polyunsaturated fatty acids, particularly in membranes. However, there are a number of ways in which such a result would be avoided. First, membranes turn over so that damaged areas would be replaced. Secondly, the accumulation of age pigments, a presumed marker of peroxidation, seems to be linear with age rather than an accelerating process. On the other hand, Nohl and Hegner (1978a) obtained evidence that the inner membrane of rat heart mitochondria has an increased free radical output in old organisms (Chapter 5, Section

II). In addition, Sagai and Ichinose (1980) observed an increased respiratory output of hydrocarbons (representing lipid peroxidation) in aged rats (Section IV,A). Nonetheless, it is obvious that at least some degree of peroxidative damage is contained by the formation and accumulation of pigment granules, a process which can be viewed as a deletion of damaged membranes. If cross-linked membranes are removed, they are presumably replaced by normal structures—a variation of turnover. One would expect that the ability to replace unsaturated fatty acids would be much greater than the very slow loss due to pigment formation. Thus, a dramatic reduction in total polyunsaturated fatty acids in membranes of old animals should not be expected, nor is it universally found, although compositional differences have been noted. For example, Thomas *et al.* (1978) found no age-related pattern of differences in fatty acids in the membrane fraction of muscle tissue from healthy human volunteers 20–73 years old. Grinna (1977a) examined microsomal and mitochondrial membranes from liver and kidney of rats, 6 versus 24 months of age. There was a decrease in linoleic acid and an increase in docosahexenoic acid with age in liver preparations. In kidney, oleic acid increased with age. There were also changes in neutral lipid composition. The author states that other experiments carried out at other ages between 6 and 24 months gave results consistent with the data presented. Still, without data from intervening and older ages, it is difficult to judge whether these changes are truly related to senescence or whether they are simply differences found at the specified ages. For example, if one were to select results from two ages in the table of fatty acids presented by Thomas *et al.* (1978), differences could certainly be noted, but these do not show an overall, age-related trend. Further discussion of age-related changes in membrane lipids will be found in Chapter 4.

C. Comment

Much of the early concern about free radicals has been directed toward lipid peroxidation and the resulting formation of malonaldehyde and thence age pigment. However, it is important to remember that free radicals have the potential for initiating other reactions including additions, scissions, cross-linking, and aromatic hydroxylations. These reactions, in turn could result in many deleterious processes including mutagenic and carcinogenic effects, transcription errors, DNA damage, and protein inactivation. In general, although a few searches for cross-linked products have been made, particularly in DNA (Chapter 6, Section VII), there is little unequivocal evidence that shows that these reactions are major events in the aging process. However, the idea of a small but steady burden on the tissues cannot be excluded. In fact, the idea of a relationship between free radicals and aging is one of the few examples of a theory of aging that has not only survived but seems to be gaining ground.

III. AGE PIGMENTS

Since before 1900 it has been known that with age, fluorescent pigments accumulate in cells. In fact, the material, known generally as lipofuscin (there are other pigments such as ceroid), is widespread, having been observed in the form of yellow, autofluorescent, membrane-bound granules. The pigments have been found in vertebrates and invertebrates alike. Although many studies have involved postmitotic cells, particularly nerve cells, age pigments have been identified in practically all tissues. They are not evenly distributed in all cells and they do not accumulate at the same rate in all tissues. In fact, pigment may accumulate differently in different cell types, even in the same organ. The material may be aggregated in a few clumps or may show a scattered distribution.

For the biochemist, it hardly seems necessary to review the large mass of histological and morphological information that deals with lipofuscin. A few general references should be adequate, outside of specific aspects of lipofuscin metabolism or structure which fall under a more narrow interpretation of "biochemical" considerations. For recent general references, see Bourne, 1973; Brizzee *et al.*, 1975; Miquel *et al.*, 1977; Siakotos and Koppang, 1973; Siakotos and Armstrong, 1975; Aune, 1976.

The pigment granules are believed by many investigators to derive from subcellular organelles as a result of cross-linking of membranes. The cross-linking is thought to derive from peroxidation of lipids (Section II). Miquel *et al.* (1978) provided ultrastructural evidence for mitochondrial involvement in pigment formation in mouse testis. Other authors have also considered mitochondria as a source of lipofuscin (Gopinath and Glees, 1974; Spoerri and Glees, 1974; Travis and Travis, 1972), although this conclusion is not universally agreed upon. Cross-linked membranes and other cross-linked products, e.g., proteins, are thought to be encompassed by lysosomes. Since the cross-linked products are not readily susceptible to the action of proteolytic enzymes, they accumulate with time. In short, the granules appear to represent secondary lysosomes in which the products constituting lipofuscin have accumulated.

The amount of lipofuscin that accumulates can be quite significant.

Malkoff and Strehler (1963) calculate it to be 6% of the total intracellular volume for cardiac muscle in a centenarian. In old *Drosophila*, it may account for 50% of the cytoplasmic volume in digestive cells (Miquel *et al.*, 1974).

Although the accumulation of pigment with age has been observed histologically for many years, the relatively recent development of chemical procedures for quantitative measurement (see below) made it easier for investigators to examine the effect of various agents or experimental conditions on rate of formation. Such measurements have also permitted a more sensitive means for comparing the amounts of the material in young and old animals.

The long recognized accumulation of age pigments was brought together with lipid peroxidation by Tappel and co-workers so that the structural relationship became clear. Among the products of peroxidation of polyunsaturated fatty acids is malonadlehyde, which is formed from the three carbon atoms between the allylic double bonds: $-CH=CH-CH_2-CH=CH- \rightarrow OHC-CH_2-CHO$. Chio and Tappel (1969a) reacted leucine, valine, their ethyl esters, and hexylamine, respectively, with malonaldehyde to form Schiff bases, which then yield compounds with the structure $RNHCH=CH-CH=NR$, the internal carbons deriving from the malonaldehyde. The absorption and fluorescence spectra of the compounds were similar to those reported for lipofuscin. The authors then demonstrated that enzymes could be inactivated in the presence of peroxidizing lipids or by direct reaction with malonaldehyde (Chio and Tappel, 1969b). They incubated a number of enzymes in the presence of methyl linolenate and showed that they became inactivated as a linear function of oxygen uptake. Ribonuclease A, after incubation with ethyl arachidonate or malonaldehyde, was shown to form monomeric, dimeric, trimeric, and aggregated molecules, making it clear that cross-linking had taken place. Moreover, the spectral properties of the products were shown to be similar to those of age pigments.

Fletcher *et al.* (1973) developed a procedure for quantitative analysis of age pigments, which is based upon chloroform–methanol extraction and measurement of fluorescence. Retinal, an interfering pigment, which is also extracted, was destroyed by UV irradiation. The fluorescence and thiobarbituric acid (for malonaldehyde) procedures are stated to be extremely sensitive, especially the former. Other aldehydes and sucrose also give color with thiobarbituric acid and, therefore, should not be present during the assays. Recently, Mihara *et al.* (1980) reported that maintaining a low pH improves the method and circumvents the uncertainties in the earlier procedures. Csallany and Ayaz (1976) chromatographed chloroform–ethanol extracts of tissues on Sephadex LH-20 to remove retinol and a low molecular weight compound which interfered with the quantitative assay of lipofuscin pigments. The authors also noted that UV radiation did not entirely remove contaminating retinol and tended to cause extra fluorescence. Shimasaki *et al.* (1977) showed that the extracts included many fluorescent products separable on TLC plates. Thus, it seems that determination of total pigments may not accurately measure age-related changes. One pigment, which separated on the TLC plates, was found only in older animals. This pigment, "age-related fluorescent substance," was shown, in a subsequent report, to increase linearly with age (Shimasaki *et al.*, 1980).

As measured histologically, many investigators have observed that pigment increases with age, although not always consistently for each tissue. Biochemical analysis generally supports this finding, but with occasional unexplained complications. An *in vitro* lipid peroxidation system was utilized by Dillard and Tappel (1971) to study peroxidation of rat liver mitochondria, microsomes, and heart muscle sarcosomes. Fluorescence developed linearly with absorption of oxygen. Production of malonaldehyde, as determined by color formation with thiobarbituric acid, increased, but not in a linear fashion. Perhaps this was due to the fact that mitochondria can metabolize malonaldehyde (Recknagle and Ghoshal, 1966), or perhaps the molecule becomes involved in other reactions. Recently, peroxidation *in vivo* has been determined by measuring hydrocarbons in expired air by means of the gas chromatograph (Section IV,A). Tappel *et al.* (1973), using a total fluorescence-extraction procedure, found an increase in fluorescence with age in rat testes (0.25 versus 1.9 year). The heart also showed an increase in fluorescence with age, although the individual variation was extensive in the older animals. Csallany *et al.* (1977) also found age pigments in liver to increase with age, but not until after the animals were 12 months old. They suggested that the lack of accumulation for the first 12 months may be due to turnover (removal) of the pigments. Miquel *et al.* (1978) found that lipofuscin increased rapidly with age in mouse testis until about 24 months of age but did not increase subsequently in older animals (up to 30 months). In the brain, the rate of increase was slow, but it continued to rise linearly through all age groups. It is of interest to note that the level of pigment was not increased in neurons from the brain of a patient with progeria over that of control cells of the same age (West, 1979).

In studies of the effect of vitamin E or other antioxidants (Section IV), both control and experimental animals generally show age-related increases in pigment levels. The type of diet appears to make a considerable difference. As might be expected, diets high in polyunsaturated fatty acids increase the level of lipofuscin found in the tissues (Reddy *et al.*, 1973).

Insects are composed of postmitotic cells, which readily accumulate lipofuscin. Since *Drosophila* and the housefly (*Musca*) also have a short life-span and can be conveniently reared in the laboratory, they have been used as a model system for studying accumulation of age pigment. Miquel *et al.* (1974) and Sheldahl and Tappel (1974) showed an age-related increase in fluorescent pigment extracts of *Drosophila*. The latter authors sought to relate the life-shortening effect of increased tempera-ture to pigment formation. The large production of pigment in this organism might be related to a reduced complement of protective en-zymes. *Drosophila*, for example, has a low glutathione peroxidase level (Miquel *et al.*, 1974) compared to tissues of other species. The enzyme is believed to be involved in protection from oxygen radicals (Section II,A). Armstrong *et al.* (1978) also observed that peroxidase, as well as catalase, declines with age in *Drosophila*, whereas the peroxide content increases. Donato and Sohal (1978) showed age-related increases in lipofuscin in the housefly (head, thorax, abdomen). They then attempted to experi-mentally relate the accumulation of pigment to altered life-spans (Sohal and Donato, 1979). They observed that active, short-lived flies accumu-lated lipofuscin faster than flies which, by restriction of flight, lived more than twice as long. The amount of pigment at the end of the life-span was the same in both cases. The authors, therefore, feel that the amount of pigment present is a good marker of physiological age.

Epstein *et al.* (1972) found that pigment granules increase with age in the free-living nematode *Caenorhabditis briggsae*. Klass (1977) made a sim-ilar observation in the related *C. elegans*.

IV. ANTIOXIDANTS

If there is a net leakage *in vivo* of free radicals to sensitive parts of the cell, in spite of protective enzymatic mechanisms, then the presence of antioxidants, particularly in the lipids of the membrane, should reduce any deleterious effects. Since age pigments appear to be a product of free radical damage, antioxidants should reduce their quantity. Alter-natively, a deficiency of antioxidants should increase the amount of pigments formed. Furthermore, if there is a direct relationship between free radical damage and aging, then antioxidants might be expected to decrease both processes.

Harman (1961), following up his hypothesis that free radicals might, in part, be responsible for aging, tested free radical inhibitors (mercaptoethylamine (MEA), hydroxylamine–HCl, cysteine–HCl, diaminodiethyl sulfide) in the diet of C3H, AKR, and Swiss mice. An extension of the half-survival time of C3H female mice from 14.5 to 18.3 months was noted when the diet contained 1% MEA. Small improvements were obtained with the AKR mice and none with the Swiss mice. In a subsequent study, Harmon (1968) tested another series of antioxidants in the diet of LAF_1 mice. Among the compounds tested (0.25–1% of the diet) were MEA and di-t-butylhydroxytoluene (BHT), which, Harmon concluded, had a beneficial effect on the mortality rate of LAF_1 mice. Mean survival time was increased 13 and 29% when the former was added to the commercial (control) diet at 0.5 and 1%, respectively, but only if deaths before 15 months of age were ignored. Maximal life-span was not significantly increased.

Along similar lines, Comfort et al. (1971) reported that 0.5% ethoxyquin in the diet extended the mean life-span of male and female C3H mice by about 15%.

Three years after Harmon's experiments, Kohn (1971) again tested the effects of both MEA and BHT on the longevity of C57Bl/6J mice. The criterion used was that an agent would be considered to be effective only if it increased the maximum life-span beyond that of the most long-lived controls. This standard eliminated the substantial variation normally observed in mean life-span, which may be a result of environmental and nutritional factors. When survival of control mice was optimal, the antioxidants had no effect on life-span, although the treated animals were of lower weight. In one experiment in which the control animals had a shorter life-span, MEA and BHT were beneficial in extending the life-span and 50% survival time, but never more than optimal control values. Kohn concluded that the antioxidants do not inhibit the processes that determine maximal life-span. He suggested that the antioxidants may inhibit some harmful environmental or nutritional factor.

The problem of antioxidant effect on life-span was recently tackled once again by Clapp et al. (1979). The effects of BHT on BALBc barrier-reared mice was examined as follows: BHT (0.75%) was given for a period of 8–11 weeks of age, for the entire life-span starting at 8 weeks, and for the entire life-span starting at 11 weeks. All BHT-treated groups showed improved mean survival times (206 days maximum extension in males and 174 days in females) over controls, due mostly to a reduction in early deaths. Maximal life-span was not increased. These results differ from those of Kohn (1971) in that under *optimal* environmental conditions (these were barrier-reared animals) an extension of the mean survival time is obtained. They differ from the results of Harman in that early deaths need not be eliminated from the calculations. The authors suggest that the improvement in mean life-span could be due to prevention of spontaneous tumors or other disease. One should keep in mind that antioxidants may act indirectly, that is, not as antioxidants, but as inducers of certain microsomal enzymes (Walker *et al.*, 1973, also see Chapter V, Section III). Presumably, higher levels of these enzymes could affect metabolism so as to alter mean life-span. In fact, treated animals showed an increase in liver size.

In other experiments, Harman and Eddy (1979) using Swiss mice, fed antioxidants to mothers for a 40-day period between conception and weaning and then determined the life-span of the offspring. The most effective antioxidant was MEA, which at 0.5% resulted in a small increase in the average life-span of male offspring (15%) and females (8%). Effects on maximal life-span were indeterminate. Other antioxidants tested were Santoquin (1,2-dihydro-6-ethoxy-2,2,4-trimethylquinoline), sodium hypophosphite, BHT, vitamin E, and butylated hydroxyanisole. The body weight of the offspring showed differences throughout their life-span, depending on the antioxidant that had been fed to the mothers. In a subsequent paper, Harman (1980) reported on the effect of antioxidants on the immune system, using New Zealand Black mice, a strain which loses T cell suppressor function early in life. A rather substantial increase in the average life-span (32%) was obtained using 0.25% Santoquin in the diet.

Harman (1971) utilized another approach to explore the role of lipid peroxidation in aging by feeding C3H/HeJ and Swiss male mice different degrees of unsaturated fatty acids (5, 10, or 20% of diet by weight) and determining the effects on life-span. The diets included lard, olive oil, corn oil, safflower oil, and menhaden oil. Vitamin E was provided in excess of calculated requirements. There was a substantial degree of species-specific difference. The C3H mice showed a decreased mean life-span with increased unsaturation or amount of dietary fat, partly because of increased mammary carcinoma. Body weight was not affected significantly. In Swiss male mice, the life-span was not significantly related to the amount of dietary fat or degree of unsaturation, but body weight increased with increased dietary fat.

In subsequent experiments, Harman *et al.* (1976) fed Sprague-Dawley rats semisynthetic diets containing different amounts of unsaturated fats from shortly after weaning to old age. Mortality was not affected. Performance in a maze was lower in senescent and mature rats according to the amount and degree of unsaturated fat in the diet. The authors take the modification of learning behavior to support the idea of damage by lipid peroxidation. Porta *et al.* (1980a; 1980b) also examined the effects of age of highly saturated versus unsaturated fat in the diet of Wistar rats. At 24 months, they observed little evidence for lipid peroxidation in brain in either case.

It is hard to evaluate the meaning of the reported results. The improvements in mean life-span noted by some investigators appear to be species-related rather than general. In some cases, there were differences in weight. If the test animals ate less, there would certainly be long-term effects. Perhaps the antioxidants simply improve the basic diet provided. After all, diet is an important determinant in the life-span of rodents (Chapter 2, Section II). Perhaps, as already mentioned, induction of enzymes is a factor; perhaps hormone balance is affected; perhaps tumor incidence is reduced. What role antioxidants or extra unsaturates play in moderating the aging process (if any) is yet to be determined.

A. Vitamin E

Vitamin E (α-tocopherol) was discovered in 1923 by H. M. Evans and K. S. Bishop. It is required for development of the fetus in female rats, but no fertility properties were found for humans. After searching in vain for a function for the vitamin, researchers have focused on its role as a biological antioxidant. α-Tocopherol is fat soluble and is found in membranes, interdigitated with the phospholipid structure. That there may be a relationship between the vitamin, a known antioxidant, and protection of polyunsaturated fatty acids from peroxidation, seems obvious. For this reason, many investigators have sought to prove that the vitamin acts as an antioxidant *in vivo*. They have further attempted to demonstrate a beneficial sequence of events arising from large doses of the vitamin—that is, that less peroxidation results in less age pigment and a longer life-span. The possible reverse effects of deficiency, namely, increased peroxidation and increased pigment formation, have also been explored. For a recent review of vitamin E effects on free radical antioxidant reactions, see Witting (1980).

Dillard and Tappel (1971) fed an α-tocopherol-deficient diet to a group of rats (Sprague-Dawley) and supplemented the diet of two other groups with 10.5 and 45 mg/kg of the vitamin, starting at 4 weeks of age. The control group received a basal chow diet (16 mg/kg of α-tocopherol). Microsomes and mitochondria were prepared and peroxidation of membranes was measured in an O_2 atmosphere. The rate of oxygen uptake and production of fluorescence were in the following order: deficient, 10.5, and 45 mg/kg dosages. The same order was noted for production of color with thiobarbituric acid (malonaldehyde production), but with some inconsistencies related to the time course of peroxidation when measured this way. Reddy *et al.* (1973) found that several tissues of rats fed diets containing 10% lard and 1% cod liver oil without vitamin E for 4 months contained twice as much fluorescent pigment as animals also provided with 45 mg/kg of vitamin E. Adipose tissues of animals fed cod liver oil (highly unsaturated) had 3 times more fluorescence than animals fed corn oil. Pigment formation was generally inverse to the vitamin E concentration. Subsequently Tappel *et al.* (1973) checked the effect of feeding a combination of antioxidants (α-tocopherol, BHT, ascorbic acid, DL-methionine, and sodium selenite) to mice in various proportions. The diet with the highest levels of antioxidants gave the greatest reduction in the amount of fluorescent products in testes and heart. Mortality rate was unaffected. Drawn to its logical conclusion, these experiments suggest that reduction of age pigments does not affect life-span. In agreement with Tappel *et al.* (1973), Jager (1972) did not find any relationship between mortality and vitamin E content of food, in rats (Wistar) up to 22 months. Porta *et al.* (1980a) also found that low levels of vitamin E (2 mg%) and high levels (200 mg%) in diets with varied fatty acid composition had no effect on the maximal life span of a strain of Wistar rats. However, in one diet high in polyunsaturated fatty acids, the high level of vitamin E improved the 50% survival time beyond that of the other dietary groups.

Blackett and Hall (1981) make the findings unanimous. Long-term studies of two strains of mice (C3H/He and LAF_1) fed vitamin E supplements (0.25%) showed no effect on maximal longevity. The vitamin supplement, however, did increase the number of animals surviving to 24 months.

At least until maturity, vitamin E also does not appear to affect learning function. Freund (1979) observed a retarded accumulation of pigment in the brains of mice by increased levels of vitamin E in the diet, although all animals received adequate amounts of the vitamin. The test groups received ethanol or vitamin E supplements (7 mg/day). Brain lipofuscin increased steadily with age, as measured by fluorescence of lipid extracts. The vitamin E reduced fluorescence by about 19% for the chow and ethanol-containing diet at 10 months of age. When tested for learning ability, the mice (C57BL/6J) receiving vitamin E supplements showed no differences from the controls, suggesting that, at least at 8 months of age, the pigments were not deleterious to brain function.

An interesting discovery is that lipid peroxidation results in the formation of volatile hydrocarbons. Exhalation of these products can be measured by gas chromatography. Tappel (1980) and Tappel and Dillard (1981) in recent reviews, discuss the production of ethane and pentane.

Studies show that vitamin E in the diet reduces hydrocarbon production and unsaturated fat increases it. Sagai and Ichinose (1980) reported that hydrocarbon expiration increased with age with an exponential relationship, i.e., there is an increase in the level of peroxidation reactions with age. Thiobarbituric acid assays for malonaldehyde in serum did not show this relationship.

There are many other reports of vitamin E reduction of pigment formation, although these studies in general do not deal with senescent animals (Chow and Tappel, 1972; Takeuchi *et al.*, 1976; Mihara *et al.*, 1980; Sylven and Glavind, 1977; see Leibovitz and Siegel, 1980, for a brief review).

It would seem safe to conclude that animals given high doses of vitamin E are at least partly protected from lipid peroxidation especially as absence of vitamin E or feeding increased amounts of unsaturated fatty acids increases pigment formation. However, these findings are by no means unanimous. For example, the results reported by Grinna (1976) are, in part, in contrast to the generally observed pattern. She found that young rats (11 weeks old) became deficient 7 weeks after removal of vitamin E from the diet, based upon a test for red blood cell hemolysis. Older animals (47 weeks) required 16 weeks; at 67 weeks of age, no deficiency could be generated. Thus, there seems to be a declining need for the vitamin with increasing age. Accumulation of vitamin E appeared to depend upon the amount fed. In this study, although pigments increased with age, there was no correlation between amount of fluorescence and dietary levels of vitamin E in either microsomes or mitochondrial membranes. Moreover, in microsomes, the amounts of unsaturated fatty acids (18:1, 18:2, 20:4, and 22:6) did not vary with the dietary level of vitamin E for a given age-group, although composition did change with age. (For age-related changes in the lipid composition of membranes, see Chapter 4, Section V.) If vitamin E deficiency results in increased lipid peroxidation, then one might expect that the level of unsaturated fatty acids would be reduced. Since this is not the case, it must be assumed either that peroxidized molecules are replaced or that basal levels of the vitamin are adequate. Csallany *et al.* (1977) also obtained inconsistent results from studies on the effects of vitamin E on pigment formation. The authors examined several tissues of mice after 18 months of diets which contained varying amounts of vitamin E (deficient, 30, and 300 ppm). They found no significant differences in the pigment levels of heart, spleen, kidney, brain, lungs, or uterus. Liver showed a reduction with increasing vitamin E. Most of the increase of lipopigment in liver occurred between 12 and 18 months of age. Pigments were analyzed by the column procedure of Csallany and Ayaz

(1976). Therefore, the results are not due to technical problems involving total lipid extracts (Section III). In a similar vein, Porta *et al.* (1980b) recently reported that up to 24 months, vitamin E had no effect on the amount of pigment in brain cells of Wistar rats fed diets containing different amounts of polyunsaturated fatty acids.

Grinna (1976) pointed out that the lack of change in fatty acid composition and lack of change in certain membrane-associated enzymes does not support the idea that vitamin E is a general membrane stability factor in microsomal membranes. Yeh and Johnson (1973) concluded that random lipid peroxidation is not responsible for the respiratory decline in deficient mitochondria. They believe that, for protection, a cytosolic factor is necessary in addition to the vitamin.

Most of the above experiments do not deal with truly senescent animals. Unfortunately, the data must suffice until more information becomes available.

Although vitamin E has not been shown to extend the maximal life span of rats or mice, presence of the quinone form (α-tocopheroquinone) in the medium gives a 19% increase in life-span of the free-living nematode *Caenorhabditis briggsae* (Epstein and Gershon, 1972). The compound causes a delay in the accumulation of age pigment granules in the intestinal cells. Although this effect may appear to have a direct relationship to increased longevity, one should bear in mind that there could be a beneficial effect on the complex medium in which the organisms are grown. The vitamin also appears to delay age-related changes in the levels of certain enzymes in the nematode *Turbatrix aceti* (Bolla and Brot, 1975). It has also been reported to increase the life-span of the rotifer (Enesco and Verdone-Smith, 1980).

In sum, the evidence is not consistent as to whether or not vitamin E is required for some sort of membrane stabilization or preservation of polyunsaturated fatty acids *in vivo*. On the other hand, there is little reason to doubt that it functions as an antioxidant. However, its role in preventing aging effects is dubious. Certainly, there is little evidence that greater than the normal levels of the vitamin perform any real function in this respect. In fact, the studies with vitamin E suggest that age pigment accumulation is neither life-threatening or even seriously debilitating.

B. Vitamin C

Ellery *et al.* (1979) fed ascorbic acid to guinea pigs. In spite of high levels of the material in tissues of the experimental animals, no effect on fluorescent levels of brain extracts was observed. Vitamin C is involved in the nonenzymatic formation of free radicals, which may explain why low levels appear to stimulate lipid peroxidation *in vitro* (Section II,A), whereas high levels are claimed to prevent these effects. Although there is little work on the direct effects of vitamin C on aging, a brief review of its possible role is presented by Leibovitz and Siegel (1980).

C. Centrophenoxine

Centrophenoxine has been claimed to remove lipofuscin *in vivo* and *in vitro*. Bourne (1973) reported a reduction of age pigments in nerve cells of squirrel monkeys to which the drug had been fed. Riga and Riga (1974) observed reductions of pigment in rat central nervous system. Nandy and Bourne (1966) and Hasan *et al.* (1974) obtained a reduction in neuronal tissues of senile guinea pigs. Spoerri and Glees (1975) report that the removal of pigment from the anterior hypothalmus is carried out by phagocytic cells under the influence of centrophenoxine. Spoerri and Glees (1974) observed removal of pigment from cultured cells of spinal ganglia after treatment with the drug.

In the free-living nematode *Caenorhabditis briggsae,* centrophenoxine in the medium lessens the increase in osmotic fragility and specific gravity shown by aged organisms. Lipofuscin accumulation is also retarded (Kisiel and Zuckerman, 1978).

D. Comment

It is reasonable to conclude that lipofuscin is formed by cross-linking reactions of malonaldehyde generated by lipid peroxidation reactions and that the pigment accumulates with time. It seems equally reasonable to conclude that the accumulation does not seriously damage cells, though it may add to their burden. This conclusion may be derived from the fact that reduction in the amount of pigment in cells of animals fed high doses of vitamin E does not produce concomitant effects on lifespan. It would seem that age pigments represent a mechanical accumulation of undigestible material, a result rather than a cause of aging.

V. METABOLISM OF LIPIDS

Studies of lipids with respect to age deal with compositional changes as well as with metabolism per se. Thus, there is much literature that reports on changes in level of cholesterol, triglycerides, free fatty acids, phospholipids, etc. in serum, membranes (Chapter 4), or in various tissues. Serum levels have been emphasized particularly in humans, because of the ostensible relationship of lipid components to cardiovascular disease. The observed changes, however, even where they show a clear, age-related direction, have not been successfully tied to underlying metabolic events. With the exception of studies of cholesterol metabolism in which some of the enzyme activities involved have been measured, the literature in this respect is sparse indeed.

A. Turnover

Little work is reported on the effect of age on lipid turnover. There is some evidence that cholesterol turnover slows with age (Section V,B). Beyond that, the most detailed paper is that of Grinna (1977b). The author utilized a double isotope procedure in which [^{14}C]glycerol was injected, and 4 days later, [^{3}H]glycerol was injected. The rate at which the [^{14}C]glycerol has been degraded will be reflected in the ratio of ^{3}H:^{14}C. The procedure was derived from that developed for determining protein turnover. Studies of total lipids in rats of 6, 12, and 24 months of age yielded, in each case, heterogeneous turnover rates for various fractions of liver: soluble phase > whole liver homogenates > microsomal membranes > mitochondrial membranes > plasma membranes. The ratio generally was the same at 6 and 24 months, that is, turnover rates were about equal, both of these ages being slower than at 12 months. However, given the substantial standard error, the true changes in rate may be minimal. Data were given in terms of ^{3}H:^{14}C ratios. Half-lives were not calculated. Phospholipids and neutral glycerides showed the same age-dependent pattern of isotope ratios (6<12>24 months) in microsomes and mitochondria. Relative rates of turnover were similar to the values for total lipids in these fractions. Of individual phospholipids, phosphatidylcholine showed no age-related change. Phosphatidylethanolamine, cardiolipin, and "other" showed the same 6>12< 24 month pattern observed for other lipids. The turnover rate did vary with the membrane type (microsomes and mitochondria). It should be noted that lipid composition of membranes has been reported, although not unanimously, to change with age. (Chapter 4, Section II). Whether such compositional changes reflect an altered metabolism or replacement of damaged components with different products is not known. In attempting to evaluate such changes, it should be noted that dietary differences can cause greater changes than are observed as a result of aging (Chapter 4, Section II). Not only diet, but even sex, may cause substantial differences in lipid composition of tissues. Eddy and Harman (1977) reported that the ability to incorporate dietary 22:6ω3 into brain lipids is impaired by 12 months of age in male, but not female, rats.

B. Cholesterol

1. Tissue Levels

Studies of cholesterol levels are particularly numerous, as high serum cholesterol has been associated with an increased risk of heart disease in humans. Whether cholesterol levels are truly related to aging rather than to cardiovascular difficulties, which may cause premature death, is a legitimate question. Heart attacks or atherosclerosis are not a problem in most other vertebrates or even in some groups of humans, and certainly bear no relationship to aging in invertebrates. Nonetheless, age-related changes in cholesterol levels, if they exist, must reflect changes taking place in rates of synthesis, utilization, or oxidation. Unfortunately, little is known of these aspects of cholesterol metabolism.

As pointed out in a review by Kritchevsky (1980), the serum levels of cholesterol in humans appears to rise with age, although the pattern of increase varies in different reports. Thus, Keyes *et al.* (1950) found that cholesterol levels in men increased between 17 and 55 years of age (177 versus 248 mg/dl, respectively) and, subsequently, leveled off with an actual drop occurring in the 60- and 78-year group. Werner *et al.* (1970) found that in men, cholesterol levels increased steadily from under 12 to 69 years of age and leveled off between 70 and 79. The rate of increase was quite slow after age 39. Women showed a similar pattern, but there was a substantial jump in concentration between 49 and 59 years. After 50 years of age, their level was consistently higher than that for men of equivalent ages (285 mg/dl for women versus 259 mg/dl for men in the 60- to 69-year age-group). Heiss *et al.* (1980) also observed a steady increase in serum cholesterol with age, the most rapid rise occurring between 20 and 39 and leveling off at 45 years of age. Although there are clearly variations in the patterns observed by different investigators, there is a consistency in the trend toward an increase with age.

In rats, cholesterol also appears to increase with age, although the reported results are by no means consistent. Carlson *et al.* (1968) found that liver and plasma levels increased with age in Sprague-Dawley rats between 1 and 18 months. Lacko and David (1979) also found that plasma cholesterol increased with age in Sprague-Dawley rats, but observed no change between 2 and 12 months. A substantial increase occurred by 24 months. Carlisle and Lacko (1981) observed no change up to 24 months in the same strain of rats. In agreement, Kritchevsky (1980) and Story and Kritchevsky (1978) reported a small increase up to 12 months, a large increase by 18 months, and no subsequent change by 24 months. A somewhat similar pattern was observed by Dupont et al. (1972), who showed increasing cholesterol levels with age, the rise being gradual between 3 and 12 months, and then doubling between 12 and 18 months of age in rats of the CFE strain (Sprague-Dawley derived). A somewhat different pattern was obtained by Uchida et al. (1978). They noted that in Sprague-Dawley rats, a large increase in serum cholesterol came somewhat earlier than reported by Story and Kritchevsky (1978) and Lacko and Davis (1979), occurring between 7–23 and 48–51 weeks. Older animals were not studied.

In the Fischer 344 rat, the pattern seems to be one of no increase or slow increase between 2 and 18 months, but a substantial increase by 24 months (Story and Kritchevsky, 1978; Story et al., 1976; Kritchevsky, 1980; Carlisle and Lacko, 1981). Story and Kritchevsky (1978) also estimated liver pools of cholesterol. Combined with serum values, there is a steady increase of cholesterol with age for both Fischer and Sprague-Dawley rats, the levels being much higher in the latter. Liepa et al. (1980) obtained a different age pattern for serum cholesterol in Fischer rats. They observed a sharp increase between 12 and 18 months, leveling off between 18 and 24 months, and decreasing by 27 months. This last change is particularly interesting because it deals with the oldest group for which data are available. Food restriction (Chapter 11) sharply reduced and delayed the degree of increase.

Uchida et al. (1978) also investigated age-related changes in Wistar rats (male, 7–106 weeks). Serum cholesterol increases with age from 82 to 147 mg/dl at ages of 17–23 and 100–102 weeks, respectively. Porta et al. (1980a), however, observed no significant age-related difference in the serum cholesterol levels of Wistar rats up to 24 months of age, even though diets of varying lipid composition were utilized. Similarly, no consistent changes were noted in the brain (Porta et al., 1980b).

Unmistakably, the pattern of change for cholesterol levels in rats varies with the strain. Perhaps some of the differences are related to weight. Unlike Sprague-Dawleys, Fischer rats do not gain weight beyond 6–9 months. Even beyond strain differences, it is apparent that results show substantial variation from laboratory to laboratory. These differences in patterns observed by various investigators may perhaps relate to diet, season, exercise, or for that matter, to the time of day at which experiments are conducted in the various laboratories. Even from the same laboratory, considerable disparities may be noted. For example, though the age-related effects are the same, the absolute values for serum cholesterol obtained by Story *et al.* (1976), in two sets of experiments carried out at different times, differed by over 100% in Fischer rats. In this regard, Cohen (1979) pointed out that nutritional studies are difficult to compare with one another even when "identical" diets are used. Perhaps nutrition provides a clue to the variations observed, because diet can dramatically affect lipid composition (Witting, 1980).

In individual tissues, Hrachovec and Rockstein (1959) showed an accumulation with age of cholesterol in brain, lungs, liver, and muscle of rabbits. Hegner and Platt (1975) reported an increase in cholesterol content in aged rat liver membranes, but Dupont *et al.* (1980) did not observe increase in the total carcass cholesterol content of aging rats (9–21 months), male or female. In aging humans, Crouse *et al.* (1972) found increases with age of cholesterol and cholesteryl esters in muscle, skin, adipose tissue, and connective tissue. However, Thomas *et al.* (1978) found no difference in the level of cholesterol in skeletal muscle membrane in healthy, nonobese, human males from 22 to 73 years of age. Levels in primates increased only slightly (Lacko and Davis, 1979), but high-density lipoprotein increased and low-density lipoprotein decreased substantially with age.

2. Metabolism

Although patterns of change vary with strains, the evidence makes it clear that cholesterol levels increase with age in rats as well as humans. The cause must be related to rates of synthesis and utilization, the latter process encompassing oxidation, conversion to other metabolites, and elimination. Therefore, age-related changes in the enzymes involved in cholesterol synthesis and metabolism should reflect the observed changes in cholesterol levels. Unfortunately, there are few studies of cholesterol at the metabolic level. Of these existing, most deal with "mature" rather than with senescent animals, about 2 years being the upper limit of age. Story *et al.* (1976) found that, in liver slices of Fischer rats, cholesterol synthesis from acetate dropped sharply between 2 and 12 months, but rose slightly at 18 and 24 months. A similar pattern was reported by Story and Kritchevsky (1978) and by Kritchevsky (1978; 1980). In Sprague-Dawley rats, there was a rather steady decline. With mevalonate as substrate, there was a steady decline in synthetic ability after 6 months in both Fischer and Sprague-Dawley rats. Liver microsomal 3-hydroxymethylglutaryl-CoA reductase varied erratically in both species. None of the changes in enzyme levels bore any direct relationship to the picture shown by total cholesterol, namely, that it increases with age.

As to the catabolism of cholesterol, the enzyme cholesterol 7α-hydroxylase is involved in the synthetic pathway to bile acids, which are eventually excreted. Story and Kritchevsky (1978) showed that the hydroxylase levels in liver microsomes bear no relationship to age. The patterns of change with age were dramatically different for Fischer and Sprague-Dawley rats. Uchida *et al.* (1978) found that biliary secretion of cholesterol and bile acids decreased with age in Wistar and Sprague-Dawley rats on a per kilogram basis, but remained unchanged on a per rat basis. Liver cholesterol (mg/g) was found to increase with age. Kritchevsky (1980) noted little change in this parameter.

Several other aspects of cholesterol metabolism appear to change with age. Turnover was reported to decrease in papers by Hruza and co-workers (Hruza and Zbuzkova, 1973), but the "old" rats were 13–14 months of age. Dupont *et al.* (1980) provided experimental evidence for increased retention of labeled cholesterol in CFE (Sprague-Dawley) rats between 9 and 21 months of age, but there was either little change or an actual decline after 15 months of age both in the liver and in the carcass. Liver mitochondrial preparations from young Wistar rats oxidize more [26-^{14}C]cholesterol to $^{14}CO_2$ than do old preparations (Story and Kritchevsky, 1974).

Both synthesis and hydrolysis of cholesteryl esters increase with age in acetone powders of rat aortas (Kritchevsky *et al.*, 1973). The serum levels of cholesteryl esters also increase with age in rats (Story and Kritchevsky, 1978). The finding that the activity of lecithin–cholesterol acyltransferase, an important enzyme in plasma lipid metabolism, increases with age in both Fischer and Sprague-Dawley rats perhaps explains this observation. However, in Sprague-Dawley rats, Lacko and Davis (1979) saw no such enzyme increase between 12 and 24 months of age. In fact, Carlisle and Lacko (1981) reported a substantial age-related *decrease* in activity between 12 and 14 months.

In general, enzyme studies have been too sparse to provide a coherent picture in cholesterol metabolism, and they do not explain the observed rise in serum levels.

C. Fatty Acid Metabolism

In humans, triglycerides rise through the middle years and start declining at about 50 years for males and 60 for females (Bierman, 1978). In rats, the pattern observed has not been consistent with age. Old Wistar rats show an increase in serum triglycerides. In Fischer rats, one set of experiments showed that an increase occurred between 12 and 18 months, but a second set showed no change with age, although the concentrations were 2 times as high in the latter experiment (Story *et al.*,

1976). Liepa *et al.* (1980) observed a sharp increase between 12 and 18 months and an equally sharp decline through 27 months. According to these authors, free fatty acids in serum also decline with age in Fischer rats (12 through 27 months). The decline is reflected in individual fatty acids. Porta *et al.* (1980a) found a decreasing trend in serum triglycerides between 3 and 18 months in Wistar rats, a sharp rebound occurring at 24 months with some of the test diets. In a general way, depending on the diet, brain tissue showed a small increase up to 18 months and a decrease by 24 months, although the changes were generally small and, when averaged for the various dietary groups, were not statistically significant (Porta *et al.*, 1980b).

In general, beyond cataloging longitudinal effects of age on triglyceride levels, few studies involving metabolism have been made. The rate of lipolysis in adipocytes decreases between 12 and 24 months and subsequently rises somewhat in Fischer rats (Bertrand *et al.*, 1980; Masoro *et al.*, 1980). As to fatty acid composition, changes in membranes have been investigated in some detail (Chapter 4, Section II).

The oxidation of fatty acids in aged rat hearts has been studied by Abu-Erreish *et al.* (1977). They used isolated, perfused hearts (male Fischer 344) and measured oxidation of [^{14}C]palmitate under various work loads. Since mitochondria from old hearts have lower rates of fatty acylcarnitine metabolism, lipid metabolism may perhaps be impaired in this tissue. On the basis of whole hearts, the old organs (5 versus 24 months) showed only slightly reduced levels of oxygen consumption and palmitate oxidation at high work loads. On the basis of per gram of dry heart tissue, the rates were significantly lower in old hearts at each time period measured. Tissue levels of ATP, ADP, AMP, and creatine phosphate were similar. The small changes in coenzyme A were not considered of importance. The level of carnitine was reduced in 12- and 24-month-old animals. The old hearts proved as capable as young in energy production on an organ basis.

Kritchevsky (1978) recorded the enzyme activity of hepatic acetyl–CoA carboxylase and fatty acid synthetase in Fischer 344 and Sprague-Dawley rats at various ages from 2 to 24 months. Fatty acid synthetase increases in old Fischer but not in old Sprague-Dawley rats. Acetyl–CoA carboxylase is thought to be rate limiting for fatty acid synthesis. It appears to increase with age in both species. Heger (1980) also reported an age-related increase in the specific activity of acetyl–CoA carboxylase in rat liver homogenates, although the increase occurred between 4 and 15 months with no further change by 30 months. The author also noted differences in the kinetic properties of the "old" enzyme. As is the case for cholesterol, several published reports on age-related changes in fatty acid metabolism are not considered here as they deal with comparisons between young and mature animals and not with aging.

D.   Other Lipids

There have been a number of determinations made of the effect of age on levels of various phospholipids in tissues. For example Liepa *et al.* (1980) report a steady increase of serum phospholipids with age in Fischer rats. Kritchevsky (1980) provided data on the levels of several liver phospholipids in Fischer and Sprague-Dawley rats at 3, 12, and 18 months of age. Most phospholipids and the enzymes sphingomyelinase and choline kinase did not appear to change with age. However, small changes would not be seen as statistics allow for considerable variation. Porta *et al.* (1980a) did not find age-related increases in the serum phospholipids of Wistar rats, although with some test diets there seemed to be an increase between 18 and 24 months. Brain phospholipids rose at 18 months and declined at 24 months (Porta *et al.*, 1980b). In general, these results do not give us any information that can be applied in a biochemical context. One could go on detailing reports of changes or no changes in this and that component of tissues, but it is painfully clear that such results, until they attain more depth and consistency, can only be of marginal value in understanding the role of lipids in the aging process.

E.   Comment

It is clear that studies of the lipid content of tissues have not pointed toward any particular age-related metabolic misfunction. Kritchevsky (1980) concluded that as animals age, they become increasingly unable to metabolize lipid, synthesizing less and degrading and excreting less, so that there is a net accumulation. The contention may well be correct, but there is as yet not nearly enough evidence to prove it. Species differences make it difficult to distinguish between deleterious "aging" effects and normal metabolic patterns. The variation in results from different laboratories makes it particularly uncertain that a firm base for future work has been established.

Studies of enzymes involved in lipid metabolism have been minimal, and information regarding regulatory mechanisms is nonexistent. Moreover, many references to "aging" changes in lipid metabolism refer to mature animals only. In some cases, the "young" animals are selected at such early ages that they are still developing and therefore provide an exaggerated base for comparison.

The most negative view of lipid metabolism with respect to aging is that it simply continues unchanged except for minor or nonthreatening fluctuations brought about by changes in some other region of the metabolic scheme or by diet. Given the fact that the composition of dietary fat can bring about large and presumably harmless changes in the lipid composition of tissues, it is difficult to know whether modest changes due to aging are matters of concern. The research so far accomplished has provided few clues and indicated few roads for new researchers to follow. Perhaps the most notable accomplishment to date is to demonstrate that with increasing age, dramatic changes do not occur in lipid metabolism.

VI. TISSUE CULTURE

Very little work has been performed on the biochemistry of lipids in cells in culture. Packer and Smith (1974) caused a flurry of excitement by reporting that human diploid fibroblasts (WI-38) in tissue culture showed a large increase in the number of population doublings the cell would undergo if vitamin E was added to the culture medium. However, the authors subsequently found that when a new batch of fetal calf serum was used in the preparation of the medium, the startling effect of vitamin E disappeared (Packer and Smith, 1977). Balin *et al.* (1977) also reported that the vitamin did not extend the doubling potential of the cells.

Human glia cells accumulated pigment when kept for long periods in stationary phase (Brunk *et al.*, 1973). Spoerri and Glees (1974) also used cultured neurons and satellite cells for studies of lipofuscin.

The presence of increased partial pressures of oxygen decreased WI-38 cell doublings (Balin *et al.*, 1977). Vitamin E exerted no protective effect. This result and the fact that decreased oxygen levels did not extend the life-span are interpreted by the authors as evidence that free radical reactions do not play a significant role in determining the doubling potential cells in culture. Nonetheless, age pigments can be formed *in vitro*.

Kritchevsky and Howard (1970) reported little change in the lipids of late-passage WI-38 cells except for phospholipid profiles. Total lipid content increased, but fatty acids showed few significant differences except for a lowering of myristic acid in the late-passage cells. Polgar *et al.*

(1978) also found no change in cholesterol, phospholipid, or neutral fat content of late-passage human embryo lung fibroblasts. The effect of adding various unsaturated acids on growth of cells at different passage numbers of human fibroblasts (IMR-90) was examined by Lynch (1980).

REFERENCES

Abraham, E. D., Taylor, J. R., and Lang, C. A. (1978). *Biochem. J.* **174**, 819–825.
Abu-Erreish, G. M., Neeley, J. R., Whitmer, J. T., Whitman, V., and Sanadi, D. R. (1977). *Am. J. Physiol.* **232**, E258–E262.
Armstrong, D., Rinehart, R., Dixon, L., and Reigh, D. (1978). *Age* **1**, 8–12.
Aune, J. (1976). *Interdiscip. Top. Gerontol.* **10**, 44–61.
Baird, M. B. (1980). *Biochem. Biophys. Res. Commun.* **95**, 1510–1516.
Balin, A. K., Goodman, D. B. P., Rasmussen, H., and Cristofalo, V. J. (1977). *J. Cell Biol.* **74**, 58–67.
Barrett, M. C., and Horton, A. A. (1975). *Biochem. Soc. Trans.* **3**, 124–126.
Bertrand, H. A., Masoro, E. J., and Yu, B. P. (1980). *Endrocinology* **107**, 591–595.
Bierman, E. L. (1978). *Fed. Proc. Fed. Am. Soc. Exp. Biol.* **37**, 2832–2836.
Blackett, A. D., and Hall, D. A. (1981). *Gerontology* **27**, 133–134.
Bolla, R., and Brot, N. (1975). *Arch. Biochem. Biophys.* **169**, 227–236.
Bourne, G. H. (1973). *Prog. Brain Res.* **40**, 187–201.
Brizzee, K. R., Kaack, B., and Klara, P. (1975). In "Neurobiology of Aging" (J. M. Ordy and K. R. Brizzee, eds.), p. 463. Plenum, New York.
Brunk, V., Ericsson, J., Ponten, J., and Westermark, B. (1973). *Exp. Cell Res.* **79**, 1–4.
Carlisle, S. I., and Lacko, A. G. (1981). *Comp. Biochem. Physiol.* **70B**, 753–758.
Carlson, L. A., Froberg, S. O., and Nye, E. R. (1968). *Gerontologia* **14**, 65–79.
Chio, K. S., and Tappel, A. L. (1969a). *Biochemistry* **8**, 2821–2827.
Chio, K. S., and Tappel, A. L. (1969b). *Biochemistry* **8**, 2827–2832.
Chow, C. K., and Tappel, A. L. (1972). *Lipids* **7**, 518–524.
Christophersen, B. O. (1969). *Biochem. Biophys. Acta* **176**, 463–470.
Clapp, N. K., Salterfield, L. C., and Bowles, N. D. (1979). *J. Gerontol.* **34**, 497–501.
Cohen, B. J. (1979). *J. Gerontol* **34**, 803–807.
Cohen, G., and Cederbaum, A. I. (1980). *Arch. Biochem. Biophys.* **199**, 438–447.
Comfort, A., Youhotsky-Gore, I., and Pathmanathan, K. (1971). *Nature (London)* **229**, 254–255.
Crouse, J. R., Grundy, S. M., and Ahrens, E. H. (1972). *J. Clin. Invest.* **51**, 1292–1296.
Csallany, A. S., and Ayaz, K. L. (1976). *Lipids* **11**, 412–417.
Csallany, A. S., Ayaz, K. L., and Su, L. C. (1977). *J. Nutr.* **107**, 1792–1799.
DelMaestro, R. F., Thaw, H. H., Bjork, J., Planker, M., and Arfors, K. E. (1980). *Acta Physiol. Scand., Supp.* **492**, 43–57.
Dillard, C. J., and Tappel, A. L. (1971). *Lipids* **6**, 715–721.
Donato, H., and Sohal, R. S. (1978). *Exp. Gerontol.* **13**, 171–179.
DuPont, J., Mathias, M. M., and Cabacugan, N. B. (1972). *Lipids* **7**, 576–589.
DuPont, J., Mathias, M. M., Spindler, A. A., and Janson, P. (1980). *Age* **3**, 19–23.
Eddy, D. E., and Harman, D. (1977). *J. Am. Ger. Soc.* **25**, 220–229.
Ellery, P. M., Hughes, R. E., and Jones, E. (1979). *Exp. Gerontol.* **14**, 49–50.
Enesco, H. E., and Verdone-Smith, C. (1980). *Exp. Gerontol.* **15**, 335–338.
Epstein, J., and Gershon, D. (1972). *Mech. Ageing Dev.* **1**, 257–264.

Epstein, J., Himmelhock, S., and Gershon, D. (1972). *Mech. Ageing Dev.* **1**, 245–255.
Fletcher, B. L., Dillard, C. J., and Tappel, A. L. (1973). *Anal. Biochem.* **52**, 1–9.
Freund, G. (1979). *Life Sci.* **24**, 145–152.
Fridovich, I. (1975). *Ann. Rev. Biochem.* **14**, 147–159.
Gopinath, G., and Glees, P. (1974). *Acta Anatomica* **89**, 14–20.
Grinna, L. S. (1976). *J. Nutr.* **106**, 918–929.
Grinna, L. S. (1977a). *Mech. Ageing Dev.* **6**, 197–205.
Grinna, L. S. (1977b). *Mech. Ageing Dev.* **6**, 453–459.
Grinna, L. S., and Barber, A. A. (1973). *Biochem. Biophys. Res. Commun.* **55**, 773–779.
Harman, D. (1956). *J. Gerontol.* **11**, 298–300.
Harman, D. (1961). *J. Gerontol.* **16**, 247–254.
Harman, D. (1968). *J. Gerontol.* **23**, 476–482.
Harman, D. (1971). *J. Gerontol.* **26**, 451–457.
Harman, D. (1980). *Age* **3**, 64–73.
Harman, D., and Eddy, D. E. (1979). *Age* **2**, 109–122.
Harman, D., Hendricks, S., Eddy, D. E., and Seibold, J. (1976). *J. Am. Geriat. Soc.* **24**, 301–307.
Hasan, M., Glees, P., and Spoerri, P. E. (1974). *Cell Tissue Res.* **150**, 369–375.
Hazelton, G. A., and Lang, C. A. (1980). *Biochem. J.* **188**, 25–30.
Heger, H. W. (1980). *Mech. Ageing Dev.* **14**, 427–434.
Hegner, D., and Platt, D. (1975). *Mech. Ageing Dev.* **4**, 191–200.
Heiss, G., Tamir, T., Davis, C. E., Tyroler, H. A., Rifkind, B. M., Schonfeld, G., Jacobs, D., and Frantz, I. D. Jr. (1980). *Circulation* **61**, 302–315.
Hrachovec, J. P., and Rockstein, M. (1959). *Gerontologia* **3**, 305–326.
Hruza, Z., and Zbuzkova, V. (1973). *Exp. Gerontol.* **8**, 29–37.
Hughes, B. A., Roth, G. S., and Pitah, J. (1980). *J. Cell Physiol.* **103**, 349–353.
Jager, F. C. (1972). *Nutr. Metab.* **14**, 1–7.
Kellogg, E. W., and Fridovich, I. (1976). *J. Gerontol.* **32**, 405–408.
Keys, A., Mickelsen, O., Miller, E. V. O., Hayes, E. R., and Todd, R. L. (1950). *J. Clin. Invest.* **29**, 1347–1353.
Kisiel, M. J., and Zuckerman, B. M. (1978). *Age* **1**, 17–20.
Klass, M. R. (1977). *Mech. Ageing Dev.* **6**, 413–429.
Kohn, R. R. (1971). *J. Gerontol.* **26**, 378–380.
Kritchevsky, D. (1978). *Fed. Proc. Fed. Am. Soc. Exp. Biol.* **38**, 2001–2006.
Kritchevsky, D. (1980). *Proc. Soc. Exp. Biol. Med.* **165**, 193–199.
Kritchevsky, D., and Howard, B. V. (1970). *In* "Aging in Cell and Tissue Culture" (E. Holeckova and V. J. Cristofalo, eds.), pp. 57–82. Plenum, New York.
Kritchevsky, D. J., Genzano, J. C., and Kothari, H. V. (1973). *Mech. Ageing Dev.* **2**, 345–347.
Lacko, A. G., and Davis, J. L. (1979). *J. Am. Geriatr. Soc.* **27**, 212–217.
Leibovitz, B. E., and Siegel, B. V. (1980). *J. Gerontol.* **35**, 45–56.
Lesko, S. A., Lorentzen, R. J., and Ts'o, P. O. P. (1980). *Biochemistry* **19**, 3023–3028.
Liepa, G. V., Masoro, E. J., Bertrand, H. A., and Yu, B. P. (1980). *Am J. Physiol.* **238**, E253–257.
Lippman, R. D. (1980). *Exp. Gerontol.* **15**, 339–351.
Lippman, R. D. (1981). *Exp. Gerontol.* **36**, 550–559.
Lippman, R. D. (1982). *In* "Luminescent Assays: Perspectives in Endocrinology and Clinical Chemistry" (M. Serio and P. Puzzagli, eds.), *in press.* Raven, New York.
Lippman, R. D., Agran, A., and Uhlen, M. (1981a). *Mech. Age. Develop.* **17**, 275–281.
Lippman, R. D., Agren, A., and Uhlan, M. (1981b). *Mech. Age. Develop.* **17**, 283–287.

Loschen, G., Fluke, L., and Chance, B. (1971). *FEBS Lett.* **18**, 261–264.

Lynch, R. D. (1980). *Lipids* **15**, 412–420.

Malkoff, D. B., and Strehler, B. L. (1963). *J. Cell Biol.* **16**, 611–616.

Masoro, E. J., Yu, B. P., Bertrand, H. A., and Lynd, F. T. (1980). *Fed. Proc. Fed. Am. Soc. Exp. Biol.* **39**, 3178–3182.

McCay, P. B., Gibson, D. D., Fong, K. L., and Riger, K. (1976). *Biochem. Biophys. Acta* **431**, 459–468.

McCay, P. B., Gibson, D. D., and Hornbrook, R. (1981). *Fed. Proc. Fed. Am. Soc. Exp. Biol.* **40**, 199–205.

McCord, J. M., and Fridovich, I. (1969). *J. Biol. Chem.* **244**, 6049–6055.

Mead, J. F. (1976). *In* "Free Radicals in Biology" (W. A. Pryor, ed.), pp. 51–68. Academic Press, New York.

Mihara, M., Uchiyama, M., and Fukuzawa, K. (1980). *Biochem. Med.* **23**, 302–311.

Miquel, J., Tappel, A. L., Dillard, C. J., Herman, M. M., and Bensch, K. G. (1974). *J. Gerontol.* **29**, 622–637.

Miquel, J., Oro, J., Bansch, K. G., and Johnson, J. E. Jr. (1977). *In* "Free Radicals in Biology" (W. Pryor, ed.), Vol. III, pp. 133–182. Academic Press, New York.

Miquel, J., Lundgren, P. R., and Johnson, J. E. Jr. (1978). *J. Gerontol.* **33**, 5–19.

Nandy, K., and Bourne, G. H. (1966). *Nature (London)* **210**, 313–314.

Nohl, H. (1979). *Z. Gerontologie* **12**, 9–18.

Nohl, H., and Hegner, D. (1978a). *Eur. J. Biochem.* **82**, 563–567.

Nohl, H., and Hegner, D. (1978b). *FEBS Lett.* **89**, 126–130.

Nohl, H., and Kramer, R. (1980). *Mech. Ageing Dev.* **14**, 137–144.

Nohl, H., Hegner, D., and Summer, K. H. (1979). *Mech. Ageing Dev.* **11**, 145–151.

Oberly, L. W., Oberley, T. D., and Buettner, G. R. (1980). *Med. Hypoth.* **6**, 249–268.

Packer, L., and Smith, J. R. (1974). *Proc. Natl. Acad. Sci. U.S.A.* **71**, 4763–4767.

Packer, L., and Smith, J. R. (1977). *Proc. Natl. Acad. Sci. U.S.A.* **74**, 1640–1641.

Pedersen, T. C., and Aust, S. D. (1972). *Biochem. Biophys. Res. Commun.* **48**, 789–795.

Pfeifer, P. M., and McCay, P. B. (1972). *J. Biol. Chem.* **247**, 6763–6769.

Player, T. J., Mills, D. J., and Horton, A. A. (1977). *Biochem. Biophys. Res. Commun.* **78**, 1397–1402.

Polgar, P., Taylor, L., and Brown, L. (1978). *Mech. Ageing Dev.* **7**, 151–160.

Porta, E. A., Joun, N. S., and Nitta, R. T. (1980a). *Mech. Ageing Dev.* **13**, 1–39.

Porta, E. A., Nitta, R. T., Kia, L., Joun, N. S., and Nguyen, L. (1980b). *Mech. Ageing Dev.* **13**, 319–355.

Recknagel, R. O., and Ghoshal, A. K. (1966). *Lab. Invest.* **15**, 132–146.

Reddy, K., Fletcher, B., Tappel, A., and Tappel, A. L. (1973). *J. Nutr.* **103**, 908–915.

Reiss, U., and Gershon, D. (1976). *Biochem. Biophys. Res. Commun.* **73**, 255–262.

Riga, S., and Riga, D. (1974). *Brain Res.* **72**, 265–275.

Sagai, M., and Ichinose, T. (1980). *Life Sci.* **27**, 731–738.

Sheldahl, J. A., and Tappel, A. L. (1974). *Exp. Gerontol.* **9**, 33–41.

Shimasaki, H., Nozawa, T., Privett, O. S., and Anderson, W. R. (1977). *Arch. Biochem. Biophys.* **183**, 443–451.

Shimasaki, H., Veta, N., and Privett, O. S. (1980). *Lipids* **15**, 236–241.

Siakotas, A. N., and Armstrong, D. (1975). *In* "Nemobiology of Aging" (J. M. Ordy and K. R. Brizzee, eds.), p. 369. Plenum, New York.

Siakotos, A. N., and Koppang, N. (1973). *Mech. Ageing Dev.* **2**, 177–200.

Sohal, R. S., and Donato, H. Jr. (1979). *J. Gerontol.* **34**, 489–496.

Spoerri, P. E., and Glees, P. (1974). *Mech. Ageing Dev.* **3**, 131–155.

Spoerri, P. E., and Glees, P. (1975). *Exp. Gerontol.* **10**, 225–228.

Stohs, S. J., Hassing, J. M., Al-Turk, W. A., and Masoud, A. N. (1980). *Age* **3**, 11–14.

Story, J. A., and Kritchevsky, D. (1974). *Experentia* **30**, 242–243.

Story, J. A., and Kritchevsky, D. (1978). *In* "Liver and Aging—1978" (K. Kitani, ed.), pp. 193–202. Elsevier/North Holland, New York.

Story, J. A., Tepper, S. A., and Kritchevsky, D. (1976). *Lipids* **11**, 623–627.

Svingen, B. A., Buege, J. A., O'Neal, F. O., and Aust, S. D. (1979). *J. Biol. Chem.* **254**, 5892–5899.

Sylven, C., and Glavind, J. (1977). *Int. J. Vit. Nutr. Res.* **47**, 9–16.

Takeuchi, N., Tanaka, F., Katayama, Matsmiya, K., and Yamamura, Y. (1976). *Exp. Gerontol.* **11**, 179–185.

Tappel, A. L. (1980). *In* "Free Radicals in Biology" (W. A. Pryor, ed.), pp. 2–44. Academic Press, New York.

Tappel, A. L., and Dillard, C. J. (1981). *Fed. Proc. Fed. Am. Soc. Exp. Biol.* **40**, 174–178.

Tappel, A., Fletcher, B., and Deamer, D. (1973). *J. Gerontol.* **28**, 415–421.

Thomas, T. R., Londeree, B. R., Gerhardt, K. L., and Gehrke, C. W. (1978). *Mech. Ageing Dev.* **8**, 429–434.

Tien, M., Svingen, B. A., and Aust, S. D. (1981). *Fed. Proc. Fed. Am. Soc. Exp. Biol.* **40**, 179–182.

Tolmasoff, J. M., Ono, T., and Cutler, R. C. (1980). *Proc. Natl. Acad. Sci. U.S.A.* **77**, 2777–2781.

Travis, D., and Travis, A. (1972). *J. Ultrastruct. Res.* **39**, 124–148.

Turrens, J. F., and Boveris, A. (1980). *Biochem. J.* **191**, 421–427.

Uchida, K., Nomura, Y., Kadowaki, M., Takase, H., Takano, K., and Takeuchi, N. (1978). *J. Lipid Res.* **19**, 544–552.

Walker, R., Rahim, A., and Parke, D. V. (1973). *Proc. Roy. Soc. Med.* **66**, 780.

Werner, M., Tolls, R. F., Hultin, J. V., and Mellecker, J. (1970). *Z. Klin. Chem. Klin. Biochem.* **8**, 105–115.

West, C. D. (1979). *J. Comp. Neur.* **186**, 109–116.

Witting, L. A. (1980). *Free Radicals Biol.* **4**, 295–319.

Wright, J. R., Colby, H. D., and Miles, P. R. (1981). *Arch. Biochem. Biophys.* **206**, 296–304.

Yeh, Y. Y., and Johnson, R. M. (1973). *Arch. Biochem. Biophys.* **159**, 821–831.

Chapter **4**

Membranes

I. OVERVIEW

The idea that membranes are involved in the aging process is an attractive one because of their pervasive importance in cellular function and the obvious relationship of lipid-rich membranes to free radical theories involving lipid peroxidation (Chapter 3). Membranes not only mediate metabolite and ion fluxes between the interior and exterior of the cell and its organelles, but are involved in processes requiring structural elements such as respiratory assemblies, or more generally, membrane-bound enzymes and hormone receptors. Obviously then, changes in membranes could affect permeability of ions, active transport, enzyme kinetics, or complex processes such as electron transport. Whether or not changes in membranes, if they exist, initiate and develop the process of aging or are simply a result of the process, is of course, another matter and one for which we have no answer. There does seem to be some experimental evidence for age-related changes in membranes (e.g., changes in mitochondrial function, altered behavior of membrane-bound enzymes, changed lipid composition of mitochondrial and microsomal membranes), although their effect on metabolism is obscure.

One method often used in the search for age-related changes in membranes is to tackle the problem from the point of view of changes in lipid content. Thus, direct analysis of fatty acid composition, ratios of saturated to unsaturated fatty acids, molar ratio of cholesterol to phospholipids, and relative amounts of specific lipids have all been utilized. The reported results show few characteristics in common. If age-related changes occur, their nature varies with the type of membrane involved.

A consideration pursued by some investigators is that if polyunsaturated fatty acids are indeed peroxidized, then the content of polyunsaturated fatty acids in membranes should decrease with age. For a number of reasons, this approach seems to be overly simplified. For one thing, membranes turn over, and thus altered components would be replaced. Moreover, turnover appears to be unaltered by age. Therefore, in old animals, a reduction in the amount of polyunsaturated fatty acids would presumably have to come about by altered synthetic patterns or by peroxidation at a rate above that of replacement and segregation of cross-linked membranes in pigment granules. In this regard, evidence has been provided for an increased rate of free radical formation in mitochondria from old compared to young rats (Chapter 5, Section II). On the other hand, accumulation of age pigments tends to be linear with time—that is, they do not seem to form faster in old organisms. Moreover, there is some evidence that peroxidation of phospholipids *in situ* does not take place readily. Finally, the direct evidence for a decreased level of polyunsaturated fatty acids in old membranes is contradictory.

One can, as is typical in most areas of aging research, come up with counter arguments that are supported by appropriate examples of experimental evidence, but that is exactly what they are—arguments, not established facts. A reasonable point of view is that some peroxidation of membranes occurs, as proved by the continuous production of age pigments throughout the life-span of animals. Evidence that these pigments are indeed a result of peroxidation lies in the fact that several investigators have reduced their amount by feeding high levels of vitamin E, an antioxidant (Chapter 3, Section IV). Quite telling is the finding that this protection against peroxidation has no discernible effect on the life-span of mammals, although the life-span of certain invertebrates (nematodes, rotifers) appears to be lengthened. In short, the level of detectable peroxidation and also presumably, of undetected peroxidation, can be substantially reduced. Yet, physiological aging and life-span of mammals, as far as current experiments have shown, are not affected.

Besides studies of lipid composition, other approaches to age-related effects on membranes, such as changes in transport activity, properties of membrane-bound enzymes, or ultrastructure have been explored. Although no clear picture emerges, the results taken together indicate that indeed there are age-related changes in membranes. The nature of the changes, their effect on membrane architecture, the degree of functional loss, and even the types of membranes affected are yet to be ascertained.

II. LIPID COMPOSITION

With regard to the lipid composition of membranes, investigations suffer from the fault all too frequent in such generalized, but necessary, studies of aging, namely, the results obtained are sometimes contradictory. Table 4.I summarizes relevant data. As can be seen, there is a roughly consistent pattern of changes between the same type of membranes in different tissues, but no apparent relationship between different membranes (e.g., mitochondrial versus microsomal) even in the same tissue. The sparsity of data makes it premature to draw general conclusions, but the cholesterol:phospholipid ratio seems to increase with age. This change would affect the fluidity of the membranes.

Grinna (1977a) made a fairly detailed study of the lipid composition of mitochondrial and microsomal membranes from rat liver and kidney (Table 4.I) and provided a useful review (1977b). The author found that microsomal phospholipids per milligram of protein (liver and kidney) dropped about 13–17%, but cholesterol content did not change significantly at 6 versus 24 months of age. Mitochondrial membranes from the old animals showed substantial increases in cholesterol (69% in kidney) and no change in phospholipid content. In general, the microsomes contained 2 to 3 times more phospholipid and 6 to 7 times more cholesterol than the corresponding mitochondrial membranes.

As to individual phospholipids, no age-related changes were observed in the amount of phosphatidylcholine or phosphatidylethanolamine as the percentage of total phospholipids. Individual fatty acids in microsomes and mitochondrial membranes showed age related changes in both polar and neutral lipids which are listed in Table 4.II. The trends are remarkably consistent for each tissue of origin [except for arachidonic acid (20:4) in the neutral lipids of kidney mitochondria] rather than the type of membrane (microsome versus mitochondrial). This result is quite the opposite of the situation shown in Table 4.I.

In contrast to the rather substantial increases in docosahexenoic acid (22:6) reported for liver microsomes and mitochondrial membranes, change in this compound was not observed in the comparable kidney fractions. In neutral lipids, oleic acid (18:1) showed an increase with age in all membranes; in polar lipids, only kidney fractions showed an increase.

Hawcroft and Martin (1974) studied lipid composition of mouse liver microsomes at intervals from 3 to 22 months of age. The changes they reported are rather substantial. For example, estimated from the published graphs, loss of total phospholipids was about 57% and phosphatidylcholine, about 35% between 3 and 22 months. The latter dif-

Table 4.I. Reported Changes in Membrane Lipid Composition with Respect to Age[a]

Tissue	Age (months)	Lipids[b] Total PL	PC	PE	C	C/PL	PS	Reference[c]
Rat liver mitochondria	6 versus 24	NC	NC	NC	–	+		1,2
Rat kidney mitochondria	6 versus 24	NC	NC	NC	+	+		1,2
Rat heart mitochondria		NC						1
Rat liver microsomes	6 versus 24	–	NC	NC	NC	+		1,2
Rat kidney microsomes	6 versus 24	–	NC	NC	NC	+		1,2
Rat heart microsomes		NC						1
Mouse liver microsomes	4 versus 22	–	–	+			–	3
Rat liver plasma membrane	2.7 versus 26	(–)[d]		(–)[d]	+			4
Rat muscle sarcoplasmic reticulum	12/24/28[e]	NC	+	(–)[d]		–[f]		5
Human muscle membrane	22 → 73 (years)[g]	NC			NC			6

[a] PL, Phospholipids; PC, phosphatidylcholine; PE, phosphatidylethanolamine; C, cholesterol; PS, phosphatidylserine. (+), increases with age; (–) decreases; NC, no change.

[b] Based on milligrams of protein.

[c] Key to references: (1) Grinna and Barber, 1972; (2) Grinna, 1977a; (3) Hawcroft and Martin, 1974; (4) Hegner and Platt, 1975; (5) Bertrand et al., 1975; (6) Thomas et al., 1978.

[d] Difference not statistically significant. These experiments are based on percentage of total lipids.

[e] Younger animals also studied.

[f] Low at 12 months, higher at 24 months, and low again at 28 months.

[g] Arrow indicates intervening age-groups.

Table 4.II. Age-Related Changes in the Fatty Acid Composition of Membranes

Tissue	Age (months)	Percentage change with age in unsaturated fatty acid content[a,b]						Reference
		In polar lipids				In neutral lipids		
		18:1	18:2	20:4	22:6	18:1	20:4	
Rat liver microsomes	6 versus 24	NC	−66	NC	+320	+22	NC	Grinna, 1977a
Rat liver mitochondria	6 versus 24	NC	−20	NC	+159	+19	NC	Grinna, 1977a
Rat kidney microsomes	6 versus 24	+50	NC	NC	NC	+27	NC	Grinna, 1977a
Rat kidney mitochondria	6 versus 24	+26	NC	NC	NC	+27	+63	Grinna, 1977a
		In total lipids						
Heart mitochondria (inner membrane)	3 versus 24[c]	NC	+20	−19	−17			Nohl, 1979
Human muscle membrane	22 → 73 (yr.)[d]	NC	NC	NC				Thomas et al., 1978
Rat liver microsomes	3 versus 17	NC	−23	NC	+58			Grinna, 1976
Mouse liver microsomes	4 versus 22	—	—	—				Hawcroft and Martin, 1974
Rat liver plasma[e] membrane	3 versus 24	+32	−13	−8	+127			Hegner, 1980

[a] Calculated as the percentage increase or decrease over the young value.
[b] NC, No change; (+), increases with age; (−), decreases with age.
[c] Percentage of total acids as recovered by gas chromatography.
[d] Arrow indicates several intervening age-groups.
[e] Fatty acids in lecithin.

ference is even greater between 10 or 12 months and 22 months. By contrast, Grinna (1977a) observed only a small decrease in phospholipids in liver microsomes (17% between 6 and 24 months) and very small changes between 11, 42, and 67 weeks of age in rats being tested for effects of Vitamin E (Grinna, 1976). The results of Hawcroft and Martin (1974) also are conflicting for phoshatidylcholine and ethanolamine. Unfortunately, the picture of age-related changes in lipid composition of membranes is even further complicated, as Thomas *et al.* (1978) found no significant changes in fatty acid composition of the membrane fraction of human muscle tissue obtained by biopsy of groups of subjects aged 20–30, 31–40, 41–50 and 65–73 years, respectively. Moreover, they were a composite of nuclear, mitochondrial and microsomal membranes. The authors point out that the oldest subjects used in the study were relatively active for their ages and may not, therefore, present a typical "old" pattern. On the other hand one might view the lack of change as militating against the existence of a common mechanism of membrane change with age. In this regard, Bertrand *et al.* (1975) found few age-related differences in the sarcoplasmic reticulum of rat skeletal muscle using animals at a number of ages from 1 through 28 months. Phospholipid content did not change in adult animals. Phosphatidylcholine increased somewhat between 12 and 28 months and phosphatidylethanolamine decreased. However, there was enough fluctuation between age groups that all of the changes appear to be within normal physiological range. Thus, the changes noted would not seem to be of great significance to aging.

From the above results, there appear to be age-related changes in the lipid composition of certain membranes, but these changes are not consistent either as to membrane type or to tissue of origin. Given the wide variation of results, too much emphasis at this time should not be placed upon quantitative aspects of the changes. The information available does not support the idea that peroxidation in old animals is so extensive as to reduce the amount of polyunsaturated fatty acids as general phenomenon, although perhaps it exists in specific places, such as the inner mitochondrial membrane as proposed by Nohl (1979) (Table 4.II). In fact, Grinna (1977a) reported a very large increase in highly unsaturated docosahexenoic acid (22:6) with age in liver (but not kidney) microsomes and mitochondria and an increase in arachidonic acid (20:4) in the neutral lipid of kidney mitochondria (Table 4.II). Hegner (1980) also reports an increase in docosahexenoic acid though he reports that lecithin becomes less saturated in old rat liver plasma membranes.

A large source of error undoubtedly exists in the preparation of membranes. For example, microsomes consist of a variety of components.

Moreover, age-related changes might change the susceptibility of membranes to fragmentation and thus would yield dissimilar preparations for analysis. In short, the reports of age-related changes in lipid components could be in part, a reflection of differences in the preparations. One should also bear in mind that the effect of diet can bring about as great a change in membrane composition as aging, or perhaps a greater one. In Grinna's study (1976) of the effect of vitamin E on fatty acid composition of rat liver microsomes, the stock diet animals at 11 weeks showed 5.5% of total lipids as docosahexenoic acid (22:6) versus 1.9% for the experimental diet (with or without vitamin E), a change of 65%. At 67 weeks of age, the values are 8.7 versus 5.6%, respectively. Other fatty acids showed even larger differences between the two diets. In fact, linoleic acid (18:2) showed a difference of 380% between stock and experimental diets in young rats. If "normal" differences due to diet can be so large, how important are the changes in composition reported for aging? Another dietary consideration is that Hegner and Platt (1975) found that although cholesterol levels were elevated and phospholipid levels decreased in old rat plasma membranes (2.7 versus 22 months), the feeding of phospholipids rich in dilinoleylphosphatidylcholine tended to reduce the differences. The rather unspecific changes in the lipid composition of membranes that have been observed appear to fluctuate quantitatively more with conditions unrelated to aging than with aging itself.

It should be noted that substantial changes in the lipid composition of membranes have been reported to occur at early ages, yet there appears to be no threatening loss of function—young animals are healthy—suggesting that lipid composition provides a rather insensitive probe into meaningful membrane changes. Although much of the work appears to have been carried out by well-established procedures, and the changes in lipid composition have been determined with apparent care and are statistically significant (Grinna, 1977a), there is a danger inherent in lack of comparative results from other laboratories when dealing with relatively crude subfractions of membranes such as microsomes. An idea of the difficulties can be obtained from the contradictory data obtained with mitochondria (Chapter 5, Section II), an organelle whose preparation is well established and for which the integrity of the preparations may be tested. In reality, little can be understood about the meaning of the changes provided by the pioneering studies mentioned above, standing as they do without an adequate supporting framework of information. It must suffice for the present to assume that the evidence indicates that indeed there are probably changes in the lipid composition of some membrane types with age. That these changes are due to technical differences in obtaining membranes from old tissues, i.e., greater adherence of lipids, or a different fragmentation pattern for old membranes is a real possibility. The latter situation in itself would suggest that there are changes in membrane properties in old organisms.

III. PEROXIDATION OF MEMBRANE LIPIDS

The idea that membranes are damaged by peroxidation is supported by the fact that the residual bodies that accumulate in old tissues appear to contain membranous material. However, the idea that old membranes are damaged (with the inference that young ones are not) is a different consideration altogether. Nohl (1979) reported that the ratio of unsaturated to saturated fatty acids decreases with age (from 2.51 to 1.98, a change of 21%) in the inner membrane of rat heart mitochondria. This information together with the observation of an increased net production of free radicals by old mitochondria (Nohl and Hegner, 1978) was interpreted by these authors as support for the idea of free radical damage to the old membranes. The change in membranes was demonstrated by differences in discontinuities of Arrhenius plots of several inner membrane enzymes in mitochondria of young versus old rats (Nohl, 1979). Hawcroft and Martin (1974) reported a rather suddenly appearing reversal in the ratio of unsaturated to saturated fatty acids in mouse liver microsomes. At roughly about 21–24 months of age, the percentage of saturated acids became greater than unsaturated. Examination of the data of Grinna (1977a) shows no such change in the relationship of saturated to unsaturated fatty acids in liver or kidney microsomes or mitochondrial membranes. Thomas *et al.* (1978) also found no such changes in human muscle membranes. On the basis of current data, one would, therefore, have to conclude that if there is a measurable loss of unsaturated fatty acids, it must be restricted to specific membranes, perhaps even to the inner mitochondrial membrane.

A bothersome problem with the idea that membranes are damaged as a result of increased peroxidation is the magnitude of the reported change in ratio of unsaturated to saturated fatty acids. It would seem that even subtle changes in composition should alter membrane function. The decrease of 21% in the ratio of unsaturated to saturated fatty acids (Nohl, 1979) between young and old rat mitochondrial membranes seems overwhelming, particularly since mitochondria from old rats still maintain normal function in most respects and good function in others (Chapter 5, Section II). Hegner (1980) showed an 11% decrease in the unsaturated:saturated fatty acid ratio in lecithin from rat liver plasma membranes, but there is a large increase in docosahexenoic acid (22:6) in the older animals.

Another difficulty with the idea of peroxidative damage arises from the already mentioned problem of membrane turnover. Either turnover slows with age (it does not appear to) or old mitochondrial membranes are very quickly peroxidized so as to spend their limited period of existence in damaged form. If they are irreversibly cross-linked, they seem to become sequestered in lysosomes.

IV.   TURNOVER

Membrane turnover does not appear to change with age, although the studies available are far from being comprehensive. Grinna (1977c) studied the turnover of lipid components of liver and kidney microsomal and mitochondrial membranes in rats of 6, 12, and 24 months of age. [^{14}C] and [^{3}H] glycerol were injected 3 days apart, and the animals were sacrificed 3 hours after the second injection. The ratio of ^{3}H:^{14}C in the extracted lipids provides an estimate of relative turnover rates. The procedure, originally devised to study protein turnover, is based on the adaptation to lipids of Lee and Snyder (1973). In general, no age related change was found in the turnover of phospholipid or neutral fractions or for the individual phospholipids, phosphatidylethanolamine and phospatidylcholine in either microsomes or mitochondria when comparing rats of 6 versus 24 months. The 12-month values showed a generalized faster turnover. If the experimental technique provides an accurate picture, the results mean that not only synthesis and degradation, but also transport of lipids into the membranes are unchanged with age, as many of the mitochondrial lipids are synthesized in the endothelial reticulum. From there, they must be transferred to the mitochondria. There is other evidence that suggests that age does not affect membrane turnover. Menzies and Gold (1971) found no change in the half-life of mitochondrial proteins from a variety of tissues in rats of 12 versus 24 months and in brain (Menzies and Gold 1972), although subclasses of the protein obtained by organic solvent extraction gave differing results. Comolli *et al.* (1972) found no age-related differences in the turnover of tissue fractions representing rat liver microsomes, lysosomes, and mitochondria.

V.   CHANGE IN OTHER MEMBRANE COMPONENTS

There are few reports of the effects of age on the composition of membrane components other than lipids. Barclay *et al.* (1973) reported changes in the relative amounts of membrane proteins of rat liver plasma membrane. The amount of protein extractable with salt was increased with age from 10 through 116 weeks of age.

VI. CHANGES IN FUNCTION

Bertrand et al. (1975) examined the functional ability of muscle membranes from aged rats. They measured the ATPase required for Ca^{2+} transport together with the steady state concentration of membrane-bound phosphorylated intermediate and the rate of Ca^{2+} transport. At the steady state, the phosphorylated intermediate was unchanged with age. ATPase activity fell from 2 through 12 months of age and rose at 24 and 28 months, ending at a higher level than at the young ages. Ca^{2+} transport, although it proved difficult to measure with consistent results, was not reduced with age, but appeared to increase after 12 months. In fact, the efficiency of Ca^{2+} uptake/Ca^{2+},Mg^{2+}-ATPase activity may also increase with age. It certainly does not decrease.

Other age-related changes of membrane function include the slowing of cholic acid transport into rat hepatocytes (Hegner, 1980). Thymidine transport also is reduced. The content of membrane-bound hormone receptors may also change with age (Chapter 10).

VII. CHANGES IN MEMBRANE-BOUND ENZYMES

A sensitive method for detecting changes in membranes should be the comparison of membrane-bound enzymes from young and old animals (Chapter 9, Section VIII). Even subtle changes in the membrane would most likely be reflected in changed enzyme properties. Evidence for such alterations has been reported by Nohl (1979) to occur in the inner mitochondrial membrane (Chapter 5, Section II). Grinna and Barber (1972) provided evidence for changes with age in a number of microsomal enzymes, presumable due to membrane alteration. Later, Grinna and Barber (1975) suggested that the lower specific activity that had been noted for glucose 6-phosphatase was due to the presence of fewer enzyme molecules in the old preparations. Grinna (1977d) more recently attributed the altered enzyme activity in rat kidney and liver microsomes to membrane changes; based on differences observed in Arrhenius plots. However, transport of the substrate into the microsomal vesicles may be a complicating factor in measuring this enzyme (Chapter 5, Section III). Bertrand et al. (1975) found that ATPase activity declined in rat skeletal muscle sarcoplasmic reticulum between 2 and 12 months and then rose, ending up with a higher activity at 28 months than at 2 months. Hegner and Platt (1975) observed that Mg^{2+}-ATPase and Na^+, K^+-ATPase activities in liver plasma membranes of young rats (2.7 months) are almost 2 times those for the old animals (22 months of age), a finding quite different from that described above for Ca^{2+}, Mg^{2+}-ATPase in rat sarcoplasmic reticulum membranes. The use of animals only 2.7 months old for comparison with aged animals is a matter of concern. Barclay et al. (1973) reported that Na^+, K^+-ATPase was constant in the rat liver plasma membranes from about 5 to 22 months of age. However, before 5 months of age, the value was over twice as large, showing a sharp drop between 2.5 and 5 months. The drop observed by Hegner and Platt (1975), presumably occurred in this early period. Barclay et al. (1973) also reported that the activity of 5'-nucleotidase, a membrane-bound enzyme, dropped between 16, 22, and 29 months. Changes in other membrane-bound enzymes are discussed in Chapter 9, Section VIII.

It should be borne in mind that although the work with Arrhenius plots shows differences in membrane properties, measurement of the specific activity of enzymes cannot distinguish between altered membranes, which cause reduced enzyme function, and the presence of fewer enzyme molecules. Moreover, even the Arrhenius plots are subject to age-related changes in cell structure, which may in turn, subtly alter the nature of the isolated membrane preparations.

VIII. COMMENT

There is substantial qualitative evidence for changes in various types of membranes with age shown by changes in lipid composition, in membrane-bound enzymes, in mitochondrial function, and in hormone receptors. Yet, there are disturbing contradictions and a lack of consistency in the results, from author to author, which leaves one uneasy about the quantitative aspects of the studies. The nagging realization that even modest changes in the degree of contamination of membrane preparations (perhaps resulting from age differences) would distort the analyses, further reduces any feeling of security. One must wonder whether changes would not be more subtle than a 70% increase in this fatty acid and a 40% decrease in that one. Moreover, the prevailing use of animals at only two ages does not really show that the changes occur late in the life span. At least it seems certain that whatever changes exist are not specific for any one type of membrane: no single change that affects all membranes has been described. Clarification of the meaning of age-related changes in membranes must lie in long-term study by different laboratories using the best prevailing technology for isolation and study of membranes. Given the importance of the field and the rapid advances now being made in research on membrane biology, it is unfortunate that there are so few investigators working on age-related effects.

Masoro (1975) and more recently Hegner (1980) have provided overviews of the effect of age on membrane composition and function.

IX. TISSUE CULTURE

The information available makes it clear that changes occur in the properties of the surface of cells at late passage. The changes appear to be related to alterations in glycoproteins, although the evidence is, as yet, more suggestive than conclusive. Bowman and Daniel (1975) reported changes in surface features of late-passage cells as seen by electron microscopy. Azencott and Courtois (1974) reported changes in adhesiveness of chick fibroblasts suggesting changes in the membrane surface. Yamamoto *et al.* (1977) observed a decrease in agglutination caused by concanavalin A (Con A). The authors suggest that an increase in a hyaluronic acid-protein conjugate at the cell surface is responsible for these results.

Milo and Hart (1976) reported a large decrease (67%) in the sialic acid content of Phase III compared to Phase II WI-38 cells. The analysis was on the basis of per milligrams of protein. However, since protein nearly doubles at late passage (Schneider and Mitsui, 1976; Spataro *et al.*, 1979), the real relationship of the finding to "aged" cells is not certain. Yet, sialic acid is reported to be largely responsible for the negative charge on the cell surface, and the net charge decreases with passage time (Bosmann *et al.*, 1976). Spataro *et al.* (1979) studied sialyltransferase activity (transfer of terminal sialic acid residues into glycoproteins) in middle (18–23) and late passage (40–45) WI-38 cells. Total enzyme activity increased sharply in the latter, but this result appeared to be due to the presence on old cells of more acceptors for sialic acid. The authors concluded that there is a decreased amount of membrane-bound sialic acid in "old" cells, presumably representing a decreased amount of sialic acid in the glycoproteins.

As to other changes in membrane properties, Berumen and Macieira-Coelho (1977) observed an increase in albumin uptake late in the life span of human adult lung fibroblasts, presumably due to changes in membrane permeability. Aizawa and Mitsui (1979) showed that after treatment with Con A, red blood cells adsorbed increasingly to human fetal lung cells with increasing passage number. Subsequently, Aizawa *et al.*, (1980a) extended the work to cells from heart, liver, skin, and muscle. All showed a continously increasing degree of adsorption of Con A-treated red cells with passage number. If the fibroblasts (instead of the red cells) are coated with Con A, the adsorption of red cells increases only at Phase III (Aizawa et al., 1980b). Of interest is that skin fibroblasts show increased adsorbance of Con A treated red cells with donor age, thus reflecting the situation observed in vitro. The authors suggest that the adsorption is in part related to the presence of a large protein (MW 220,000) at the cell surface (Aizawa et al., 1980c). The change in binding of Con A-treated red blood cells appears to be related to division age and not to cell size or even ability to divide rapidly. That is, even rapidly dividing cells in later passage behaved as "old" cells (Aizawa et al., 1980d). On the other hand, red cell binding to Con A-treated cells was related to nondividers.

Passage number does not appear to affect cell surface antigens. Moley and Engelhardt (1981) found no change in specific cell surface antigens in cells derived from human foreskins. Brautbar et al. (1972, 1973) had earlier reported no change, qualitative or quantitative, in HLA antigens in several human diploid cell strains.

There has been little work on chemical changes in membranes of cells in culture. Polgar et al. (1978) found no change in the fluidity of membranes of late-passage human embryo lung fibroblasts.

REFERENCES

Aizawa, S., and Mitsui, Y. (1979). J. Cell Physiol. 100, 383–387.
Aizawa, S., Mitsui, Y., Kurimoto, R., and Matsuoka, K. (1980a). Mech. Ageing Dev. 13, 297–306.
Aizawa, S., Mitsui, Y., and Kurimoto, F. (1980b). Exp. Cell Res. 125, 287–296.
Aizawa, S., Mitsui, Y., Kurimoto, F., and Nomura, K. (1980c). Exp. Cell Res. 127, 143–157.
Aizawa, S., Mitsui, Y., Kurimoto, F., and Matsuoka, K (1980d). Exp. Cell Res. 125, 297–303.
Azencott, R., and Courtois, Y. (1974). Exp. Cell Res. 86, 69–74.
Barclay, M., Skipski, V. P., and Terebus-Kekish, O. (1973). Mech. Ageing Dev. 1, 357–365.
Bertrand, H. A., Yu, B. P., and Masoro, E. J. (1975). Mech. Ageing Dev. 4, 7–17.
Berumen, L., and Macieira-Coelho, A. (1977). Mech. Ageing Dev. 6, 165–172.
Bosmann, H. B., Gutheil, R. L. Jr., and Case, K. R. (1976). Nature (London) 261, 499–501.
Bowman, P. D., and Daniel, C. W. (1975). Mech. Ageing Dev. 4, 147–158.
Brautbar, C., Payne, R., and Hayflick, L. (1972). Exp. Cell Res. 75, 31–38.
Brautbar, C., Pellegrino, M. A., Ferrone, S., Reisfeld, R. A., Payne, R., and Hayflick, L. (1973). Exp. Cell Res. 78, 367–375.
Comolli, R., Ferioli, M. E., and Azzola, S. (1972). Exp. Gerontol. 7, 369–376.
Grinna, L. S. (1976). J. Nutr. 106, 918–929.
Grinna, L. S. (1977a). Mech. Ageing Dev. 6, 197–205.
Grinna, L. S. (1977b). Gerontology 23, 452–464.
Grinna, L. S. (1977c). Mech. Age. Develop. 6, 453–459.

Grinna, L. S. (1977d). *Gerontology* **23**, 342–347.

Grinna, L. S., and Barber, A. A. (1972). *Biochim. Biophys. Acta* **288**, 347–353.

Grinna, L. S., and Barber, A. A. (1975). *Exp. Gerontol.* **10**, 319–323.

Hawcroft, D. M., and Martin, P. A. (1974). *Mech. Ageing Dev.* **3**, 121–130.

Hegner, D. (1980). *Mech. Ageing Dev.* **14**, 101–119.

Hegner, D., and Platt, D. (1975). *Mech. Ageing Dev.* **4**, 191–200.

Lee, T. C., and Snyder, F. (1973). *Biochim. Biophys. Acta* **291**, 71–82.

Masoro, E. J. (1975). *Adv. Exp. Biol. Med.* **6l**, 81–94.

Menzies, R. A., and Gold, P. H. (1971). *J. Biol. Chem.* **246**, 2425–2429.

Menzies, R. A., and Gold, P. H. (1972). *J. Neurochem.* **19**, 1671–1683.

Milo, G. E., and Hart, R. W. (1976). *Arch. Biochem. Biophys.* **176**, 324–333.

Moley, J., and Engelhardt, D. L. (1981). *J. Gerontol.* **36**, 136–141.

Nohl, H. (1979). *Z. Gerontologie* **12**, 9–18.

Nohl, H., and Hegner, D. (1978). *Eur. J. Biochem.* **82**, 563–567.

Polgar, P., Taylor, L., and Brown, L. (1978). *Mech. Ageing Dev.* **7**, 151–160.

Schneider, E. L., and Mitsui, Y. (1976). *Proc. Natl. Acad. Sci. U.S.A.* **73**, 3584–3588.

Spataro, A. C., Bosmann, H. B., and Myers-Robfogel, M. W. (1979). *Biochim. Biophys. Acta* **553**, 378–387.

Thomas, T. R., Londcree, B. R., Gerhardt, K, O., and Gehrke, C. W. (1978). *Mech. Ageing Dev.* **8**, 429–434.

Yamamoto, K., Yamamoto, M., and Ooka, H. (1977). *Exp. Cell Res.* **108**, 87–93.

Chapter 5

Mitochondria, Microsomes, and Lysosomes

I. OVERVIEW

As is the case for intact organisms, cell components, such as mitochondria and lysosomes, might be expected to display age-related changes at morphological, physiological, and molecular levels. The first of these may be observed by microscopic and perhaps histological techniques; the second by measuring functional aspects of isolated mitochondria, lysosomes, and microsomes; the third by attempting to find changes in molecular composition or enzymatic capability. Mitochondria, particularly, may have problems forced upon them by extramitochondrial events that result from aging. For example, deficiencies in synthesis or transport of mitochondrial components that are synthesized in the cytoplasm would certainly affect function of the organelles. Even small changes in membranes, either pre- or postsynthetic could have significant effects. There could also be changes internal to mitochondria, e.g., reduced protein synthesis or inadequate assembly of respiratory structures. Moreover, these considerations may change from tissue to tissue, so that it might be necessary to deal with liver mitochondria, kidney mitochondria, etc., rather than with an "aging" factor common to all mitochondria.

Considerable attention has been given to morphological changes in cell components with age. Unfortunately, like many of the biochemical studies, the data provided are difficult to assess. Evidence has been provided for age-related effects, which run the gamut from more, but smaller mitochondria, to fewer, but larger mitochondria, to no change at all. The amount of endothelial reticulum has also been claimed to increase, decrease, or not change at all. There does seem to be general agreement, although it is not unanimous, that lysosmal volume increases in cells of old animals.

One important conclusion emerging from the above studies is that there is probably a class of fragile mitochondria in old tissues, which might be lost during processing. Support for this idea has been provided by centrifugal separation of two types of organelles from old rat heart and by several studies that indicate that there is an increased susceptibility of old mitochondria to damage by osmotic shock or by storage.

There have been several detailed studies of mitochondrial oxidative capacity. From this work, it is apparent that in old animals, there is a reduction in the ability to oxidize certain but not all substrates when measured at "full speed" (State 3). At least some of the effects are probably related to membrane changes, but this conclusion is by no means proven. In this regard, evidence has been put forward to explain the deficiencies in old mitochondria on the basis of peroxidation caused by oxygen radicals generated by the organelles. Whether or not the reduced capabilities of "old" mitochondria have any physiological significance is not known.

As to the endothelial reticulum, there are conflicting reports of morphological changes with age. Certain enzyme levels in microsomes show age-related changes, but it is not known if they result from membrane changes, altered enzymes, or a reduction in the number of enzyme molecules. The last of these alternatives is most likely, because in some cases there is a change with age in the degree of increase or decrease of these enzymes in response to various drugs. Results of these studies are in good agreement, but more data are needed dealing with the period between maturity and senescence as well as more information on changes arising from differences in sex, strain, and maintenance conditions, all of which may affect hormonal balance and thus heavily influence the results. In fact, research on microsomal enzymes and their response to drugs has not yet probed deeply into the reasons for the observed age-related differences, that is, whether the response is locally initiated or a result of endocrine changes. Moreover, almost all of the work in the area has so far dealt with rodent systems—providing much too narrow a base from which to draw general conclusions.

Although a number of lysosomal enzymes have been shown to change with age, it is fair to say that no direct link has been established between changes in lysosomal activity and aging, exclusive of the role of these organelles in the formation of age pigments.

Further study of cell components can be expected to provide useful information that will broaden our knowledge of the effects of aging. Although one could argue persuasively that changes in these structures are secondary to such basic processes as those related to gene expression (chromatin behavior, protein synthesis, etc.), study of mitochondrial and microsomal metabolism in old organisms should offer a chance for findings that will point the investigator toward primary facets of aging. On the basis of current results, study of lysosomes would seem to offer less promise.

II. MITOCHONDRIA

There has been considerable research carried out on the effect of age on mitochondria. There seems to be tacit agreement that the organelles are victims of the process and not its instigators.

A. Structure

Although mitochondrial morphology may not appear to be of immediate interest to the biochemist, he must nonetheless be concerned with structural considerations that may provide a rationale for any age-related biochemical changes that may come to light. Therefore, a brief outline of work dealing with changes in mitochondrial numbers, size, and structural elements is provided.

There are several reports that the number of mitochondria decreases with age. Herbener (1976) observed this trend in mouse liver (C57BL/6J) using stereological techniques. Between 30 and 43–44 months of age, the number of mitochondria per unit of cytoplasm decreased 39%. Before that time (between 8 and 30 months), the drop was not statistically significant. Tate and Herbener (1976) also reported the same trends when they compared mouse liver at 9, 18, and 36 months of age. The decrease in the number of mitochondria per unit of cytoplasm (numerical density) was about 20% between 18 and 36 months. This value was not statistically significant (95% confidence level), and the decrease between 9 and 18 months was small. Thus, most of the decrease in mitochondrial number reported by Herbener (1976) between 30 and 40–44 months presumably occurred after 36 months in rather remarkably old animals. In heart mitochondria, Herbener (1976) found a 32% decrease in number between 8 and 30 months, subsequent changes between 30 and 43–44 months being insignificant. Tate and Herbener (1976) found the greatest decrease (about 30%) between 18 and 36 months with little change between 9 and 18 months. The combined results of Herbener

(1976) and Tate and Herbener (1976) with heart mitochondria would seem to place the loss in numbers between 18 and 30 months, whereas in liver, the largest decrease was between 30 and 44 months. The above findings are based on cytoplasmic volume, and, therefore, age-related changes in this parameter would be reflected in the values found for mitochondrial number. Meihuizen and Blansjaar (1980) observed substantial increases in the cytoplasmic volume of hepatocytes from each of three different zones of rat liver between 3 and 35 months of age.

As to other reports of mitochondrial loss, Wilson and Franks (1975) observed a reduction in mitochondrial numbers in old mouse liver (6 versus 30 months), but the difference was not statistically significant. Schmucker (1978) also suggested that in rat liver there is a net loss of mitochondria as a function of age. Volume density (in essence, the cytoplasmic volume occupied by mitochondria) and specific volume showed statistically significant decreases in both centrolobular and portal locations. Meihuizen and Blansjaar (1980) observed a small decrease in mitochondrial numbers in hepatocytes from old rats (3 versus 35 months) from the midzone, but no change in central or peripheral regions. Tauchi and Sato (1968, 1978) found a decline in mitochondrial numbers with age in human liver.

Several findings differ with the above conclusions. Tauchi *et al.* (1974) observed no change in the number of mitochondria in mouse liver between 12 and 24 months of age, although there was an increase compared to young (3-month-old) animals. Zs-Nagy and Pieri (1977) also reported no substantial change in the numerical density of rat liver mitochondria between 12 and 27 months of age. They did observe a decrease between 2 and 12 months, a period which may be considered development, not aging. In contrast to all of the above reports, Kment and Hofecker (1977) reported that there were more (36%) but smaller mitochondria in aged rat heart (5–6 versus 20–21 months), and Pieri *et al.* (1975) reported small increases in mitochondrial number per single rat hepatocyte at 27 versus 12 months, although changes per unit of volume of liver were negligible. Reference to single hepatocytes is problematic since the volume of the cells is not constant.

Not only the number of mitochondria, but the volume density (fraction of the cytoplasm occupied by the mitochondria) has been the subject of much investigation. Results are quite varied, and there seems no way to provide a cohesive summary. Herbener (1976) noted a progressive decrease in the volume density of mouse liver mitochondria that followed the age pattern reported by this author for numerical density. That is, the value for liver dropped between 30 and 44 months and for heart between 8 and 30 months. It should be noted that the average mitochondrion showed no change in volume, so that losses in volume density are due to losses in numerical density. Analogous results were obtained by Tate and Herbener (1976) with animals aged 9, 18, and 36 months, although in this case, the average mitochondrion showed a slight (16%) but not significant increase in volume. Tauchi *et al.* (1974) also observed no statistically significant change in mitochondrial size (not volume) in mouse liver cells at 12 versus 24 months of age. Schmucker (1978) reported that reduction in the volume density of male Fischer 344 rat liver mitochondria occurs at later ages (25 versus 30 months) after an earlier rise in both centrolobular and portal locations. The pattern of change with age for each of the two regions was quite distinct. In general, volume densities based on percentage of liver tissue or volume per cell rose through development from the 1 month levels and declined late in life (typically between 20–25 versus 30 months). The values for the oldest animals were not greatly different from those observed at early ages (1 or 6 months). Of the three liver regions tested, Meihuizen and Blansjaar (1980) found a significant decrease only in mitochondria from midzonal hepatocytes of 35-month-old rats compared to 3-month-old animals.

Other reports differ sharply from the general sense of a decreased mitochondrial volume density. Pieri *et al.* (1975) observed a slight increase in mitochondrial volume between adult and old animals (12 versus 27 months), although there was a considerable decrease between 1 and 12 months. These results, discussed by Zs-Nagy and Pieri (1977), follow, as expected, the pattern of numerical density found by these authors. Not only the volume density, but the size of individual mitochondria, although reported to be unchanged with age by some investigators (see above; Herbener, 1976; Tauchi *et al.*, 1974) has been found by others to alter considerably. Wilson and Franks (1975) observed a considerable increase in the mean size of liver mitochondria (60%) in old mice (6 versus 30 months). Tauchi and Sato (1968, 1978) also reported on increased size (but decreased number) of mitochondria in old human hepatic cells. Kment and Hofecker (1977) reported somewhat different results. They found no change in total mitochondrial area, but a smaller individual size of the organelles in the liver of older rats (5–6 versus 20–21 months). In short, they report an increased number of smaller mitochondria.

As to the structure of mitochondria, several studies provide evidence that age-related changes occur. Tate and Herbener (1976) observed a small, statistically significant reduction (15%) in the surface density of cristae per unit of mitochondria in mouse liver (between 18 and 36 months), but not in heart, suggesting that mitochondria from the two sources age differently. Total cristae per unit of cytoplasm dropped about 35% in both tissues between 9 and 36 months because of the reduced numbers of mitochondria. As mentioned above, Wilson and Franks (1975) found a greatly increased mean size in old mouse liver mitochondria. Structural changes included a light, "foamy" vacuolated matrix, short cristae, and a loss of dense granules. The old preparations showed considerable variability in mitochondrial size. Normal and altered mitochondria were observed side by side, so that it is apparent that all of the organelles do not show age-related changes. The changes were not so dramatic in isolated mitochondria, perhaps because the fractionation procedure leads to considerable loss of abnormal mitochondria (up to 47%) in the old preparations. Pieri *et al.* (1975) also reported the presence of swollen mitochondria in hepatocytes of old rat liver. Investigators should thus take care during the preparation of mitochondria that samples from old animals are not converted to "young" preparations by selective losses.

A number of other investigators also reported evidence that mitochondria become fragile in old animals. Inamdar *et al.* (1974, 1975) found evidence that mitochondria from skeletal muscle, heart, and liver of hamsters become fragile in older animals (18 months for muscle, 15 months for heart, and 20 months for liver). The yield of mitochondria was substantially lower and storage in the cold led to a more rapid lowering of the respiratory control ratio and of phosphorylation in the presence of substrate (pyruvate–malate). Electron micrographs showed no obvious changes in structure. On the other hand, Weindruch *et al.* (1980) observed that the yield of mitochondria from liver, brain, or spleen of aged mice (9–12 months versus 23–26 months) remained the same. Murfitt and Sanadi (1978) separated two bands of mitochondria from rat heart by isotonic gradient centrifugation from both young and old animals (8 versus 27–34 months). The lighter fraction showed little difference in properties with age, but the heavy fraction from old animals was deficient in respiratory efficiency. The authors attribute the results to the presence of partly damaged mitochondria which are perhaps in the process of being turned over.

There is other evidence for greater fragility of mitochondria in old organisms. Weinbach and Garbus (1959) observed a slightly increased sensitivity of old rat liver mitochondria (3–4 versus 24 months of age) to various treatments (e.g., freezing and thawing). Barrett and Horton (1976) observed that rat liver mitochondria from old animals, when exposed briefly (20 sec.) to hypotonic media of appropriate strength (20–50 mOsm/liter), showed an increased decline in respiratory control ratios (25–40%). At higher osmotic strength and in 250 mM sucrose, there were no age-related differences. Since there were no differences in swelling, the authors attribute the problem to changes in membrane properties. Spencer and Horton (1978) provided evidence that mitochondria from old rats (28 months), which were exposed briefly to hypoosmotic media, leak slightly more malic dehydrogenase and significantly more glutamate dehydrogenase than mitochondria from young animals (7–14 months). These data are consistent with an increased permeability of the inner mitochondrial membrane.

There are a few other reports in which no evidence was found for the presence of altered or fragile mitochondria in old animals. Based on a reduction of mitochondrial DNA and protein content, Stocco and Hutson (1978) support the view that there are less mitochondria in liver from old rats (Fischer). Total DNA and protein content of the old mitochondria based on liver weight (which was reported to change little with age) declined 35 and 44%, respectively, in 24 versus 12-month-old animals. However, the mitochondrial protein nearly doubled and DNA more than doubled between 3 and 12 months. If changes of this magnitude are in the normal range, the decline noted between 12 and 24 months does not appear to be particularly notable. The authors used a rate zonal centrifugation technique to separate mitochondrial fractions. In a search for "damaged" mitochondria, they noted that the percentage of cytochrome oxidase and lipoamide dehydrogenase distribution did not change with age when the fractions obtained from the sucrose gradients were divided into three regions. They, therefore, concluded that they were not dealing with losses due to mitochondrial fragility in old preparations. Although it seems unlikely that Stocco and Hutson (1978) lost fragile mitochondria during their relatively gentle isolation procedure, it is regrettable that the amount of cytochrome oxidase recovered in mitochondria from young and old livers was not compared to the total originally present in the respective homogenates. This determination would have shown whether there were extra losses of "old" mitochondria. In this regard, it is of interest that Baird and Massie (1976) did not find evidence for fragility of catalase-containing organelles (presumably peroxisomes) in liver and kidney in two strains of old rats. The amount of catalase in the supernatant solution from old preparations should increase if the particulate fraction containing the enzyme is more fragile. However, the percentage of enzyme activity in the particulate fraction does not decrease with age, and there is not an increase in the proportion of the enzyme that becomes soluble.

The preponderance of evidence suggests that there is a reduction in the number and volume density of mitochondria with age. The explanation for the widely varying results with regard to these and other parameters such as size is not clear. Kment and Hofecker (1977) pointed out that even from the same tissue, quite variable results may be obtained because of the relatively small size of the sample. They also suggest variability occurs during different stages of development and aging. For example, the volume density could decrease during maturation and then remain constant or increase slightly during adulthood and then fall steeply during late aging. Although this proposal might explain some of the observed differences, it does not explain all of them. Wilson and Franks (1975) pointed out that the morphology of mitochondria in different cell types of the liver varies at any given age. For example, Kupffer cells and endothelial cells have smaller mitochondria than parenchymal cells. The differences observed by Schmucker (1978) in hepatocytes from two locations emphasizes the point. In intact hepatocytes, there are definite changes in cytoplasmic volume with age, but such changes are usually disclaimed as being responsible for the reported changes in measuring mitochondrial size or volume. Perhaps more important, are changes that may be due to diet. The staple diet in each laboratory may vary in carbohydrate, lipid, and protein content and thus bring about a compensatory mitochondrial adjustment. Exercise, strain, and sex are other considerations. Technical circumstances in preparing electron micrographs and mitochondrial preparations may also be involved.

The above state of affairs leaves one with a sense of dissatisfaction. However, when all is said and done, the biochemist can afford to be somewhat unconcerned with numbers and size, but he must certainly pay attention to the fact that there is substantial evidence that a fraction of the mitochondria in old preparations have increased fragility. The difference in the conclusions reached by Stocco and Hutson (1978) and Wilson and Franks (1975) and others regarding the presence of fragile mitochondria may be explained by the fact that the latter used liver from mice at ages of 6 versus 30 months, whereas the former used rats at 12 versus 24 months. Thus, the changes noted by Wilson and Franks could have occurred after 24 months. However, as reported above, a number of other investigators have noted changes in the fragility of mitochondria in old animals. Hence, although the situation remains unresolved, the preponderance of evidence seems to support the idea of the existence of damaged mitochondria in old animals. Biochemical studies reinforce the idea that there are age-related changes in mitochondria, although these would seem to be too subtle to be manifested as changes in gross structure.

B. Function

Some of the early work on function consists of assays of mitochondrial enzymes in crude homogenates of various tissues. These results are found in reviews of enzyme changes assembled by Finch (1972) and Wilson (1973). The accuracy of such determinations is open to question. Studies of the effect of age on the metabolic abilities of mitochondria began in the late 1950s and has continued steadily since that time.

1. Respiration

A detailed study of the respiratory activity of mitochondria from heart and skeletal muscle was carried out by Chen *et al.* (1972). They used Fischer 344 rats at several ages between 3 and 28 months. In isolated cardiac muscle mitochondria, with either palmityl carnitine or mal-ate–glutamate as substrate, State 3 respiration decreased slowly till about 16 months of age, and then more sharply. The change was of the order of 18% at 12 versus 28 months. State 3 respiration, in which the rate of phosphorylation of ADP is determined, may be considered to be the maximal respiratory capacity of the mitochondria. In old muscle mi-tochondria, malate–glutamate showed a similar decreased respiratory rate, but palmityl carnitine was unaffected. For succinate and cyto-chrome *c*, there was no age-related change in either tissue. For β-hy-droxybutyrate, the rate was 20% lower in old heart mitochondria, agree-ing with the earlier findings of Weinbach and Garbus (1959) and disagreeing with those of Gold *et al.* (1968), who found no change. Chen *et al.* (1972) attribute the latter result to the use of bovine serum albumin in the assay medium. Chiu and Richardson (1980), however, observed a decrease with age in β-hydroxybutyrate metabolism even though they utilized the conditions of Gold *et al.* (1968). The results of other studies on the effect of age on State 3 respiration are listed in Table 5.I. State 4 oxidation (limiting amounts of ADP) was, in general, unchanged and ADP:O ratios were generally close to theoretical values, suggesting that the integrity of the mitochondria was maintained.

Chen *et al.* (1972) concluded, from the substrates that yielded de-creased respiration in old mitochondria, that the problem was located before NAD reduction and might involve changes in dehydrogenase activity, anion transport, or other transport-associated systems. They noted that changes of around 20% could be physiologically significant under conditions of stress.

With the paper of Chen *et al.* (1972) as the focus, let us examine other work on aging effects in mitochondria. Wilson *et al.* (1975) studied malic dehydrogenase, a matrix enzyme, and cytochrome oxidase, which is attached to the inner membrane, to see if the observed age-related changes in structure (Wilson and Franks, 1975) (Section II,A) had af-fected the level or function of these enzymes in aged (30–32 months) compared to young (3–6 months) mice. No age-related effects were

Table 5.I. Effects of Age on State Three Respiration and Enzyme Activities of Mitochondria[a]

Source	Age (months)	Enzyme or substrate	Change (%)[b]	Reference
Rat liver	12 versus 21–24	Cyt oxidase	−13 (NS)	Paterniti et al., 1980
	7–14–29 (homogenate)	Cyt oxidase	0	Abu-Erreish & Sanadi, 1978
	12 versus 21–24	Succ dehydrog	−68	Paterniti et al., 1980
	3–4 versus 24	HO-butyrate	−39	Weinbach & Garbus, 1959
	6 versus 24	HO-butyrate	−33	Grinna & Barber, 1972
	12–14 versus 24–27	HO-butyrate	0	Gold et al., 1968
	3–4 versus 24	Succinate	0	Weinbach & Garbus, 1959
	14 versus 33–35	Succinoxidase	0	Barrows et al., 1962
		KG + malonate	0	Barrows et al., 1962
		Glu + malonate	0	Barrows et al., 1962
		Malate	0	Barrows et al., 1962
		Succ cyt c reductase	−26	Grinna & Barber, 1972
Mouse liver	6 versus 24	Cyt oxidase	0	Wilson et al., 1975
	6 versus 30	Malate	0	Wilson et al., 1975
		Succinate	0	Wilson et al., 1975
	9–12 versus 23–26	Cyt oxidase	−9	Weindruch et al., 1980
		Glutamate	0	Weindruch et al., 1980
		Mal + pyr	0	Weindruch et al., 1980
		Succ + rotenone	0	Weindruch et al., 1980
		HO-butyrate	−12	Weindruch et al., 1980
Rat kidney	12 versus 21–24	Cyt oxidase	−39	Paterniti et al., 1980
	7–14–29 (homogenate)	Cyt oxidase	0	Abu-Erreish & Sanadi, 1978
	14 versus 33–35 (homogenate)	Succinoxidase	−19	Barrows et al., 1962
	12 versus 21–24	Succ dehydrog	−69	Paterniti et al., 1980
	6 versus 24	Succ cyt c reductase	−31	Grinna & Barber, 1972
		HO-butyrate	−22	Grinna & Barber, 1972
	12–14 versus 24–27	HO-butyrate	0	Gold et al., 1968
Rat skeletal muscle	9 versus 24	Cyt oxidase	0	Chen et al., 1972
		Succinate	0	Chen et al., 1972

continued

Table 5.1. *Continued*

Source	Age (months)	Enzyme or substrate	Change (%)[b]	Reference
Rat heart	12–28	Palmitylcarnitine	0	Chen et al., 1972
	12–28	Glu-mal	−19	Chen et al., 1972
	12–28	Palmityl CoA + carnitine	0	Chen et al., 1972
	5 versus 26	Cyt c oxidase	−27	Abu-Erreish & Sanadi, 1978
	3 versus 24	Cyt c oxidase	0	Nohl, 1979
	6 versus 26	Succinate	0	Abu-Erreish & Sanadi, 1978
	10 versus 24	Succinate	0	Starnes et al., 1981
	9 versus 24	Succinate	0	Chen et al., 1972
	12–14 versus 28–32	Succinate	0	Chiu and Richardson, 1980
	3 versus 24	Succinate	−30	Nohl, 1979
	9 versus 24	Succ dehydrog	−17	Nohl, 1979
	6 versus 24	Cyt oxidase	0	Nohl, 1979
	12–14 versus 24–27	Succ cyt c reductase	0	Grinna & Barber, 1972
	8 versus 24	HO-butyrate	0	Gold et al., 1968
	3 versus 24	HO-butyrate	−23	Chen et al., 1972
	12–14 versus 28–32	HO-butyrate	−19	Nohl, 1979
	6 versus 24	HO-butyrate	−55	Chiu and Richardson, 1980
	12 → 28	HO-butyrate	+18	Grinna & Barber, 1972
	6 versus 24	Glu-mal	−18	Chen et al., 1972
	12–14 versus 28–32	Glu-mal	0	Hansford, 1978[c]
	8 versus 24	Glu-mal	−36	Chiu & Richardson, 1980
	10 versus 24	Glu-pyr	−29	Chen et al., 1972
		Glu	0	Starnes et al., 1981
	6 versus 24	Pyr-mal	0	Starnes et al., 1981
	12–14 versus 28–32	Pyr-mal	0	Hansford, 1978[c]
	8 versus 24	Pyr-mal	−27	Chiu and Richardson, 1980
		KG	0	Chen et al., 1972

12 → 28		Palmitoylcarnitine	−20	Chen et al., 1972
6 versus 24		Palmitoylcarnitine + mal	−27	Hansford, 1978a[c]
		Octanoate + mal	−37	Hansford, 1978a
		Palmitoyl-CoA + mal − carnitine	−12	Hansford, 1978
12 → 28		Palmitoyl-CoA + carnitine	−(NS)	Chen et al., 1972
6 versus 24		Acetylcarnitine + mal	−27	Hansford, 1978a[c]
		Acetate	0	Hansford, 1978a
		Pyr + malonate + carnitine	−22	Hansford, 1978a
6 versus 24		Mal	0	Hansford, 1978a[c]
		Acyl-CoA synthetase	−40	Hansford, 1978a
		Carnitine acetyltransferase	−35	Hansford, 1978a
		Carnitine palmitoyltransferase	0	Hansford, 1978a
		Acyl-CoA dehydrogenase	0	Hansford, 1978a
		3-HO-acyl-CoA dehydrogenase	−33	Hansford, 1978a
12–14 versus 28–32	Rat brain[e]	Glu	−74	Chiu & Richardson, 1980
3–4 versus 24	(N)	Succ	0	Weinbach & Garbus, 1959
12 versus 24[d]	(S)	Pyr-mal	−24	Deshmukh et al., 1980
	(N)	Pyr-mal	0	Deshmukh et al., 1980
	(S)	HO-butyrate	−18	Deshmukh et al., 1980
	(N)	HO-butyrate	−33	Deshmukh et al., 1980
	(S)	HO-butyrate	−(NS)	Deshmukh et al., 1980
	(N)	HO-butyrate dehydrogenase	−36	Deshmukh et al., 1980
	(S)	3-Keto acid-CoA transferase	−27	Deshmukh et al., 1980
	(N)	3-Keto acid-CoA transferase	−32	Deshmukh et al., 1980
	(S)	Acetoacetyl-CoA thiolase	+39	Deshmukh et al., 1980
	(N)	Acetoacetyl-CoA thiolase	+22	Deshmukh et al., 1980
	(S)	Pyruvate dehydrogenase	0	Deshmukh et al., 1980
	(N)	Pyruvate dehydrogenase	0	Deshmukh et al., 1980
12 versus 24	(N)	Glu-mal	−28	Deshmukh & Patel, 1980[d]

continued

99

Source	Age (months)	Enzyme or substrate	Change (%)[b]	Reference
(S)		Glu-mal	0	Deshmukh & Patel, 1980
(N + S)		Glutamine dehydrog	0	Deshmukh & Patel, 1980
(N + S)		Aspartate aminotransferase	0	Deshmukh & Patel, 1980
Rat brain	7-14-29	Cyt c oxidase[f]	-16	Abu-Erreish & Sanadi, 1978
	12–14 versus 28–32	Succinate	0	Chiu and Richardson, 1980
		Glu-mal	0	Chiu and Richardson, 1980
		Pyr-mal	0	Chiu and Richardson, 1980
		Glutamate	-30	Chiu and Richardson, 1980
Mouse brain	9-12 versus 23-26	Cyt c oxidase	0	Weindruch et al., 1980
		Glutamate	0	Weindruch et al., 1980
		Mal + pyr	0	Weindruch et al., 1980
		Succ + rotenone	0	Weindruch et al., 1980
Mouse spleen		Cyt c oxidase	-26 (NS)	Weindruch et al., 1980

[a] Intact mitochondria are used unless otherwise noted.
[b] NS, Not statistically significant.
[c] Measurements were made at 25°C. A set of results was also reported for 37°C.
[d] Three months also.
[e] (S), Synaptic; (N), nonsynaptic.
[f] Homogenate.

observed based on milligrams of protein in the mitochondria; the investigators did not find cytochemical evidence for changes in location of the enzymes with age. They also found no change in State 4 ("resting") respiration and no change in State 3 respiration, using succinate as substrate, both findings in agreement with those of Chen *et al.* (1972). Of course, the loss of the enlarged mitochondria from livers of old mice reported by Wilson and Franks (1975) would tend to leave "young" mitochondria in the final preparations used for assay. In this regard, Inamdar *et al.* (1974) found no changes of activity in young versus old mitochondria from heart, liver, or skeletal muscle of guinea pigs, using glutamate–malate, succinate, pyruvate–malate, or β-hydroxybutyrate as substrates. These authors observed that the "old" mitochondria were less stable to storage than the young preparations, and the yield was lower. A loss of the most fragile organelles during preparation might explain the lack of difference in the surviving mitochondria.

More recent work by Murfitt and Sanadi (1978) indicates that the age-related change reported for State 3 respiration is associated with a heavy fraction of rat heart mitochondria, which can be isolated by isotonic density gradient fractionation. The heart mitochondria from both young and old Fischer 344 rats (8 months versus 27–34 months) separated into upper and lower bands. The upper band showed no significant differences between young and old mitochondria in State 3 respiration (glutamate–malate). In the lower band, State 3 and State 4 respiration rates and respiratory control were sharply depressed in the old preparations. The authors postulated that the lower band mitochondria represent organelles that are being degraded during normal turnover, and in older animals they become more severely damaged before being eliminated. Whatever the case, these mitochondria appear to be responsible for the lower State 3 respiration noted earlier by Chen *et al.* (1972) in old heart muscle mitochondria.

Paterniti *et al.* (1980) examined a number of heme-containing and other enzymes with respect to age in rats of 1–2, 12, and 22–24 months of age. Liver and kidney mitochondria as well as microsomal cytochrome *P*-450 and soluble fractions were assayed. The interest in heme-containing enzymes was inspired by the previous finding that δ-aminolevulinic acid synthetase, which is the key enzyme in heme biosynthesis, decreased 50–70% in liver, heart, and brain of the aging rat, but did not decrease in kidney (Paterniti *et al.*, 1978). Liver mitochondria showed a small decline (not statistically significant) in cytochrome oxidase activity between 12 and 21–24 months. However, there was a 68% drop in the oxidation of succinic acid. In kidney, the losses were 39 and 69%, respectively, for the two activities. The values rose steeply between 1–2 and 12 months, so that the subsequent decrease between 12 and 24 months bring the "old" values relatively close to those for the 1 to 2-month-old animals. These reported changes are rather dramatic. Wilson *et al.* (1975) using mouse liver mitochondria and several other workers using mitochondria from other tissues found no age-related differences in succinate oxidation (Table 5.I). Barrows *et al.* (1962) had reported a small (19%) age-related drop in succinoxidase activity in kidney homogenates. One should bear in mind that measurements of succinate oxidation reflect the most limiting constituent of the total process. Assay of succinic dehydrogenase measures only this enzyme.

Hansford (1978a) compared the metabolism of heart mitochondria in young (6-month) and old (24-month) rats with particular respect to lipid oxidation and observed a generalized loss of the enzymes associated with fatty acid oxidation. In agreement with the report of Chen *et al.* (1972), the State 3 rate of oxidation of palmitylcarnitine was found to be reduced (30%) in old organelles. However, he found no difference in glutamate–malate oxidation, contrary to the results of Chen *et al.* (1972) and Abu-Erreish and Sanadi (1978). In explanation, Hansford (1978a) noted that the absolute rate of oxidation he obtained was lower and suggested that the higher level of phosphate used in his experiments might limit entry of the substrate into the mitochondria, thus preventing the reaction from achieving a high enough rate to show an age-related difference. Acetate metabolism did not decline but acetyl carnitine showed a 27% loss of activity. A reduced ability to oxidize octanoate in old mitochondria (37%) was attributed to a decline in the activity of acyl-CoA synthetase.

Enzyme assays of disrupted mitochondria showed 30–40% lower levels in old preparations of acyl-CoA synthetase, carnitine acetyltransferase, and 3-hydroxyacyl-CoA dehydrogenase. There was no change in carnitine palmitoyltransferase or acyl-CoA dehydrogenase. There was a reduction in the exchange of carnitine and acetylcarnitine across the mitochondrial membrane and this appears to be a result of a decreased pool of exchangeable carnitine. The reduced exchange was not believed to be responsible for the lowered oxygen utilization of the substrates tested. Chiu and Richardson (1980) also examined heart mitochondria from rats of several ages. Comparing ages of 12–14 versus 28–32 months, their results are in agreement with those of Chen *et al.* (1972) with respect to lack of change in the State 3 oxidation of succinate and lowered oxidation of β-hydroxybutyrate and glutamate–malate. The reduction found for pyruvate–malate (-27%) differs from other reports (Table 5.I).

Recently, Starnes *et al.* (1981) studied mitochondria isolated from perfused, working rat hearts at 10 versus 24 months. In their control organs (no perfusion), they found no age-related change in the State 3 respiration of succinate. They also observed no change in pyruvate–malate or glutamate (Table 5.I). With low-intensity perfusion (180 beats/min) the latter two substrates showed reduced respiration (22 and 31%, respectively) in the old samples. With high intensity perfusion (420 beats/min), both young and old respiratory rates fell substantially, but reductions for the latter were greater. From these results, it seems clear that "old" mitochondria responded less well to stress. Respiratory control and ADP:O ratios were unchanged with age in all cases, suggesting that once NADH is produced, oxidative phosphorylation proceeds normally in old rats. The authors suggested, as did Chen *et al.* (1972), that there is an age-related decline in the function of pyridine-linked dehydrogenases.

Deshmukh *et al.* (1980) isolated nonsynaptic and synaptic mitochondria from rat brain at 3, 12, and 24 months of age. The organelles appeared to have good integrity. The authors observed that State 3 oxidation of pyruvate–malate declined continuously with age in the former, being lowered by 24% between 12 and 24 months. In synaptic mitochondria, a decline occurred only between 3 and 12 months. β Hydroxybutyrate declined only slightly in nonsynaptic mitochondria, but considerably (33%) in synaptic organelles between 12 and 24 months. The level of 3-hydroxybutyrate dehydrogenase and 3-keto acid–CoA transferase decreased with age, but other mitochondria enzymes (acetoacetyl–CoA thiolase, pyruvate dehydrogenase complex) showed no change. Similarly prepared nonsynaptic mitochondria showed a drop in the State 3 respiration of glutamate–malate not observed in synaptic mitochondria (Deshmukh and Patel, 1980). The results may be due to decreased transport. Mitochondria from neither source showed age-related changes in CO_2 production from glutamate or α-ketoglutarate. No changes were found in the activity of glutamic dehydrogenase or aspartate aminotransferase obtained from the mitochondria. Chiu and Richardson (1980) also examined rat brain mitochondria but found no change in State 3 oxidation of pyruvate–malate, succinate, or glutamate–malate from animals 12–14 versus 28–32 months of age. Glutamate oxidation was reduced by 30%.

Weindruch *et al.* (1980) found little age-related change in respiration in mouse liver, brain, or spleen mitochondria between 9–12 and 23–26 months of age when activity was based on milligrams of mitochondrial protein. Glutamate, malate–pyruvate, β-hydroxybutyrate, and succinate–rotenone were tested. Slightly altered results were obtained if calculations were based on the cytochdrome *c* oxidase content of the mitochondria.

Study of Table 5.I shows that with a few exceptions, there is rather substantial agreement on age-related changes in mitochondrial metabolism, although there are a few differences in the reported behavior of old mitochondria derived from different tissues. It is difficult to tell for certain if these are the result of varying experimental conditions or are genuine examples of differing patterns of aging that depend on the tissue of origin. Undoubtedly, State 3 oxidation of a number of substrates declines in the mitochondria of old animals. State 4 respiration and ADP:O ratios are generally reported to be unchanged. Indirect support for this conclusion was provided by Brouwer *et al.* (1977), who found no age-related decline in the respiratory control ratios of mitochondria in intact hepatocytes from young (3 months), mature (12 months), and very old (36 months) rats. Insofar as this measurement reflects mitochondrial integrity, there seems to be no dramatic change.

2. *Cytochrome Content*

Several investigators have reported changes in the cytochrome content of old mitochondria (Table 5.II). Gold *et al.* (1968) performed initial investigations with kidney and heart organelles. Subsequently, Abu-Erreish and Sanadi (1978) studied changes in cytochrome concentration of rat heart mitochondria at ages from 5 through 26 months. Cytochrome c oxidase, solubilized with deoxycholate, dropped steadily with age between these ages, losing about 27% of its activity. Whole homogenates of heart tissue showed the same level of decrease, so that the results were not due to the breakage of fragile mitochondria during isolation from the old tissue. The heart muscle enzyme showed no change of turnover number with age, so that the loss in activity is not due to less effective oxidase molecules, but to a reduced amount of the enzyme in the mitochondria. Other cytochromes in heart mitochondria also showed reduced concentrations (nmole/mg heart mitochondrial protein) as determined spectrophotometrically. Cytochromes a-a_3, b, and c-c_1 dropped substantially between 15 and 26 months of age. There was no change between 5 and 15 months. The mitochondria used were shown to be structurally in good condition by testing the rate of NADH oxidation and by measuring respiratory control ratios. Since the loss of the different cytochromes remains proportional, it appears that the number of respiratory assemblies decreases with age in heart mitochondria. Abu-Erreish and Sanadi (1978) pointed out that the rate-limiting step in mitochondrial respiration is not in the cytochrome region, but at the substrate end of the chain. Thus, the observed decrease in cytochrome content with age may not be responsible for the observed decrease in State 3 rates observed by several investigators (Table 5.I). Indeed, the

Table 5.II. Age-Related Changes in Mitochondrial Cytochrome Content

Tissue source	Age (months)	Cytochrome[a,b]	Change (%)[c]	Reference
Rat liver	12 versus 21–24	b	−6(NS)	Paterniti et al., 1978
		c-c_1	−8(NS)	Paterniti et al., 1978
		a-a_3	−37	Paterniti et al., 1978
	12–14 versus 24–27	a-a_3	0	Gold et al., 1968
Rat kidney		b	NS	Gold et al., 1968
		c-c_1	+42	Gold et al., 1968
		a-a_3	−72	Gold et al., 1968
		a-a_3	−12	Gold et al., 1968
Rat heart	5 versus 26	b	−41	Abu-Erreish and Sanadi, 1978
		c-c_1	−25	Abu-Erreish, 1978
		a-a_3	−22	Abu-Erreish, 1978
	12–14 versus 24–27	a-a_3	−17	Gold et al., 1968

[a] nmole/mg protein.
[b] Protein content may change with age and thus could affect results.
[c] NS, Not statistically significant.

authors suggested that changes in membrane composition and membrane-associated enzymes may be responsible. Paterniti *et al.* (1980) measured cytochrome content of liver and kidney mitochondria in rats of 12 versus 21–24 months. They found no differences in the content of cytochromes b or c-c_1 in liver, but a-a_3 dropped by 37%. In kidney, the level of cytochrome b dropped, but not significantly, c-c_1 increased 42%, and a-a_3 showed a remarkable drop (-72%). These results do not show the proportionality observed by Abu-Erreish and Sanadi (1978) for heart mitochondria.

3. *Peroxidative Damage*

It seems likely that the changes in function reported for old mitochondria are related to alterations of the membrane. Support for this thesis has been provided by a number of investigations. Nohl *et al.* (1978) and Nohl (1979) reported that in old rats, several enzymes bound to the inner mitochondrial membrane had altered properties. Arrhenius plots of ATPase, succinic dehydrogenase, succinate oxidase, β-hydroxybutyrate dehydrogenase, and cytochrome oxidase were obtained from 3 versus 24-month-old rat heart mitochondria. In all but cytochrome oxidase, discontinuities showed significant differences in the young versus old preparations. Glutamate oxidase, which is not part of the inner membrane, was unchanged.

The discontinuities in the Arrhenius plots presumably represent changes in the physical state of the membrane to which the enzymes are attached. When the mitochondria were broken up and solubilized in detergent (Triton X-100), succinic dehydrogenase, succinate oxidase, and ATPase lost their discontinuities and became identical whether they were derived from young or from old animals. β-Hydroxybutyrate dehydrogenase was inactivated by the procedure. These experiments provide evidence for the existence of changes in the mitochondrial membrane, which in turn affects the enzymes. There is other more or less direct evidence for alterations in mitochondrial membranes with age. Nohl and Kramer (1980) reported that translocation of adenine nucleotides is reduced by 40% in 30-month compared to 3-month-old rat heart mitochondria. Since the number of sites involved is unchanged, the authors attribute the change in transport activity to changes in the physiochemical state of the membrane lipids. Direct analysis of phospholipid and fatty acid composition also shows differences, particularly in a reduction of negatively charged phospholipids.

Nohl and Hegner (1978), Nohl *et al.* (1978), Nohl and Kramer (1980), and Hegner (1980) have marshaled evidence to support the idea that the membrane changes are brought about by increased oxygen radical pro-

duction in old mitochondria. It has long been known that intact mitochondria can produce small amounts of peroxide and superoxide radical, which are believed by Turrens and Boveris (1980) to be generated by
NADH dehydrogenase in the mitochondrial membrane and in the ubiq-
uinone–cytochrome *b* area. In this regard, Nohl and Hegner (1978) investigated formation of free radicals by heart mitochondria and submitochondrial particles in young (3-month) and old (23-month) rats. The
authors found that the submitochondrial particles from old rats, freed
from the protective effect of superoxide dismutase, generated 25% more
superoxide radical than the equivalent "young" preparations. Using
intact mitochondria, a small amount of superoxide radical was produced, about 20% of that of the submitochondrial fraction, representing
the degree to which the radicals escaped the action of superoxide dismutase. Again, production in the old preparations was about 25% greater than in young preparations. In support of this thesis, Nohl, *et al.*
(1978) reported that superoxide radical was formed at a rate of 1.90
(± 0.09) nmole/minute/mg in young versus 2.54 (± 0.22) for old mitochondria. For H_2O_2, the rates were 0.76 versus 1.02, respectively. Moreover, the authors subjected "young" heart mitochondria to a radical-
generating situation (xanthine–xanthine oxidase) and observed a loss in
respiratory activity of 3-hydroxybutyric acid and of respiratory control
similar to that found for 24-month-old mitochondria. The P:O ratios
were less affected. Protection was afforded by addition of superoxide
dismutase and catalase. As in the normal "old" mitochondria, glutamate–malate was unaffected by the presence of radicals. These experiments provide suggestive evidence that peroxidation is the cause of the
changes noted for old mitochondria. The degree of lipid peroxidation
was, however, much higher in the externally treated than in old
mitochondria and did not affect respiration to a comparable degree.

Nohl and Jordan (1980) subsequently considered peroxide metabolism
and noted the protective effect of mitochondrial superoxide dismutase,
catalase, and glutathione peroxidase. H_2O_2 itself apparently does not
affect mitochondrial function, but may generate hydroxyl radicals,
which are an extremely reactive and presumably dangerous species
(Chapter 3, Section II). Interestingly, the authors ascribe the major de-
struction of H_2O_2 to catalase with GSH peroxidase activity accounting
for 15% of intra-mitochondrial peroxide metabolism, GSH concentrations being important to this latter function.

Presumably, the free radicals that are generated by mitochondria
would attack membrane lipids. Thus, Nohl (1979) suggested that in old
rats, some unsaturated membrane lipids are lost by peroxidation. Gas
chromatographic analysis showed a 30% decrease with age (3 versus 24

months) in the ratio of unsaturated to saturated fatty acids. A similar analysis by Nohl and Kramer (1980), but with animals of 3 versus 30 months of age showed a similar decline in the saturated:unsaturated fatty acid ratio. However, there were substantial differences in the two reports. For example, in the first study (Nohl, 1979), linoleic acid (18:2), arachidonic acid (20:4), and docosohexenoic acid (22:6) changed +20, −19, and −17% respectively, in the two age-groups. In the later study (Nohl and Kramer, 1980), the respective values were −37, +19, and +21%. Although the unsaturated:saturated ratio for the fatty acids was the same in both studies, the large differences in the concentration of the major unsaturated acids and the reversal in the direction of change are disturbing. One might explain that these differences in results occur between 24 and 30 months, the gap between the "old" animals in the two experiments. However, the oldest animals gained in the most highly unsaturated fatty acids. Under these circumstances, one must surely wonder about the role of peroxidation.

Although the evidence put forward for lipid peroxidation in mitochondria from old organisms is persuasive, it is not unequivocal. As is unfortunately the case for most of the research in aging, there are a number of other considerations that should be kept in mind. One is that Abu-Erreish and Sanadi (1978), Chen *et al.* (1972), and others (Table 5.I) did not observe any difference in succinate oxidation in old rat heart mitochondria, although Nohl (1979) reports a 30% reduction in the activity of succinoxidase and 17% in succinic dehydrogenase. Second, the use of 3-month-old animals by Nohl and co-workers as the sole basis for comparison is questionable. Several obvious examples are given above, of changes between 3 and 12 months, which are different in direction from those occurring after 12 months, 24 months, or older (Chen *et al.*, 1972; Paterniti *et al.*, 1980). One should also bear this consideration in mind when examining the data given in Table 5.I. Still other concerns are the fact that some peroxide and O_2^- radicals escape to the outside of the mitochondria where they can be coped with by cytoplasmic protective enzymes. Is there an "old–young" differential in the amount left behind that could attack the mitochondrial membrane? In any case, the question is not how much oxidative damage is done, but whether there is an age-related increase in damage. In this respect, Nohl and Hegner (1978) reported a 20% increase in superoxide radical production by intact old mitochondria. However, components of the organelles are constantly being replaced and mitochondrial turnover does not appear to change with age [although membrane proteins appear to turn over more slowly than other mitochondrial proteins (Walker *et al.*, 1978)]. Therefore, peroxidative damage would have to occur quite quickly in newly synthesized organelles if physiological effects are to occur—otherwise, the damage would be "repaired" by replacement. Furthermore, it should be noted that rat brain mitochondria (Sorgato *et al.*, 1974) and mitochondria from tumor cells (Dionisi *et al.* 1975), unlike mitochondria from liver, kidney, and heart, apparently do not generate any oxygen radicals and, thus, do not form peroxide.

One should bear in mind that the membrane changes invoked by investigators to explain their findings need not be related to lipid peroxidation. They could simply be the result of subtle programmed changes in lipid composition. Alternatively, they might be related to the conclusion of Siliprandi *et al.* (1979) that mitochondria aged *in vitro* become deficient in ion transport functions because of oxidation of vicinal thiol groups in the inner membrane.

4. Insects

Mitochondria play a direct role in providing energy under conditions of high demand in insect flight muscle. Therefore, they have been the subject of much study in this tissue. The fact that old insects have a reduced flight ability has provided an age-related rationale for such work. In contrast to the normal loss and replacement process in mammalian tissues, mitochondria do not appear to turn over in flight muscle (Maynard-Smith *et al.*, 1970; Tribe and Ashhurst, 1972). For this reason, it may well be that age-related alterations in insect mitochondria are not directly comparable to the changes observed in mammalian systems.

There have been a number of ultrastructural changes reported for mitochondria in aging insects. An extensive listing is provided by Baker (1976). Sacktor and Shimada (1972) examined the ultrastructure of flight muscle in the blowfly *Phormia regina* and found that some of the mitochondria from old organisms contained myelin-like whorls, which appear to derive from the inner membrane. In 31 to 33-day-old organisms, the structures appear in about 25% of the mitochondria. Histochemically, cytochrome oxidase is absent in these areas, although the enzyme is present in the intact cristae. Studies of ultrastructure in other insects show only a low incidence of this phenomenon, but there are many examples of changes in size, uniformity, or other changes with age (Webb and Tribe, 1974; Baker, 1976; Sohal, 1976; Sohal and Bridges, 1977; Miquel *et al.*, 1980). Bulos *et al.* (1972) examined the biochemical effects of aging on mitochondria in the blowfly *Phormia regina*. They found a decreased rate of State 3 oxidation for pyruvate-proline and α-glycerophosphate in mature (7–8 days) versus senescent (31–33 days) flies. State 4 changes were not significant, but the values for the respiratory control ratios were lower in senescent flies for both substrates (11%

for α-glycerophosphate and 39% for pyruvate–proline). No change was observed in the ADP:O ratios. Subsequently, Bulos et al. (1975) attempted to localize the lesion in the State 3 oxidation of these substrates. Uncoupled oxidation showed an age-related decline (20 and 39%, respectively), and this was similar to the values for coupled oxidation. Thus, the impairment should not be related to problems in phosphorylation, but to the electron transport chain. However, none of the partial reactions of the chain showed age-related differences, and the cytochrome content of the mitochondria was unchanged. The authors suggested possible explanations for the reduced State 3 oxidation, including altered uptake of substrates and changes in tricarboxylic acid cycle enzymes.

In agreement with the finding of Bulos et al. (1972) in *P. regina*, Tribe and Ashhurst (1972) reported a decrease in the respiratory control ratio in the blowfly *Caliphora erythrocephala*, although the change was of larger magnitude. However, these authors also reported a decrease in the P:O ratios. It should be noted that their results showed substantial variation from experiment to experiment.

Wohlrab (1976) studied age-related changes in mitochondria from the blowfly *Sarcophaga bullata*. He found that the respiratory control ratio does not decrease once muscle development is complete and that the ability to oxidize α-glycerophosphate does not decline after 9 days of age. These results are in contrast to those found in *P. regina* by Bulos et al. (1972) and Bulos et al. (1975). The only age-related change between mature and old mitochondria observed by Wohlrab (1976) was a decreased oxidation of proline–pyruvate. Even this change could be overcome by use of ATP–NaHCO$_3$ in the reaction medium, perhaps replenishing an oxaloacetate deficiency via pyruvate carboxylase. He found no change in cytochrome content, in agreement with the conclusion of Bulos et al. (1975). Swelling studies also showed no significant age-related differences. The isolated mitochondria showed excellent integrity, the NADH oxidation rate being 100 nat/minute/mg protein in mature and senescent organelles. Interestingly, no age-related differences in H$_2$O$_2$ production were observed between mature and old mitochondria. It should be noted that there were substantial differences between young and mature mitochondria, that is, in the period up to nine days of age.

It is clear that there are considerable metabolic differences between the mitochondria in *Phormia* and *Sarcophaga*. There are also differences in structure: About 25% of the senescent mitochondria of *Phormia* contain myelin-like whorls compared to less than 2% for *Sarcophaga*. However, dissimilarity between species may not be entirely responsible for the

metabolic differences observed. Hansford (1978b) also studied mitochondria from young and aged (as well as mature) *Phormia* and like
Wohlrab's findings with *Sarcophaga*, observed no age-related changes in
function. Unlike Bulos *et al.* (1972), Hansford found no change with age
in the State 3 oxidation of pyruvate. The studies differ in that Bulos *et al.*
used proline as a source of oxaloacetate. This system may have been
limiting as Hansford found that pyruvate, in the presence of bicarbonate
and ATP (see Wohlrab, 1976), provided nearly double the overall reac
tion rates. However, since Hansford did not observe an age-related
inability to metabolize proline, he suggests that the difference in results
may lie in the preparation of the mitochondria and in the incubation
procedures. Hansford also observed no change in State 4 rates, suggesting that there is no change in mitochondrial membrane integrity with
age, in contrast to the report of Tribe and Ashhurst (1972).

Vann and Webster (1977) working with mitochondria from *Drosophila
melanogaster* reported an increasing oxidation of pyruvate–malate, ATP
synthesis, P:O ratios, and ATPase activity until the flies became old—
usually about 32 days, when all of these parameters dropped in value.
The authors found evidence for an inhibitor of oxidative phosphorylation in the cytoplasm of old flies. However, the respiratory control ratio
of the mitochondria was low to start with, and statistical data are not
given.

The information available from studies of insect mitochondria indicates that changes indeed occur with age, but that they may vary in
severity and nature in different species. With such differences, it seems
unlikely that flight muscle mitochondria as a general class would represent a particularly useful model for studying mammalian systems. The
situation surely is further aggravated by the fact that mitochondria in
flight muscle are not replaced. Therefore, the first component which
"wears out" will result in a breakdown of function. In higher animals,
turnover, that is, replacement, occurs even in old animals.

C. Comment

Assuming that some, if not all of the reported alterations in mitochondrial properties are indeed real and age-related, how do these changes
come about? Since turnover of mitochondria is unaltered with age in a
variety of rat tissues (Menzies and Gold, 1971, 1972), it would seem
unlikely that the organelles become damaged simply by surviving for
unacceptable long periods in old cells. Parenthetically, the old cells we
are considering are nonmitotic in nature, otherwise, we might be dealing with new cells in old organisms. Either replacement components of

mitochondria in old organisms are mis-synthesized so that the organelles lack proper enzyme complements or possess subtly deficient membranes, or damage is rapidly initiated by the milieu of the old tissues so that at any given time, in spite of being continuously refurbished, a goodly proportion of the mitochondria are in damaged condition. The observation of Wilson and Franks (1975) that liver mitochondria in old mice show a range of forms, from normal to large sized, does not tell us which of these two possibilities is being realized. One would suspect that of the two alternatives, normal mitochondria become altered. Otherwise, the mechanism responsible for the damage would have to allow simultaneous production of both normal and defective mitochondria.

Actually, there would seem to be more than one mechanism responsible for the alterations so far detected. These include membrane changes, differing amounts of certain enzymes transported into the mitochondria, and reduced levels of cytochromes which are synthesized by the mitochondria themselves. Of course, if the original program for producing a new mitochondrion is faulty, other changes might follow as a normal consequence. Changes such as peroxidative damage after synthesis of the mitochondrion as postulated by Nohl and others are certainly a possibility. However, there is as yet no assurance that mitochondria from different tissues follow identical patterns of change.

A point, which applies to liver mitochondria, is that this organ consists of several tissue types—mainly parenchymal cells, Kupffer cells, and endothelial cells. The numbers of these various cell types change substantially with age (Knook and Sleyster, 1976). Since the mitochondrial size varies with the type of cell [endothelial and Kupffer cells have smaller mitochondria than parenchymal cells (Wilson and Franks, 1975)], perhaps some of the metabolic differences reported for mitochondria from young versus old liver may result from this situation.

It does seem clear that there are age-related deficiencies in mitochondrial metabolism. Their nature and to what degree they are primarily due to aging as against a response to an age-related change in metabolic pathways is not known.

III. MICROSOMES

The particles known to biochemists as microsomes derive from the endoplasmic reticulum (ER). Therefore, it is worthwhile to take a brief look at the results of morphological studies of the intact ER as well as data concerning metabolic aspects of microsomal preparations.

A. Endoplasmic Reticulum

Ultrasturctural changes in the ER with age, as is the case for mitochondria, are inconclusive because of contradictory reports. Pieri *et al.* (1975) reported that the surface density of smooth ER in rat liver drops between 1 and 12 months and increases again by 27 months. A sharp, age-related decrease (75%) in rough ER was reported. Total ER was also lower with age. On the other hand, Schmucker (1978) and Schmucker *et al.* (1978) also reported a decrease in total liver ER between mature (10 months) and old rats (30 months), but found this to be due to a decrease in smooth ER. The amount of rough ER was little changed in old animals. Meihuizen and Blansjaar (1980), although they found no change in rough ER (in agreement with Schmucker) observed an increase rather than a decrease in smooth ER in hepatocytes of rats of 3 versus 35 months of age in three zones of the liver lobule, in agreement with Pieri *et al.* (1975). If one takes into consideration that the amount of ER first increases with age before decreasing (Schmucker *et al.*, 1978), then the difference between early and late ages is small. The measurements by Meihuizen and Blansjaar (1980) at 3 versus 35 months would miss any decrease occurring between maturity and senescence.

Except for the report of Garg *et al.* (1979) that certain enzyme inducers bring about an age-related increase in smooth ER, there are few other reports of age-related changes in ER structure which are of direct interest to biochemists.

B. Microsomal Function

Microsomes can be considered to be a crude biochemical counterpart of the ER. The effect of age has been investigated with an eye to changes in membrane composition, changes in enzyme activity, and induction by various drugs. Studies of the effect of aging on the lipid composition of microsomal membranes are dealt with in Chapter 4, Section II.

1. Enzymes

Microsomal enzymes are membrane-bound, so that changes in activity could result from alterations in the membranes, a change in the number of enzyme molecules, or alteration of the enzyme molecules themselves. Another consideration of importance, and one which tends to be ignored, is that microsomal preparations form closed vesicles and certain enzymes may be on the inside. Thus, the rate of entry of substrate could become a factor. Many studies deal not only with basal microsomal enzyme levels in young versus old animals, but also with the relative extent to which components of the enzyme systems are induced or repressed after drug administration.

Table 5.III lists the effects of age on basal levels of microsomal enzymes and cytochromes. The data generally show a reduced function with age of the mixed oxygenase system and other drug-metabolizing enzymes. However, the results are not always consistent. In an early study of the effect of age on enzyme levels, Grinna and Barber (1972) examined three microsomal and mitochondrial membrane-bound enzymes in rat kidney, liver, and heart, and obtained results which varied with each tissue. Kidney microsomes showed the greatest degree of change. In this tissue, glucose 6-phosphatase and NADPH cytochrome c reductase were reduced by 41 and 47%, respectively, in the old animals (6 versus 24 months). NADH cytochrome c reductase activity was also reduced in kidney (34%), although this enzyme appeared to increase in liver and heart microsomes. An increase in liver was also noted by Gold and Widnell (1974). Grinna and Barber (1972) attributed the differences in enzyme activity to age-related changes in membrane conformation, although changes in the number of enzyme molecules present could not be ruled out. In a latter paper, Grinna and Barber (1975) concluded that the age-related drop in the specific activity of glucose 6-phosphatase in liver and kidney microsomes was indeed due to the presence of fewer enzyme molecules. In this case, the authors made use of a detergent (Triton X-100) to permit direct measurement of the enzyme without consideration of transport of the substrate. Subsequently, Grinna (1977a) by means of Arrhenius plots, presented evidence that membrane changes had occurred in old preparations of the enzyme.

Birnbaum and Baird (1978a) noted a decline in drug demethylation and hydroxylation in liver microsomes from old rats (12 versus 27 months). No significant decline in basal levels of NADPH cytochrome c reductase, NADPH cytochrome P-450 reductase, or cytochrome P-450 was noted. McMartin *et al.* (1980) using Wistar rats in different age-groups (7 versus 31 months and 4, 12, and 36 months of age, respectively) reported a decrease in liver microsomal NADPH cytochrome P-450 reductase and in cytochrome P-450, but a slight increase in cytochrome b_5. In kidney microsomes, the same effects were noted, but cytochrome b_5 decreased. Schmucker and Wang (1980a) reported a steady decrease in glucose 6-phosphatase activity in rat liver microsomes between 1, 16, and 27 months. They point to the anomaly that although the concentration of microsomal protein was unchanged in the old animals, stereological analyses showed a substantial decline in total

Table 5.III. Effect of Age on Microsomal Enzymes and Cytochromes[a]

Source (rat)	Age (months)	Percentage change compared to "young" values						Reference[c]
		Glucose 6-phosphatase[b]	NADH-cyt c reductase	NADPH-cyt c reductase	NADPH-cyt-450 reductase	Cyt P-450	Cyt b_5	
Liver	3 versus 24	−47	+78[d]	−23				1
	3 versus 28–30		NC	NC		−20	+13	2
	2 versus 24			NC[e]				3
	12 versus 27[f]		+(NS)	−(NS)	NC	−(NS)	−42	4
	6 versus 24	−29		−(NS)				5
	12 versus 36[f]				−14[g]	−17[g]	NC[g]	6
	16 versus 27[f]	−19						7
	16 versus 27[f]			−56		−50		8
	12 versus 24[f]					NC		9
	3 versus 20			−23[g] (Male) −29[g] (Female)		−30[g] (Male) −16[g] (Female)		10
Liver	10 versus 20 versus 600			−(NS)		−(NS)		11
Kidney	6 versus 24	−41	−34	−47	−34[g]		−34[g]	5
	7 versus 31					−42[g]		6
Heart	6 versus 24		+(NS)	NC				5

[a] NC, No change with age; NS, not statistically significant.

[b] It has recently been discovered that the enzyme is on the inside of microsomal vesicles. There is a transferase for the substrate which may change with age or drug administration. The results may reflect both transport and glucose 6-phosphatase activity and not simply the latter.

[c] Key to references: (1) Gold and Widnell, 1974; (2) Birnbaum and Baird, 1978a; (3) Adelman, 1971; (4) Birnbaum and Baird, 1978b; (5) Grinna and Barber, 1972; (6) McMartin et al., 1980; (7) Schmucker and Wang, 1980a; (8) Schmucker and Wang, 1980b; (9) Paterniti et al., 1980; (10) Kato and Takanaka, 1968a; (11) Kato and Takanaka, 1968b.

[d] Males (females showed +15%).

[e] Time lag.

[f] Paper also deals with younger animals.

[g] Approximated from graphs.

endoplasmic reticulum with age (Schmucker *et al.*, 1978). Schmucker and Wang (1980b) also observed a sharp decrease in NADPH cytochrome *c* reductase and cytochrome *P*-450 in liver microsomes of Fischer 344 rats between 16 and 27 months of age as did Kato and Takanaka (1968a). Paterniti *et al.* (1980), however, found no change in cytochrome *P*-450 levels in aged Sprague-Dawley rat liver microsomes (12 versus 24 months).

The effects of aging on microsomal enzyme levels, as summarized in Table 5.III, appear to be in general agreement. Care should be taken in interpretation of results using young animals (e.g., 1–3 months) as changes may be substantial well before 12 months is reached. Although several of the original reports include results from quite young animals, comparisons in Table 5.III are made only between mature versus old samples, wherever feasible. The listed age-related changes are, therefore, not always the same as reported by the authors or subsequent reviewers who base their conclusions on the entire life-span measured. Where the data were provided in more than one way, activity per milligram of microsomal protein has been utilized. Liver and kidney weights did not change significantly with age between maturity and old age, as noted by several investigators (Schmucker and Wang, 1980b; Gold and Widnell, 1974).

2. *Enzyme Induction*

A number of investigators have examined the effect of age on the induction or repression of microsomal enzymes after administration of various drugs. A summary of results is given in Table 5.IV. Gold and Widnell (1974) observed substantial age-related differences in the effect of hormonal microsomal enzyme inducers. Chronic low doses of triamcinolone nearly doubled the specific activity of glucose 6-phosphatase in old animals, but had no effect on young ones. Conversely, phenobarbital did not affect this enzyme in old animals, but caused a 50% increase in young ones. Triiodothyronine increased the specific activity in both young and old animals. All three compounds depressed NADH–cytochrome *c* reductase in both young and old rats. The authors point out that for both young and old animals, the enzyme activities could be either increased or decreased, but were always within the range of the highest and lowest specific activities for both ages. Therefore the age-related differences observed were ascribed to differences in endocrine states. The use of animals at 3 versus 24 months of age makes the results difficult to put into the context of "aging." Indirect support for the idea of distal rather than local control is given by Baird *et al.* (1976). They

Table 5.IV. Amount of Induction or Repression of Microsomal Enzymes with Age[a,b]

Source (rat)	Age (months)	Drug[c]	Cyt P-450	Cyt b_5	NADPH-cyt-450 reductase	NADPH-cyt c reductase	G6Pase[d]	Reference[e]
Liver	16 versus 27[f]	PB	−			−	NC[d]	1,2
	12 versus 36[f]	PB	−	NC	NC	−	−	3
	3 versus 24	PB					−	4
	7 versus 31	BNF	NC	NC	NC			3
	3 versus 24	TA				−	+	4
	3 versus 24	TI				−	NC	4
	12 versus 27[f]	PCB	+			NC	NC	5
	10 versus 19[f]	PB	−			−	NC	6
Kidney	7 versus 31	BNF	−	−	NC			3

[a] −, Lower degree of change; +, increased degree of change; NC, no change with age.

[b] Metabolism of other drugs is also measured in some of the references.

[c] PB, phenobarbital; BNF, β-naphthoflavone, PCB, polychlorinated biphenyls; TA, triamcinolone; TI, triiodothyronine.

[d] See footnote b in Table 5.III. Induction and disappearance of induced enzyme are slower.

[e] Key to references: (1) Schmucker and Wang, 1980a; (2) Schmucker and Wang, 1980b; (3) McMartin et al., 1980; (4) Gold and Widnell, 1974; (5) Birnbaum and Baird, 1978b; (6) Kato and Takanaka, 1968b

[f] Effect also measured in younger animals.

found that newly generated liver in old rats (about 31 months), following partial hepatectomy, continued to metabolize the drug "Zoxazolamine" at the normally slower rate of old animals. The authors suggest that since the liver now contains "young" cells, the slowed rate of metabolism must be due to external factors. However, there is no guarantee that newly synthesized cells in old liver behave like "young" cells.

Schmucker and Wang (1980a) reported that although phenobarbital did not greatly change the degree of repression of glucose-6-phosphatase, the pattern of loss and recovery is slower in the old animals as has been noted for hormonal effects on enzyme induction (Chapter 10, Section XIII). NADPH-cytochrome c reductase and cytochrome P-450 are induced by phenobarbital, but less so in senescent compared to mature rats (Schmucker and Wang, 1980b). McMartin *et al.* (1980) also noted a reduced induction of cytochrome P-450. In contrast, induction of this cytochrome showed no age-related difference after administration of β-napthoflavone. Birnbaum and Baird (1978b) found no age-related (3 versus 28–30 months) impairment of the monoxygenase system in rat liver microsomes in response to induction by drugs. These conclusions are contrary to the earlier findings of Kato and Takanaka (1968b). Among the problems are substantial differences according to sex, presumably because of sex hormone levels (Kato and Takanaka, 1968a). However, the Kato and Takanaka papers utilized animals only up to about 19 months of age. The differences between 10 and 19 months are quite small (1968b). As with basal enzyme levels, one should be careful about the age periods utilized for studies of enzyme induction because changes often are more dramatic between earlier than later ages. It should also be noted that "old" microsomes often start with a lower enzyme level. Therefore, comparing the degree of induction is complicated. Moreover, the rate of glucose 6-phosphatase metabolism appears to be affected by a carrier that mediates the entry of the substrate into the microsomal vesicle. The carrier itself may be induced. For example, in microsomes from young rats (3 months) both enzyme and carrier levels are increased after administration of triiodothyronine, whereas in old animals (24 months), only the enzyme is increased, although to a much greater degree. Triamcinolone increased only the carrier component in young animals, whereas both components increased in old animals (K. Dobrosielski-Vergona and C. C. Widnell, University of Pittsburgh, private communication). It is possible that other substrates are also affected in an analogous manner. Schmucker (1979) and Schmucker and Wang (1980c) have recently reviewed age-related changes in drug metabolism.

C. Comment

The mechanism whereby changes in microsomal enzymes occur has not been explained. Whether the enzymes or membranes are altered or whether there is simply less (or more) enzyme is not known for certain. Schmucker and Wang (1980c) feel that the changes in enzyme levels may be in part, related to the reduced amount of ER that occurs in liver between maturity and senescence. However, the continuous loss of glucose 6-phosphatase activity starting from early ages does not parallel the loss of ER. Adelman (1975) described isotopic evidence for *de novo* synthesis of NADPH cytochrome *c* reductase after induction by phenobarbital.

Several investigators have noted changes in the lipid composition of microsomal membranes. Such changes could affect either the activity or the location of enzymes on the membrane. However, the real picture is probably more complex. Since induction of microsomal enzymes does occur in old animals, even if in reduced amount, it is clear that synthesis of new molecules can take place. Therefore, it would seem reasonable to assume that the basal levels in microsomal enzymes are a result of increased or decreased synthesis.

As proposed by Gold and Widnell (1974), changes in enzyme level are probably a reflection of endocrine states or reduced response to stimuli. It is a little disconcerting to find a 78% increase of NADH–cytochrome *c* reductase in liver microsomes from old male rats, as compared to an increase of 15% for females (Gold and Widnell, 1974). Kato and Takanaka (1968a) also observed substantial sex-related differences in the metabolism of several drugs. Since the basal differences between male and female at a given age are sometimes larger than the age-related differences, one must suppose that the latter indeed reflect hormone levels or endocrine response. As to the differences in results among various investigators, these may in part be due to diet, circadian rhythm, environmental conditions, or strain differences, in addition to sex. Another consideration of note is that few measurements have been made at more that two points in the period between maturity and old age. Thus, there is a danger that changes may be different in the mature-senescent period. A last comment is that the rat system has been utilized almost exclusively. We, therefore, cannot be sure that the changes observed are not rat-specific and, thus, not common to the aging process. Although the oldest animals were only 50 weeks of age, Kato *et al.* (1970), in fact, observed quite different results in mice compared to rats. They found no age-related change in the mouse liver mixed-function oxidase system or in its induction.

IV. LYSOSOMES

The lysosome has been regarded with considerable suspicion by the biologist interested in aging. Its package of lytic enzymes poses an implicit threat to the cell. Leakage of these enzymes could result in damaging reactions, which could be responsible for age-related changes, real or postulated. RNA would be split, thus affecting translation of proteins; DNA might be "nicked," if not destroyed, resulting in DNA deletions and mutations; cell proteins would be lost; membranes would be damaged. It sounds somewhat like a disaster movie. Of course, only nonmitotic cells would reflect such damage, as distressed mitotic cells would be routinely replaced. The real victims of internal damage would be heart cells, brain cells, and muscle cells, as examples of nonmitotic tissues. In spite of this plausible scenario, there is no evidence that old cells, for example, from rat heart, are seriously deficient in any obvious way (Abu-Erreish et al., 1977). Rather, subtle changes in metabolic capabilities seem to occur. In this regard, one should bear in mind that aging is a smooth, continuous process. If lysosomes are involved, then damage must occur between 20 and 30 years of age as well as between 30 and 40, 40 and 50, etc. Yet, the aging process does not appear to accelerate as might be expected if damage were cumulative. One might argue that the problem lies in damage repair, a process that might occur more readily in younger tissues, but which then suggests that aging is in a realm (control of repair) that is not lysosomal in origin. In fact, there appears at the present time to be no good evidence that lysosomes are directly involved in the aging process. More likely, they, like other subcellular units, are its victims. Therefore, it is important to identify and characterize age-related changes in lysosomal function, if they exist.

The first problem is that the function of lysosomes is not clearly defined. The organelles are believed by many investigators to be responsible for the degradation of cellular proteins. They are also believed to be involved in autolysis of cells and digestion of cell organelles such as mitochondria in the process of turnover. Insofar as aging is concerned, there seems to be general although not unanimous, agreement that lysosomes are involved in the accumulation of cell materials which compose "age pigments" (Chapter 3, Section III).

A. Ultrastructure

Studies of lysosomes have included measurement of the lysosomal content of cells. Knook et al. (1975) reported an increase in lysosomal volume in aging rat liver. Schmucker (1978) also reports an age-dependent increase in volume density of 150 and 175% in 30-versus 6-month-old rats in centrolobular and portal hepatocytes, respectively. The increase in specific volume (volume per cell) was 81 and 22% respectively, for the two zones. However, the reported increase really appeared between 6 and 16 months and thereafter, relatively little change occurred between the period from maturity to senescence. Thus, although the changes are "age-related," they do not appear only to reflect "aging." Meihuizen and Blansjaar (1980) observed substantial increases in the volume density of lysosomes of rat hepatocytes (3 versus 35 months of age) from central and midzonal areas. The increase in the peripheral area was not statistically significant. Pieri *et al.* (1975), in contrast to the above reports, found no significant changes in lysosomal volume in the liver of rats up to 27 months of age.

B. Enzymes

Changes of lysosomal enzymes with age do not provide a consistent pattern. They appear to change independently of one another. Moreover, there is no consistency of change from tissue to tissue or even in different cell types from the same tissue. More recent data are summarized in Table 5.V.

1. Liver

Other than ultrastructural examination of lysosomes, most studies of age-related effects have been concerned with changes in the levels of lysosomal enzymes. Summaries of earlier work involving such changes have been included in general compilations of enzyme changes by Finch (1972) and Wilson (1973) (Chapter 9, Section II).

These early results are, however, undependable, being based generally on enzyme activities in crude homogenates from animals raised under varying conditions. They vary considerable and are even, in some cases, contradictory. A problem may lie in altered protein levels of old tissue. Hence, changes per milligram of protein would have little accuracy. Another problem may lie in the fact that in liver tissue, there are different levels of lysosomal enzymes in the different cell types (see below). If the proportion of cells of each type changes with age, then lysosomal enzyme levels would reflect this fact.

That enzyme levels do differ in the various liver cell types has recently been shown by Knook and Sleyster (1980). They compared the activities of nine lysosomal enzymes in parenchymal, Kupffer, and endothelial cells, respectively, from 3-month-old rats. About two-thirds of the liver consists of parenchymal cells. The other one-third is made up mostly of

Table 5.V. Changes in Lysosomal Enzymes with Age[a]

Source	Tissue	Age (Months)	Enzyme							
			AP	AS	β-gal	β-Gluc	Ampep	Cath D	Glucam	Reference[b]
Animal										
Rat	Nonparenchymal cells	3 versus 35	-8 -(NS)	-10 (NS)	-10 (NS)	-57	+12	+35		1
	Kupffer cells	3 versus 35	NC	-33	NC	-53	+37	+39		1
	Parenchymal cells	12 versus 30–35	NC	NC	+112[c]			+203		2
	Nonparenchymal cells	12 versus 30–35	NC	NC	NC			-12 (NS)		2
	Liver	2½ versus 24				+30		+60	+15	3[d]
	Brain	2½ versus 24				NC		NC	+73	3
	Liver	16 versus 27	-			+				4
	Heart	17 versus 27[e]	-30 (NS)	+25		NC			+35	5
	Lung	18–30	NC			-				6
	Lung	15–20	NC	NC		+			+	7

122

Days

Tissue	Comparison						Ref
Insects							
Drosophila	7–102	–				+	
	1 versus 35	NC				NCf	8
							9
Tissue culture							
Man	Passage Number						
Liver	15 versus 20	+15 (NS)	+24	–71	–21		10
Fetal lung (WI-38)	Phase II versus III				+35	+42	11
Fetal lung (WI-38)	37–42 versus 49–54	–23g					12
Fetal lung (WI-38)	25–33 versus 33–45	+14					13
Fetal lung (WI-38)	26–35 versus 36–50	+29g			+19g		14

[a] Key to abbreviations: AP, acid phosphatase; AS, arylsulfatase; β-gal, β-galactosidase; β-Gluc, β-glucuronidase; Ampep, aminopeptidase; Cath D, Cathepsin D; glucam, acetylglucosaminidase; NS, not significant NC, no change.

[b] Key to references: (1) Knook & Sleyster, 1978; (2) Knook & Sleyster, 1976; (3) Platt et al., 1973; (4) Schmucker and Wang, 1979; (5) Traurig and Papka, 1980; (6) Wilson et al., 1972; (7) Traurig, 1976; (8) Miquel et al., 1974; (9) Webster and Webster, 1978; (10) LeGall et al., 1979; (11) Milisauskas and Rose, 1973; (12) Wang et al., 1970; (13) Cristofalo et al., 1967; (14) Cristofalo and Kabakjian, 1975

[c] Large change due to sharp drop at 12 months.
[d] Other enzymes tested.
[e] Reported for lysosomal fractions. Total activity also increased.
[f] Specific activity increased by about 90%, but total activity was unchanged. The amount of protein decreased in the lysosomal fractions.
[g] Earlier passages also given.

Kupffer and endothelial cells. On a per cell basis, parenchymal cells show the highest enzyme activity. However, because of their much greater protein content, they show the lowest specific activity (i.e., on a per milligram of protein basis). Moreover, the different cell types show different changes in lysosomal enzymes with age. With respect to age-related changes, Knook and Sleyster (1976) separated parenchymal and nonparenchymal cells from young, mature, and old rats (3, 12, 24, and 30–35 months of age) and determined the activity of the lysosomal enzymes, acid phosphatase, β-galactosidase, cathepsin D, and arylsulfatase B. The most substantial change was a tripling of cathepsin D activity in parenchymal cells between 12 and 30–35 months. In nonparenchymal cells, there was no significant change in this age span, but this is because there was a sharp drop between 12 and 24 months with a subsequent increase by 30–35 months. Indeed, there were occasional decreases in several of the enzymes between a given age group. (e.g., 3 versus 12 months or 12 versus 24 months), which did not conform to a linear pattern of increase and decrease. Overall, the changes in enzyme activities in the two cell types did not match. The specific activity of the enzymes in nonparenchymal cells was much greater in all cases. In addition to the changes in enzyme activity, the authors noted several ultrastructural differences, the Kupffer cell lysosomes being variable in density and size and in endothelial cells being more uniform. Knook and Sleyster (1978) subsequently examined lysosomal enzymes in preparations of Kupffer cells (90% minimum purity) and nonparenchymal cells from rats of 3 versus 33–35 months of age. Kupffer cells are of importance because they are heavily involved in endocytotic processes, and such processes have been reported to decline with age. Because lysosomes are responsible for digestion of material that is endocytosed, age-related changes in enzyme composition should be of particular interest. On a per milligram of protein basis, cathepsin D and aminopeptidase B increased about 60% in the older samples, whereas arylsulfatase B and β-glucuronidase dropped 33% and 53%, respectively. Acid phosphatase and β-galactosidase were unchanged. Results were similar in nonparenchymal cells, but the magnitude of the changes in aminopeptidase B and arylsulfatase B was much smaller. Similar conclusions, but with differing magnitudes, were obtained if activities were calculated on a per cell basis. (The protein content of Kupffer cells increases significantly in old rats.) The fact that all enzymes did not change activity in the same direction makes it clear that the results were not simply due to an age-related increase in the number of lysosomes.

Schmucker and Wang (1979) reported quite different results using rat liver homogenates. They observed an increase in acid phosphatase during maturation (1 versus 16 months) and loss during senescence (16–27 months). β-Glucuronidase remained unchanged at first and then rose during senescence (also noted by Comolli, 1971). Levels of the latter enzyme were raised by phenobarbital administration, although less so in old than young rats. Even considering that changes in enzyme activities measured in whole homogenates of liver are reflections of differing activities in the various cell types, these findings cannot readily be rationalized with those reported above.

2. Heart

A number of lysosomal enzymes in heart tissue have been examined. Wildenthal *et al.* (1977) observed age-related alterations in lysosomal enzymes of rat and rabbit heart. For the purposes of their experiments, they compared young with mature rather then with aging animals. For the record, they observed an increase (75%) in cathepsin D in rat heart (1–2 versus 10–14 months), no change in acid phosphatase, and a 17% decrease in glucosaminidase. Rabbit heart yielded comparable results. Traurig and Papka (1980) assayed a number of lysosomal enzymes from mouse hearts ranging in age from 2.5 to 27 months and found that the patterns varied at different age intervals. β-Glucuronidase increased after 11 months, arylsulfatase rose steadily after 5 months, N-acetyl-β-D-glucosamidinase declined and then rose for a net gain. Acid phosphatase did not change significantly.

3. Other Tissues

As to other tissues, Brun and Hultberg (1975) found no age-related difference (<50 versus >50 years of age) in the activities of a number of lysosomal enzymes in the frontal cortex region of human brain. However, the data showed considerable scatter. Platt *et al.* (1973) found no change in β-glucuronidase and cathepsin D, but found a large increase (about 70%) in N-acetyl-β-glucosaminidase in old compared to young rat brain (2½ versus 24 months of age). Traurig (1976) noted no change in acid phosphatase, but noted increases in mouse lung for β-glucuronidase and glucosaminidase between adult animals and those 20 months of age. Wilson (1972) found a sharp decrease for the former enzyme in mouse lung at 6 versus 18 months and an increase between 18 and 30 months. Acid phosphatase showed little change in agreement with most results with other tissues.

4. Insects

Insofar as effects of age on lysosomal enzymes in insects is concerned, work so far has been sketchy. Miquel *et al.* (1974) demonstrated the presence of acid phosphatase in the dense bodies present in *Drosophila*, relating lysosomes to age pigments in this organism. They also found that β-acetylglucosaminidase and acid phosphatase increase with age from 7 to 40 days. Webster and Webster (1978) found age-related increases in the specific activities of acid phosphatase and cathepsin D in a lysosomal fraction. However, a considerable drop in the protein content of this fraction with age results in a large exaggeration of the specific activity of the enzymes in old organisms, based on milligrams of protein. The total activity did not change with age. This observation does not necessarily apply to determinations made in whole homogenates, but it illustrates the care that should be taken in such types of measurement. Sohal and Donato (1979) examined the levels of β-acetylglucosaminidase and β-glycerophosphatase in the heads of houseflies and, similarly, found no significant age-related differences, although there was an increase in the former enzyme through 14 days of age. They found no relationship between the lysosomal enzyme levels and the flight activity of the flies being tested, although the flight activity was related to the life-span'.

C. Comment

There seems little point in further detailing reported changes in individual lysosomal enzymes. The fact that many investigators observed substantial changes at early ages as well as later, lends credence to the belief that lysosomal enzyme activities vary considerable for whatever the metabolic need (DeMartino and Goldberg, 1978). The changes, therefore, are not necessarily a concomitant of aging, but are part of an ongoing process of metabolic regulation. They seem to vary with everything from sex to tissue of origin. It does not help that the role of lysosomes in cellular metabolism is not clearly understood. Suffice to say that changes in lysosomal enzyme levels occur, but the significance and relation to aging, if any, is obscure. Unfortunately, there is little information available on such important subjects as lysosomal turnover except for an early paper by Comolli *et al.* (1972), which , although based on crude preparations, reports a significantly increased half-life of lysosomal fractions in aged rats.

V. TISSUE CULTURE

A. Mitochondria

Lipetz and Cristofalo (1972) observed a decreasing number of mitochondria with completely transverse cristae in late-versus early-passage WI-38 cells, although the number of organelles was unchanged. Houben and Remacle (1978) observed that the mitchondrial enzyme sulfite cytochrome c reductase did not show changes in heat sensitivity in late-passage WI-38 cells.

B. Lysosomes

An increase in lysosomes with increased passage number was reported several years ago and seems to be a common feature of cells in culture. Brock and Hay (1971) observed such increases in chick fibroblasts. Robbins et al. (1970) observed increases in lysosomes in several human tissue types in culture. Subsequently, Lipetz and Cristofalo (1972) observed an increase in lysosomes and autophagic vacuoles in late-passage WI-38 cells. Similar observations were made by Brandes et al. (1972). Brunk et al. (1973) noted an increase in residual bodies in human glial cells, either in late-passage or maintained at confluence for long periods of time.

If lysosomal numbers increase with age, then one would expect an increase in the level of the associated enzymes. Cristofalo and Kabakjian (1975) observed an increase in acid phosphatase and β-glucuronidase activities with passage number in WI-38 cells. An increased proportion of the enzyme activity was found in the supernatant fraction of late-passage cells, suggesting an increased fragility of the lysosomes. Addition of hydrocortisone, which extends the number of passages the cells may undergo (Chapter 2, Section III) appears to have some stabilizing effect. Other increases in lysosomal enzyme activities in late-passage WI-38 cells have been shown by Wang et al. (1970) and Milisauskas and Rose, (1973) (Table 5.V).

Recently, LeGall et al. (1979) determined the activity of ten lysosomal enzymes at early and late passage, of several lines of adult human liver cells. The cells attained a level of about 20 passages over 5–7 months, the number varying somewhat because of the age of the donors (25–55 years of age). With increasing passage number, a gradual increase in the size and number of secondary lysosomes (residual body type) occurred five to six passages before death. In general, four to six cell lines were assayed for each lysosomal enzyme. No substantial changes occurred at any passage level until a decline in the growth rate set in (five to six passages before death). Therefore, only "growing" and "senescent" cells were measured. Arylsulfatase A, β-galactosidase, and β-glucuronidase were lower and arylsulfatase B, higher in senescent cells. Acid phosphatase was higher and hexoseaminidase decreased, but the changes were not statistically significant. α-Fucosidase and β-glucosidase did not change. One difficulty with the results is that if the number of lysosomes increase in senescent cells, why do not all of the lysosomal enzyme levels also increase? LeGall *et al.* (1979) reported an increase in secondary lysosomes of the residual variety. These may very well have lost certain enzymes and retained others. However, the particularly large loss of β-galactosidase (71%) is difficult to explain. The results obtained by LeGall *et al.* (1979) for human liver cells in culture are considerably at variance with those reported for liver cells isolated from old rats (Knook and Sleyster, 1978) and from those reported for WI-38 cells (Table 5.V).

Cells in culture at early and late passage show at least one characteristic in common with cells in intact animals, namely, that the amounts of lysosomal enzymes can vary on an individual basis. The fact that LeGall *et al.* (1979) observed no differences until after cell division slowed, whereas Cristofalo and Kabakjian (1975) obtained a steady increase (early-versus middle versus late-passage) in their enzyme levels may perhaps be ascribed to the different tissues utilized, much as differences appear in different animal tissues.

Lysosomal enzymes do not appear to become altered in late-passage cells. Houben and Remacle (1978) observed no change in the heat sensitivity of *N*-acetyl-β-D-galactosaminidase, *N*-acetyl-β-D-glucosaminidase, and β-D-glucosidase in WI-38 cells (passage 22–25 versus 42–25).

It must be concluded that the relationship of aging to lysosomal function or change of function is no more clear in tissue culture than in intact animals. The hazy picture emerging suggests that perhaps there is no such relationship.

REFERENCES

Abu-Erreish, G. M., Neeley, J. R., Whitmer, J. T., Whitman, V., and Sanadi, D. R. (1977). *Am. J. Physiol.* **232**, E258–E262.
Abu-Erreish, G. M., and Sanadi, D. R. (1978). *Mech. Ageing Dev.* **7**, 425–432.
Adelman, R. C. (1971). *Exp. Gerontol.* **6**, 75–81.
Adelman, R. C. (1975). *Fed. Proc. Fed. Am. Soc. Exp. Biol.* **34**, 179–182.
Baird, M. B., and Massie, H. R. (1976). *Exp. Gerontol.* **11**, 167–170.
Baird, M. B., Zimmerman, J. A., Massie, H. R., and Pacilio, L. V. (1976). *Exp. Gerontol.* **11**, 161–165.
Baker, G. T. III (1976). *Gerontology* **22**, 334–361.
Barrows, C. H. Jr., Roeder, L. M., and Falzone, J. A. (1962). *J. Gerontol.* **17**, 144–147.
Barrett, M. C., and Horton, A. A. (1976). *Biochem. Soc. Trans.* **4**, 64–66.
Birnbaum, L. S., and Baird, M. B. (1978a). *Exp. Gerontol.* **13**, 299–303.
Birnbaum, L. S., and Baird, M. B. (1978b). *Exp. Gerontol.* **13**, 469–477.
Brandes, D., Murphy, D. G., Anton, E. B., and Barnard, S. (1972). *J. Ultrastruct. Res.* **39**, 465–483.
Brock, M. A., and Hay, R. J. (1971). *J. Ultrastruct. Res.* **36**, 291–311.

Brouwer, A., VanBezooijen, C. F. A., and Knook, D. L. (1977). *Mech. Ageing Dev.* **6,** 265–269.

Brun, A., and Hultberg, B. (1975). *Mech. Ageing Dev.* **4,** 201–213.

Brunk, U., Ericsson, J. L. E., Ponten, J., and Westermark, B. (1973). *Exp. Cell Res.* **79,** 1–14.

Bulos, B., Shukla, S., and Sacktor, B. (1972). *Arch. Biochem. Biophys.* **149,** 461–469.

Bulos, B., Shukla, S. P., and Sacktor (1975). *Arch. Biochem. Biophys.* **166,** 639–644.

Chen, J. C., Warshaw, J. B., and Sanadi, D. R. (1972). *J. Cell Physiol.* **80,** 141–148.

Chiu, Y. J. D., and Richardson, A. (1980). *Exp Gerontol.* **15,** 511–517.

Comolli, R. (1971). *Exp. Gerontol* **6,** 219–225.

Comolli, R., Ferioli, M. E., and Azzola, S. (1972). *Exp. Gerontol.* **7,** 369–376.

Cristofalo, V. J., and Kabakjian, J. (1975). *Mech. Ageing Dev.* **4,** 19–28.

Cristofalo, V. J., Parris, N., and Kritchevsky, D. (1967). *J. Cell Physiol.* **69,** 263–272.

DeMartino, G. N., and Goldberg, A. L. (1978). *Proc. Natl. Acad. Sci. U.S.A.* **75,** 1369–1373.

Deshmukh, D. R., and Patel, M. S. (1980). *Mech. Ageing Dev.* **13,** 75–81.

Deshmukh, D. R., Owen, O. E., and Patel, M. S. (1980). *J. Neurochem.* **34,** 1219–1224.

Dionisi, O., Galeotti, T., Terranova, T., and Azzi, A. (1975). *Biochim. Biophys. Acta* **403,** 292–300.

Finch, G. E. (1972). *Exp. Gerontol.* **7,** 53–67.

Garg, B. D., Kourounakis, P., and Tuchweber, B. (1979). *Gerontology* **25,** 314–321.

Gold, P. H., Gee, M. W., and Strehler, B. L. (1968). *J. Gerontol.* **23,** 509–512.

Gold, G., and Widnell, C. C. (1974). *Biochim. Biophys. Acta* **334,** 75–85.

Grinna, L. S. (1977a). *Gerontology* **23,** 342–349.

Grinna, L. S. (1977b). *Gerontology* **23,** 452–464.

Grinna, L. S., and Barber, A. A. (1972). *Biochim. Biophys. Acta* **288,** 347–353.

Grinna, L. S., and Barber, A. A. (1975). *Exp. Gerontol.* **10,** 319–323.

Hansford, R. G. (1978a). *Biochem J* **170,** 285–295.

Hansford, R. G. (1978b). *Comp. Biochem. Physiol.* **59B,** 37–46.

Hegner, D. (1980). *Mech. Ageing Dev.* **14,** 101–118.

Herbener, G. H. (1976). *J. Gerontol.* **31,** 8–12.

Houben, A., and Remacle, J. (1978). *Nature (London)* **275,** 59–60.

Inamdar, A. R., Person, R., Kohnen, P., Duncan, H., and Mackler, B. (1974). *J. Gerontol.* **29,** 638–642.

Inamdar, A. R., Person, R., and Mackler, B. (1975). *J. Gerontol.* **30,** 526–530.

Kato, R. (1978). *In* "Liver and Aging-1978" (K. Kitani, ed.), pp. 287–299. Elsevier/North Holland, New York.

Kato, R., and Takanaka, A. (1968a). *Jap. J. Pharmacol.* **18,** 381–388.

Kato, R., and Takanaka, A. (1968b). *J. Biochem. (Tokyo)* **63,** 406–408.

Kato, R., Takanaka, A., and Onoda, K. I. (1970). *Jap. J. Pharmacol.* **20,** 572–576.

Kment, A., and Hofecker, G. (1977). *In* "Liver and Ageing" (D. Platt, ed.), pp. 65–74. Schattauer Verlag, Stuttgart.

Knook, D. L., and Sleyster, E. Ch. (1976). *Mech. Ageing Dev.* **5,** 389–397.

Knook, D. L., and Sleyster, E. Ch. (1978). *In* "Liver and Aging—1978" (K. Kitani, ed.), pp. 241–252. Elsevier/North Holland, New York.

Knook, D. L., and Sleyster, E. Ch. (1980). *Biochem. Biophys. Res. Commun.* **96,** 250–257.

Knook, D. L., Sleyster, E. Ch. and Van Noord, M. J. (1975). *Adv. Exp. Med. Biol.* **53,** 155–169.

LeGall, J. Y., Khio, T. D., Glaise, D., LeTreut, A., Brissot, P., and Guillouzo, A. (1979). *Mech. Ageing Dev.* **11,** 287–293.

Lipetz, J., and Cristofalo, V. (1972). *J. Ultrastruct. Res.* **39,** 43–56.

Maynard-Smith, J., Bozcuk, A. N., and Tebbutt, S. (1970). *J. Insect. Physiol.* **16,** 601–613.

McMartin, D. N., O'Connor, J. A. Jr., Fasco, M. J., and Kaminsky, L. S. (1980). *Toxicol Appl. Pharmacol.* **54**, 411–419.

Meihuizen, S. P., and Blansjaar, N. (9180). *Mech. Ageing Dev.* **13**, 111–118.

Menzies, R. A., and Gold, P. H. (1971). *J. Biol. Chem.* **246**, 2425–2429.

Menzies, R. A., and Gold, P. H. (1972). *J. Neurochem.* **19**, 1671–1683.

Milisauskas, V., and Rose, N. R. (1973). *Exp. Cell Res.* **81**, 279–284.

Miquel, J., Tappel, A. L., Dillar, C. J., Herman, M. H., and Bensch, K. G. (1974). *J. Gerontol.* **29**, 622–637.

Miquel, J., Economos, A. C., Fleming, J., and Johnson, J. E. Jr. (1980). *Exp. Gerontol.* **15**, 575–591.

Murfitt, R. R., and Sanadi, D. R. (1978). *Mech. Ageing Dev.* **8**, 197–201.

Nohl, H. (1979), *Z. Gerontologie* **12**, 9–18.

Nohl, H., and Hegner, D. (1978). *Eur. J. Biochem.* **82**, 563–567.

Nohl, H., and Jordan, W. (1980). *Eur. J. Biochem.* **111**, 203–210.

Nohl, H., and Kramer, R. (1980). *Mech. Ageing Dev.* **121**, 137–144.

Nohl, H., Breuniger, V., and Hegner, D. (1978). *Eur. J. Biochem.* **90**, 385–390.

Paterniti, J. R. Jr., Lin, C. P., and Beattie, D. S. (1978). *Arch. Biochem. Biophys.* **191**, 792–797.

Paterniti, J. R. Jr., Lin, C. I. P., and Beattie, D. S. (1980). *Mech. Ageing Dev.* **12**, 81–91.

Pieri, C., Zs.-Nagy, I., Mazzufferi, G., and Giuli, C. (1975). *Exp. Gerontol.* **10**, 291–304.

Platt, D., Hering, H., and Hering, F. J. (1973). *Exp. Gerontol.* **8**, 315–324.

Robbins, E., Levine, E. M., and Eagle, H. (1970). *J. Exp. Med.* **131**, 1211–1222.

Sacktor, B., and Shimada, Y. (1972). *J. Cell Biol.* **52**, 465–477.

Schmucker, D. L. (1978). *In* "Liver and Aging" (K. Kitani, ed.), pp. 21–34. Elsevier/North Holland, New York.

Schmucker, D. L. (1979). *Pharmacol. Rev.* **30**, 445–456.

Schmucker, D. L., and Wang, R. K. (1979). *Age* **2**, 93–96.

Schmucker, D. L., and Wang, R. K. (1980a). *Exp. Gerontol.* **15**, 7–13.

Schmucker, D. L., and Wang, R. K. (1980b). *Exp. Gerontol.* **15**, 321–329.

Schmucker, D. L., and Wang, R. K. (1980c). *Proc. Soc. Exp. Biol. Med.* **165**, 178–187.

Schmucker, D. L., Mooney, J. S., and Jones, A. L. (1978). *J. Cell Biol.* **78**, 319–337.

Siliprandi, N., Siliprandi, D., Zoccarato, F., Toninello, A., and Rugolo, M. (1979). *Bull. Molec. Biol. Med.* **4**, 1–14.

Sohal, R. S. (1976). *Gerontology* **22**, 317–333.

Sohal, R. S., and Bridges, R. G. (1977). *J. Cell Sci.* **27**, 273–287.

Sohal, R. S., and Donato, H. Jr. (1979). *J. Gerontol.* **34**, 489–496.

Sorgato, M. C., Sartorelli, L., Loschen, G., and Azzi, A. (1974). *FEBS. Lett.* **45**, 92–95.

Spencer, J. A., and Horton, A. A. (1978). *Exp. Gerontol.* **12**, 227–232.

Starnes, J. W., Beyer, R. E., and Edington, D. W. (1981). *J. Gerontol.* **36**, 130–135.

Stocco, D. M., and Hutson, J. C. (1978). *J. Gerontol.* **33**, 802–809.

Tate, E. L., and Herbener, G. H. (1976). *J. Gerontol.* **31**, 129–134.

Tauchi, H., and Sato, T. (1968). *J. Gerontol.* **23**, 454–461.

Tauchi, H., and Sato, T. (1978). *In* "Liver and Aging-1978" (K. Kitani, ed.), pp. 3–19. Elsevier/North Holland, New York.

Tauchi, H., Sato, T., and Kobayashi, H. (1974). *Mech. Ageing Dev.* **3**, 279–290.

Traurig, H. H. (1976). *Gerontology* **22**, 419–427.

Traurig, H. H., and Papka, R. E. (1980). *Exp. Gerontol.* **15**, 291–299.

Tribe, M. A., and Ashhurst, D. E. (1972). *J. Cell Sci.* **10**, 443–469.

Turrens, J. F., and Boveris, A. (1980). *Biochem. J.* **191**, 421–427.

Vann, A. C., and Webster, G. O. (1977). *Exp. Gerontol.* **12**, 1–5.

Walker, J. H., Burgess, R. J., and Mayer, R. J. (1978). *Biochem. J.* **176**, 927–932.

Wang, K. M., Rose, N. R., Bartholemew, E. A., Balzer, M., Berde, K., and Foldvary, M. (1970). *Exp. Cell Res.* **61,** 357–364.

Webb, S., and Tribe, M. A. (1974). *Exp. Gerontol.* **9,** 43–49.

Webster, G. C., and Webster, S. L. (1978). *Exp. Gerontol.* **13,** 343–347.

Weinbach, E. C., and Garbus, J. (1959). *J. Biol. Chem.* **234,** 412–417.

Weindruch, R. H., Cheung, M. K., Verity, M. A., and Walford, R. L. (1980). *Mech. Ageing Dev.* **12,** 375–392.

Wildenthal, K., Decker, R. S., Poole, A. R., and Dingle, J. T. (1977). *J. Molec. Cell. Cardiol.* **9,** 859–866.

Wilson, P. D. (1972). *Gerontologia* **18,** 36–54.

Wilson, P. D. (1973). *Gerontologia* **19,** 79–125.

Wilson, P. D., and Franks, L. M. (1975). *Gerontologia* **21,** 81–94.

Wilson, P. D., Hill, B. T., and Franks, L. M. (1975). *Gerontologia* **21,** 95–101.

Wohlrab, H. (1976). *J. Gerontol.* **31,** 257–263.

Zs-Nagy, and Pieri, C. (1977). *In* "Liver and Ageing" (D. Platt, ed.), pp. 43–58. Schattauer Verlag, Stuttgart.

Chapter **6**

DNA

I. OVERVIEW

DNA is a distinctive target for studies of the mechanism of aging. Any change, either brought about by programming or by random error, would have major effects on the function of a given organism. One would expect, therefore, a heavy concentration of research designed to determine whether or not changes in DNA or chromatin structure occur with aging. Similarly, there should be an increasing amount of attention given to research on metabolic functions closely associated with DNA, e.g., gene expression, repair mechanisms, DNA replication, and transcription. Indeed, the amount of work in this area that can be related specifically to aging has been substantial. However, it generally lacks the sophistication of current research in molecular biology and other specialized fields. This conclusion is exemplified by the fact that the chapter dealing with repeated genes in eukaryotes in the 1980 volume of *Annual Reviews of Biochemistry* contains 409 references with no mention of aging. In fact, there is no mention of "aging" in the entire index, although there are four chapters dealing with various aspects of DNA structure and metabolism. The omission implies that DNA research in aging is not at a level where it catches the eye of experts in the field of nucleic acids.

The investigations generated by the obvious possibilities of a relationship between nucleic acids and aging has fallen into several categories involving structure, function, and metabolism. None of this work has permitted unequivocal conclusions to be made. Early work focused on generalized changes, such as the amount of histone and non-histone proteins in chromatin, binding strength of associated proteins, and some initial attempts to show qualitative differences in some of these components. Postulated age-related changes in DNA or chromatin structure have involved searches for evidence of cross-linking, nicks, changed template activity, melting temperature, etc. Increasingly sophisticated attempts are now being made to evaluate the level of gene expression, which as the work now stands, provides some evidence that expression of rRNA genes may decrease in old age. Metabolic approaches have included studies of such parameters as activity and fidelity of DNA polymerases, DNA repair, and histone modification. Results tend to be contradictory, probably in many cases, for technical reasons. One conclusion that seems safe to make is that the fidelity of DNA polymerase is unchanged with age. One intriguing speculation has been placed in doubt: newly completed studies do not confirm the nice relationship observed between ability for excision repair and lifespan.

Work with cells in culture has more or less paralleled studies *in vivo.* Comparisons of results from the two systems are difficult to make as there are so many conflicting data within each system.

It takes little study of the material in this chapter to conclude that research in the area of DNA has not brought to light any clearly interpretable, age-related changes. The research efforts have not uncovered a "smoking pistol" in the hand of the criminal; no fingerprints have been found that implicate the culprit; no mysterious spots appear on chromatograms to focus the gaze of the eager sleuth on some guilty facet of nucleic acid metabolism. The available information does not even tell us where *not* to look, since conflicting data do not permit dismissal of negative results. Although contradictions in early work on chromatin structure should perhaps be excused as being due to problems involving chromatin isolation, it is more difficult to rationalize differences emerging from recent studies, such as those in which contrary conclusions were reached regarding age-related changes in nucleosome structure.

In spite of the lack of undisputed results, one cannot dismiss the investigations into the role of DNA in aging as unimportant. The ideas behind the explorations are, to a large degree, still valid. It is clear that we simply need more research performed with more precise techniques; that if age-related changes occur, they are subtle ones and not easily detectable.

II. DNA CONTENT

Although the DNA content of cells should be constant, it is not always so. Liver cells, for example increase in polyploidy with age (Collins, 1978), so that there is an increased DNA content per cell in old animals

(Castle *et al.*, 1978). The situation is much more complex on a weight basis, as cell size, water content, or cell density may vary with age. A good example is the DNA content reported for samples of human muscle tissue. The young tissue contained 196 μg/g wet weight, whereas the old tissue gave a value of 299 μg/g (Steinhagen-Thiessen and Hilz, 1976). Similarly, the amount of DNA in tissue of rat testes appeared to double at 30 versus 18 months, after an initial decrease between 3 and 20 months (Liu *et al.*, 1978). In this case, the increase may be due to the presence of testicular tumors. These examples are sufficient to indicate that even using DNA as a standard against which to measure age-related changes in the content of other components, e.g., RNA or protein, is not always safe.

III.   CHANGES IN DNA AND CHROMATIN

Although there is little hard data to support them, a number of theories—perhaps it is better to call them speculations—regarding the involvement of DNA in aging have been put forth. Underlying these theories lies the unresolved concept as to whether aging is caused by random error and its cumulative effects, or by a specific genetic program. In dealing with either concept, one must consider not only changes in DNA itself, but possible alterations in chromatin structure that might alter gene expression. The idea that control of the aging process may lie in changes in histones or non-histone proteins (NHP) is so simple and direct that understandably, it has great appeal. In the main, earlier work revolved around a search for such changes in the chromatin of young versus old animals; the effect of such changes, if any, on the thermal stability or template activity of the chromatin; the possibility of differential dissociation of proteins at various salt concentrations; the possibility of qualitative changes in the proteins associated with chromatin; the possibility of cross-linking of DNA to itself or to other types of molecules. To date, the best efforts of various investigators have provided us with little unequivocal data. In fact, the information available is contradictory and confusing. For example, several studies show an age-dependent increase in the thermal stability of chromatin, whereas others show no change. Various investigators have shown template activity to increase, decrease, and show no change. These results do not appear to be associated with changes in the DNA itself, as there seems to be some agreement that removal of proteins from the chromatin eliminates age-related differences.

The most reasonable explanation for the contradictory findings is that the preparations of chromatin differ significantly from laboratory to laboratory. Other technical problems may exist. For example, measurement of template activity is typically carried out using RNA polymerase from *E. coli*. This enzyme can transcribe exogenous RNA (Zasloff and Felsenfeld, 1977). Thus, RNA contamination of the various chromatin preparations could result in apparent differences in template activity. Moreover, as Gaubatz *et al.* (1979a) point out, determination of histones is subject to artifactual results unless appropriate precautions are taken.

Although much of the early work may be considered to be obsolete, it is still frequently quoted and, therefore, it deserves to be summarized in some detail. Studies of age-related changes in DNA began with the work of von Hahn and Verzar (1963) and von Hahn (1964–65). DNA from bovine thymus, containing equal amounts of histone, was found to give higher denaturing temperatures when prepared from old animals (2 months versus 10 years). Similar observations were later made with preparations from young and old rat liver (von Hahn and Fritz, 1966). In the latter study, the DNA preparations were deproteinized so as to leave varying amounts of histone (1–6%). When heated in solutions of low ionic strength, the "old" preparations containing less histone (1–2%) possessed melting temperatures equal to "young" preparations containing much more histone (6%). From these early results, it would appear that the higher T_m observed for "old" preparations is not a result of increased histone content, but perhaps is due to a selective retention of certain histones with a greater binding ability.

Since the reports of von Hahn and co-workers, a number of investigators have studied the effect of age upon the histone or protein content of various chromatin preparations. Often, the temperature stability and/or the template activity of the chromatins were also determined. The results of these investigators are given in order of publication in Table 6.I. In general, most investigations observed no change in the total histone content, whereas the results for T_m, template activity, and protein content are contradictory. Pyhtila and Sherman (1968) studied nucleoproteins derived from bovine thymus (2 months versus 10 years). They used two procedures to isolate the material—that of Zubay and Doty (1959) and that of Bonner and Huang (1963). In the first preparation, the T_m of the chromatin was greater in the old animals. In the second preparation, no age-related change was observed, and no differences were observed in the DNA itself. Use of the purification procedure of Bonner and Huang (1963) also eliminated the age-related increase in RNA content observed in the first preparation. Samis and Wulff (1969) also pointed out that template activity of chromatin can be considerably modified by shearing, heat denaturation, and extraction with sodium chloride. In

Table 6.I. Chronological Listing of Earlier Reports on the Effects of Age on Chromatin Properties[a]

Animal	Tissue	Ages	Note	Protein	Histone	NHP	RNA	T_m	Template activity[b]	Reference[c]
Mouse	Brain	Newborn to 30 months		− +				− +		1
Cow	Thymus	2 months versus 10 years	See[d]	NC			+	+	NC	2
			See[e]	NC			NC	NC	NC	
			DNA					NC		
Mouse	Prostate	6 versus 30 months		−		−f			−	3
				−		−				
				NC					NC	
Rat	Kidney	4–33 months		−		−			NC	4
	Liver			−		−			NC	
Rat	Liver	6.7, 17, 30 months		NC			−		NCg	5
				NC					NCh	
Dog	Cardiac muscle	1–16 years		NC			NC		NC	6
Rat	Liver	1–30 months		NC	NC		−	+	−	7
Mouse	Liver	3.6, 13.7, 21.4 months		NC	+	−	−		NCi	8
Rat	Submandibular gland	2 versus 12 months		−	NC	−			+	9
	Liver		DNA	NC	NC	NC			NC	
				NC					NC	

Animal	Tissue	Age				Ref.
Mouse	Brain	Average 3.4; 11.6; 18.7 months	NC –	–		10
Mouse	Liver	6, 20, 31 months	NC[i]	NC[k] +	NC	–[l]	11
	Kidney		NC[i]	NC +	NC	–
	Brain		NC[i]	NC +	NC	
Rat	Liver	2, 25, 29 months			+	12
	Regenerating liver	In situ			NC

[a] (+) Refers to an increase with age; (–), a decrease; (NC), no change; (– +) means a decrease from young to mature, then an increase; (NC +) means no change from young to mature, then an increase, etc.

[b] RNA polymerase from E. coli.

[c] Key to references: (1) Kurz and Sinex, 1967; (2) Pyhtila and Sherman, 1968; (3) Mainwaring, 1968; (4) Pyhtila and Sherman, 1969; (5) Samis and Wulff, 1969; (6) Shirey and Sobel, 1972; (7) Berdyshev and Zhelabovskaya, 1972; (8) O'Meara and Herrmann, 1972; (9) Stein et al., 1973; (10) Kurz et al., 1974; (11) Hill, 1976; (12) Pieri et al., 1976.

[d] Procedure of Zubay and Doty, 1959.

[e] Procedure of Bonner and Huang, 1963.

[f] "Acid proteins."

[g] The data show a significant decrease for mature and old versus young animals.

[h] The data show substantial increases though the authors state that there is no change.

[i] The template activities are measured after dissociation in various salt concentrations. Template activity at the initial points (no salt) appear to be unchanged.

[j] Slight increase at 31 months.

[k] Statistical data are not given.

[l] Using homologous RNA polymerases, "young" enzyme worked best with "young" chromatin and vice versa. "Young–young" and "old–old" combinations were comparable.

137

addition, they showed that the concentration of RNA polymerase used was a factor in determining template activity. They found no change in template activity in crude chromatin or sheared chromatin of rat liver at 203, 510, and 905 days of age. As one would expect, these experiments reflect the fact that chromatin preparations may yield different results depending upon the isolation technique.

Somewhat more recent work on the effect of age on transcription (Hill, 1976) deals with liver, kidney, and brain chromatin from mice. The mice (a strain of C57BL) were 6, 20, and 31 months of age. The author studied transcription using not only RNA polymerase from *E. coli*, but also homologous polymerase. Unsheared chromatin was prepared using a Triton X-100 washing procedure. No significant differences were found in the ratios of histone/DNA, NHP/DNA, or RNA/DNA in the 6- versus 20-month samples of the respective tissues. A 10–25% increase was observed in the NHP:DNA ratio for the old (31-month) animals, but no statistical analysis is provided as to its significance. Conditions selected for determining template activity using the *E. coli* polymerase permitted initiation, but no reinitation of RNA chains. The template activity of chromatin from 6- versus 31-month-old mice was determined for all three tissues as a function of increasing chromatin concentration and found to be 15–20% lower in all three tissues of the old mice. With excess polymerase, the rate of labeled UTP incorporation into RNA was 17% greater per microgram of DNA, for "young" liver chromatin. Since the lengths of the RNA chains formed were comparable (about 650 nucleotides), the "young" chromatin must have more RNA chains to account for the increased incorporation of labeled UTP. That is, "young" chromatin possessed more initiation sites.

Use of homologous RNA polymerase extracted from both "young" and "old" livers showed a surprising age-related specificity: "young" polymerase was more active with "young" chromatin than with "old" chromatin; the reverse was true when polymerase from "old" livers was used. However, "young" polymerase in conjunction with "young" chromatin, and "old" polymerase in conjunction with "old" chromatin, showed equal template activities.

In comparing the results shown in Table 6.I, it should be kept in mind that a variety of animals and tissues are being compared. If the basic cause of aging is the same in all mammals and that cause derives from changes in chromatin, then one could argue that these changes should be seen, at least to some degree, in all tissues of all animals. An opposite view is that aging results from a programmed decline in the expression of certain genes only in one or two key tissues and the resulting decline in function would in turn affect other tissues. A more reasonable view might be that most, if not all tissues age, although at different rates, gradually increasing the burden on all systems. Thus, it could happen that brain chromatin, for example, might show little change with age, whereas liver chromatin might be considerably altered. Both would not have to show common patterns of change. An example of this situation, excluding the possibility of experimental error, might be the findings of Stein *et al.* (1973) in which age-related changes occur in submandibular chromatin, but do not appear in liver chromatin (Table 6.I). In short, in comparing different tissues one need not expect a common result. Even so, the array of conflicting results even from the same tissue (Table 6.I) is rather forbidding. There is little assurance that these data can be used with any confidence to support a point of view regarding the role of chromatin changes in aging. Needless to say, age-related alterations in chromatin that are too small to detect by the rather broad methods thus far applied may occur. The amount of scatter in much of the published data would make such small changes impossible to detect.

As to reasons for the lack of agreement, the authors of the various papers put forth few explanations. Shirey and Sobel (1972) stress the method of preparation. They point out that chromatin can bind non-selectively to various components. Some tissues require special treatment, for example, use of a dense sucrose medium to remove tissue contaminants from nuclei. Use of Triton X-100 to remove the outer nuclear membrane is advantageous. Hill (1976) also notes that many systems and techniques were utilized by the various investigators. It should also be recognized that proteases and nucleases are present in nuclei and must be diligently eliminated or calculated into the results. As mentioned above, RNA polymerase from *E. coli* can transcribe RNA. Since most investigators use this enzyme for determination of template activity, chromatin contaminated with RNA could give erroneous results.

Another approach used by early investigators to study chromatin is extraction with salt of various concentrations or with detergents in an attempt to determine if proteins are removed differentially according to the age of the donors. Most of the work discussed below comes from papers referred to in Table 6.I. However, it seems advantageous to deal with these experiments as a separate group.

The initial work of von Hahn and Verzar (1963) and von Hahn (1964–65) with bovine thymus suggested that there was an age-related qualitative change in DNA–histone binding. That is, "old" DNA preparations had a higher T_m than "young" preparations when both samples had the same amount of protein attached. Therefore, either the nature of the proteins differed or the tightness of binding increased with the age of the organism. Von Hahn and Fritz (1966) extended the observations to rat liver and concluded that there was a qualitative difference in the DNA–protein binding. O'Meara and Herrmann (1972) extracted mouse liver chromatin with increasing ionic strengths of sodium chloride (0.01–3.0 M) and found that the composition of the proteins changed substantially with increasing age. Moreover, protein was more easily removed from "young" chromatin (3.6 versus 21.4 months). Mature animals (13.7 months) gave intermediate results. As the salt concentration was increased (i.e., as protein was removed), the template activity for RNA synthesis increased, although not to the same degree at all ages. The "young" chromatin started sooner and the mature and old caught up at higher salt concentrations. The authors felt that their chromatin preparations retained structural integrity, as the material manifested a low template activity (less than 5–10% of that of the purified DNA). Much higher values are observed with chromatin prepared by other workers (18–35%).

Berdyshev and Zhelabovskaya (1972) also extracted rat liver chromatin from animals of various ages (1–30 months) with increasing concentrations of sodium chloride. Above 1.0 M, a decreased amount of histone was removed from "old" chromatin. Extraction at zero salt concentration caused an extraordinary decrease in NHP with increasing age: the amount of NHP remaining in chromatin from 30-month-old animals is only 51% of the young (1-month value). Template activity of the chromatin after extraction with salt at specified concentrations showed no change in animals between 1 and 12 months, but showed a decline at 30 months. It required 2.5 M salt to obtain 100% template activity for old chromatin versus 1.5 M for "mature" and "young" chromatin. There is substantial variability and a lack of consistency in these results, which is bothersome. These and other data appear in a later report (Berdyshev, 1976).

Recently, Medvedev et al. (1979) extracted NHP from young and old mouse and rat liver chromatin using water, 0.14 M, and 0.35 M sodium chloride, respectively. Based on the results of polyacrylamide gel electrophoresis, they observed that the greatest change was an increase in number and amount of high molecular weight proteins in the water-soluble group. Subsequently, Medvedev et al. (1980) observed no such differences in NHP from mouse spleen chromatin at 4 and 29 months of age.

Stein et al. (1973) studied the effects of sodium deoxycholate on chromatin from rat submandibular glands. They found that at each concentration of the deoxycholate tested (0–0.01 M), more histone was released from adult chromatin (12-month) than from young (2-month) preparations.

In contrast to the changes reported above, Samis *et al.* (1968) found no age-related differences in the template activity of salt-extracted chromatin from rat liver. Subsequently, Shirey and Sobel (1972), in the course of their work with dog cardiac muscle, could find no age-related differences in chromatin after chromatography on a Bio-Gel A50M column in 1.25 *M* sodium chloride. The eluted chromatin showed no age-related change in the protein/DNA ratio. Quantitatively, there appeared to be no proteins that remained bound specifically to the "old" DNA.

The salt extraction procedures offer no clearer answers about changes in aging chromatin than the studies of chromatin listed in Table 6.I. The extent of age-related changes reported by some authors is startling and by their very magnitude, must be suspect. The blame probably lies in the chromatin preparations that may not be very pure to start with. Moreover, in some cases, endogenous nucleases and proteases may be active. An idea of the difficulty of eliminating impurities may be drawn from the work of Salser and Balis (1972) who extracted DNA from several tissues of CFE rats of various ages. The DNA was deproteinized, washed thoroughly, and reprecipitated several times. Analysis of individual amino acids in DNA from liver, kidney, spleen, and intestinal tissues was carried out after hydrolysis in 6 *N* HCl. Total amino acids were generally about 1–5 μmol/100 mg DNA in the various organs. No consistent age-related trends were observed. Whether or not the amino acids are bound to the DNA, are picked up during the processing, or simply derive from residual proteins was not determined.

A recent approach to the possible age-related change in chromatin structure has involved attempts to show differences in nucleosome conformation. Current ideas on the structure of chromatin (McGhee and Felsenfeld, 1980) places DNA in repeating nucleoprotein subunits (nucleosomes) in which about 70% of the base pairs of DNA are wrapped around a core consisting of eight molecules of histones (two each of H2A, H2B, H3, and H4). H1 histone is attached to the complex, but is not in the core. The nucleosomes are linked by variable lengths of spacer DNA (up to 80 base pairs) which is sensitive to nuclease attack. Micrococcal nuclease digestion of chromatin, therefore, yields a series of nucleoprotein particles containing multiples of 180–200 base pairs.

With this information as a base, Gaubatz *et al.* (1979a) examined the effect of micrococcal nuclease and DNase I on nuclei from young (1-month), mature (9-month), and old (28-month) mouse liver, brain, and heart tissue. If the chromatin structure differs, for example, in the amount of protective histone, the amount of DNA that is solubilized by micrococcal nuclease should reflect this situation. The authors point out that it is critical to consider endogenous proteolysis and nucleolysis, even though the experiments are comparative in nature. That is, if "old" tissues contain more intrinsic nuclease activity, then the results, in the absence of controls, would show an apparent, but false, increase in DNA digestion for the "old" preparation after micrococcal nuclease treatment. In this study, monitoring indicated no degradative changes in histones or other proteins during the procedure. The method of preparing nuclei (Hewisch and Burgoyne, 1973) prevented autonucleolysis. Interestingly, the control experiments for liver showed substantial endogenous nuclease activity when $CaCl_2$ (necessary for micrococcal nuclease activity) was added. This endogenous activity varied from preparation to preparation by as much as 20% without any age-related relationship. When endogenous values were subtracted for each experimental run, there were no age-related differences in the susceptibility of liver tissue to micrococcal nuclease attack. Differences observed in A_{260} (7% lower for young versus mature or aged animals) disappeared when a specific DNA assay was used. The danger in using an assay based on A_{260} is that the nuclease also attacks RNA.

Heart and brain nuclei did not show endogenous nuclease activity. No significant age-related differences were found either in amounts of DNA released or in the time course of the reaction. The nucleosome core size (140 ± base pairs) did not change with age for any of the tissues, nor did the nuclease cleave the DNA at different sites for the respective sets of young, mature, and old nuclei.

DNase I has been reported to cut chromatin in such a way as to distinguish between active and inactive genes. Presumably, the conformation of active genes is such as to permit attack by the enzyme. The results obtained by Gaubatz et al. (1979a) using DNase I with brain and liver nuclei were similar to those obtained with micrococcal nuclease. No significant change was observed in the rate of attack on young versus old chromatin. In short, either there are no age-related changes in the conformation of chromatin or as the authors point out, they are too small to be detected by present methodology. These results agree with the lack of change in transcription reported earlier by Hill (1976) (Table 6.I) for these same tissues when homologous RNA polymerase was used to measure template activity.

Tas et al. (1980), using the same approach as Gaubatz et al. (1979a), obtained quite different results. They digested liver nuclei of young (7- to 11-month) and old (22- to 27-month) mice (C57BL/6 × BALB/cF, hybrid) with micrococcal nuclease. The ratio of solubilized DNA to DNA was determined after lysing the nuclear pellet. The ratio obtained from old mice was always lower. This ratio has been interpreted to reflect transcriptionally active versus inactive chromatin. In simpler terms, the rate of appearance of solubilized DNA was always a little faster and of greater magnitude in young than old liver nuclei. The nuclease-resistant pellet was of greater density from old preparations. Use of SH-reducing agents (β-mercaptoethanolamine, dithiothreitol) decreased their density to that of young animals. The density of mechanically sheared chromatin was also reduced. Pretreatment of the nuclear fractions with the reducing agents resulted in the increased release of DNA. These results are interpreted as meaning that S—S bonds are involved in the condensed structure of chromatin through an increased protein–DNA binding, and that more chromatin exists in this state in old age. Tas (1976) had earlier made such a proposal. There appears to be little evidence at present to support this view. In fact, Carter (1979) could find no evidence for increased oxidation with age in the disulfide bonds of H3 histone or of non-histone proteins of rat liver or brain chromatin.

In attempting to reconcile the age-related effect of micrococcal nuclease reported by Tas *et al.* (1980) and the lack of difference noted by Gaubatz *et al.* (1979a), it can be pointed out that there were a number of technical differences in procedure. Tas *et al.* (1980) apparently did not run controls for endogenous nuclease activity, which is considerable in mouse liver (Gaubatz *et al.*, 1979a). Moreover, they did not use spermine and spermidine, which tend to condense the chromatin and stabilize the nuclei.

It is of interest that Zongza and Mathias (1979) reported that the nuclei of different types of liver cells had different sensitivities to nucleases. The authors separated diploid stromal, diploid parenchymal, and tetraploid parenchymal nuclei by zonal centrifugation. They observed developmentally related changes between rats of 3 weeks and 2 and 4 months of age in terms of accessibility to micrococcal nuclease and DNase, namely that the DNA repeat length increases with age. Thus, in rapidly growing rats, there seem to be detectable changes in the nuclease-susceptible conformation of DNA in nucleosomes. Presumably, such changes would be seen if they occurred to the same extent during aging.

In spite of the large amount of work performed in this area, it must be concluded that no indisputable evidence has yet been provided that implicates changes in chromatin structure in the aging process.

IV. HISTONES

The early recognition of histones as a class of proteins related to chromatin structure made these compounds an obvious target of investigation for age-related differences in amount, type, or form. As mentioned above, Von Hahn and co-workers observed an age-related increase in T_m after salt extraction of DNA preparations from calf thymus and rat liver. They attributed the differences to small residual amounts of bound protein tentatively identified as histone. For this reason, Von Hahn et al. (1969) examined the composition of histones extracted from the liver of young (6 months), mature (12 months), and old Wistar rats (26–28 months). Acid-extracted histones showed a decrease in the ratio of Arg-rich:Arg-poor fractions in the "old" preparations. Subsequently, Pyhtila and Sherman (1968) showed sharply lowered amounts (40%) of arginine-rich histones in cow versus calf thymus. The substantial changes reported in these early experiments are probably artifacts of technical procedures. Carter and Chae (1975), pointed out that Von Hahn et al. (1969) fractionated their histones in 0.02 N HCl on Sephadex G-75 and might, therefore, have been observing aggregated proteins, as the order of elution of their histones is incorrect. As to the results of Pyhtila and Sherman (1968), the method used for fractionation yields considerable cross contamination of different histone fractions. For example, these authors found nearly twice the normal amount of F3 histone in their preparations.*

Shirey and Sobel (1972) extracted proteins from the chromatin of dog cardiac muscle both with acid and with 1% sodium dodecylsulfate (SDS). Acrylamide gel electrophoresis was performed on urea-treated NHP and histones and the SDS-extracted proteins. Neither NHP or histones showed qualitative age-related differences. The quantitative aspects of the experiments are difficult to accept, being extremely variable. In fact, the differences in average band intensities for three age-groups (1- to 2-year; 8- to 9-year; 11- to 16-year) vary by as much as 150% in the smaller bands.

Carter and Chae (1975) studied the histone composition of rat liver chromatin employing the polyacrylamide gel electrophoresis procedure of Panyim and Chalkley (1969). They used Osborn-Mendel, Wistar, and Sprague-Dawley rats and C57BL mice in their study. Ages varied from 3 to 25 months in the various experiments. No age-related changes were observed in the total amount of histone. Moreover, no changes in the relative proportions of the various histones were found. The authors pointed out that conditions were selected that minimize the degradation of histones by chromosomal proteases. F1, F2b, F2a1, F2a2, and F3, including acetylated histones, were the same for all age groups within

---

*The terminology for histones changed from F to H in papers subsequent to 1977. To avoid confusion, original designations are given.

the standard deviation. The work appears to have been carefully performed, and the use of three different strains of rats strengthens the results.

Medvedev *et al.* (1977) also found no obvious age-related difference in the pattern of histones extracted from liver and spleen of young (3-month) and old (26- to 27-month) rats. However, further refinement of the procedure to examine minor subfractions of F1 histone by gel electrophoresis showed age-related changes in the proportions of F1 and F1^0 for both tissues, the latter component increasing in the old animals. The authors also observed an increase in an F1 subfraction, which had earlier been reported to contain methionine (Medvedeva *et al.*, 1975). Subsequently, Medvedev *et al.* (1978) made similar observations with mouse liver and spleen. Although the overall pattern of mouse H1 histones was found to differ from that of the analogous rat tissues, the age-related differences were similar.

Gaubatz *et al.* (1979b) carefully quantitated H1, H1^0 and H4 in young, mature, and old mouse liver and brain. No differences were observed in liver. However, in brain, H4 appeared to decrease by 50% in old animals and H1^0 to increase about 20%. These differences were ascribed to experimental artifacts. The degree of contamination (basic myelin proteins) in the H4 fraction varied with each preparation and increased with age. When corrected, the values for H4 did not differ in young and old animals. The authors also suggest that the difference they observed in the quantities of H1 could result from a shuttling of the histone between the cytoplasm and the nucleus, a process which depends on the ionic strength of the buffer, presence of detergents and cytoplasmic factors.

It seems safe to say that no age-related qualitative changes in histones have been demonstrated and that quantitative changes if they exist, are small. Moreover, as related by Gaubatz *et al.* (1979b), the possibility of technical difficulties cannot be ignored. Although Medvedev *et al.* (1977) observed quantitative changes in an H1 histone subfraction, their animals were studied at 3 versus 27 months. We do not know if the reported changes perhaps occurred relatively early in life with little or no subsequent change.

V. MODIFICATION OF HISTONES AND NON-HISTONE PROTEINS

Another approach to the effect of aging on chromatin would be to study the modification of histones and non-histone proteins. It is known that histones undergo phosphorylation, acylation, methylation, and

ADP-ribosylation, and it is believed that these processes play a role in gene expression. As to NHP, there are hundreds of different molecules in chromatin, and these differ greatly in size and charge, the pattern varying from tissue to tissue. These proteins may be phosphorylated, acylated, and methylated. Phosphorylation, in particular, is thought to play a role in gene expression.

Relatively little work, particularly with older animals, has been performed to determine whether the above processes show age-related differences, either for histones or NHP. Kanungo and Thakur (1977) and Das and Kanungo (1979) briefly reported that phosphorylation and acetylation of histones decreases with age in rat brain slices using animals 2, 14, and 84 weeks of age. The latter authors report that polyamines stimulate the reaction in the young but not the older preparations. Incorporation of labeled acetate into NHP and histones H3, H4, and H2B decreased with age (Thakur et al., 1978), although Das and Kanungo (1979) reported no change with age in H3. No age-related change was observed in H1 and H2A. In a subsequent report, Kanungo and Thakur (1979) again observed a lowered incorporation of labeled acetate into histones in slices of rat cerebral cortex with increased age. However, the major drop appeared to be in the interval of 4–32 weeks, with little change at 110 weeks (statistical accuracy not provided). Thus, the changes in acetylation reported above appear to involve development rather than aging. As the authors point out, the results might also reflect differences in the transport of acetate through the cell membrane or a decreased level of acetyltransferase.

In stark contrast, Liew and Gornall (1975) had earlier reported that in brain nuclei from old mice (29 months), the label from ^{3}H-labeled acetyl–CoA is incorporated into histones to a 40% greater degree in old (29 month) than in young (2 months) animals. No age-related difference was found in the incorporation of ^{32}P, but both acetylation and phosphorylation of NHP were markedly increased with age. Sarkander and Knoll-Kóhler (1978) also observed an increase with age in acetylation of histones by nuclei of rat brain. The authors point out that various problems involving acetate metabolism and pool size are eliminated when acetyl-CoA is used as the label. Unfortunately, the "old" animals in these experiments were only 12 months of age. Since Liew and Gornall (1975) only measured acetate incorporation at 2 and 29 months, they may really have been observing an increase in acetylation at early ages (say between 2 and 12 months) which subsequently changed little. Thus, they would be observing the same maturational changes reported by Sarkander and Knoll-Kóhler (1978).

Using intact rats, Oh and Conard (1972) observed that incorporation of labeled acetate into histones in the liver decreased markedly between 2 and 6 months of age, leveled off till 12 months, and rose modestly at 12 months to a plateau till 24 months. There seemed to be no age specificity in the pattern of labeling. The authors also noted a decline in histone acylation during liver regeneration in the older animals. More recently, O'Meara and Pochran (1979) injected labeled acetate into male Sprague-Dawley rats of various ages and examined incorporation into liver histones. There appeared to be no differences in acetate pool sizes. A decline in the specific activity of labeled histones was observed up to 21–24 months of age with no further change at 27 months. There were age-dependent differences in the incorporation of acetate into H3 and H4, the histones containing nearly all of the label. However, the rate of incorporation of labeled acetyl–CoA into histones by intact liver nuclei appeared to be unchanged between 12 and 24 months of age. After a "chase" of unlabeled acetyl–CoA, the rate of deacetylation was also approximately equal at these two ages, although the younger preparations appeared to have a faster initial rate.

The difficulty with interpretation of the whole animal experiments is that results were reported for a single time after injection of label. It might be that the data would be quite different at other time intervals. For example, in the case of induction of certain enzymes, old animals achieve the same level of activity as young animals, but it may take longer (Chapter 10, Section XIII).

A summary of the above work carried out with nuclei or slices provides widely varying conclusions: acetylation of certain histones decreases, increases, and remains the same during maturation with little reference to senescence. Perhaps the contradictions lie in differences in tissue preparations and experimental conditions, such as substrate concentration and time of incubation. Certainly, the conflicting data do not provide a feeling of security in making use of the data. The situation with respect to intact animals is little better, lacking in data from intermediate ages and being questionable on technical grounds. It is obvious that much remains to be done before it can be established that acetylation or phosphorylation of histones or NHP show consistent changes that are related to the aging process. In any case, there is no assurance that one can interpret the meaning of such changes in terms of being a cause rather than an effect of aging: changes in phosphorylation or acetylation of histones or NHP may simply reflect changing maintenance requirements. If the older animals utilize a different pattern of metabolic enzymes, there will surely be differences, however subtle, in the level of active genes and presumably, in the characteristics of histones and NHP.

VI. ALTERED DNA

In contrast to the idea of programmed changes in nucleic acids, there is a school of thought that proposes that aging results from an accumulation of errors. Those errors which might arise from a change in the DNA would be of greatest importance. The effect of occasional errors in other macromolecules such as mRNA or tRNA would surely be diluted out by large numbers of normal molecules, although in theory, errors in informational molecules could become amplified. For example, if faulty transcription occurred, the resulting mRNA could give rise to faulty proteins. If any of these proteins were involved in translation or transcription, they might bring about even more errors in the next generation of proteins (Orgel, 1963, 1970). The idea of errors in the protein-synthesizing system and its ramifications are discussed in Chapter 9, Section III.

An early theory of aging involving mutations was proposed by Szilard (1959). Subsequently, Curtis (1963) based a theory of aging on somatic mutation. He related this idea to chromosomal aberrations, which were observed to be more numerous in cells of old compared to young animals. In part because it can produce life-shortening effects and also cause chromosomal aberrations, radiation was first thought to provide a useful model for aging studies. However, it is now generally accepted that the life-shortening effect of radiation is not analogous to the normal aging process.

The role of mutations in aging is difficult to assess, particularly since accurate data on mutation rate and frequency in eukaryotic organisms are not available. Even without worrying about the effects of aging, we are dealing with estimates. For human germ cells, the proposed values appear to be extremely low. The frequency of amino acid substitutions in hemoglobin is also very low (Hirsch, 1978). These facts, coupled with the existence of repair systems makes it dubious that mutation plays a primary role in the aging process. In fact, it is safe to say that there is no coherent pattern of evidence that supports such a thesis. Many writers, however, accord at least a contributory role to mutations or DNA damage from chemical or environmental factors.

One should be careful to distinguish between the terms, mutation, error, and damage. In the present context, mutation is considered to be an irreversible change in DNA, which would lead to altered transcription. Errors refer to such events as mistranslation or mistakes in transcription. The end results would be similar to those arising from mutations except that the latter would have a permanent effect. Damage refers to chemical or physical changes in the DNA or chromatin. Damage would probably, but not necessarily, result in loss of information and, therefore, a loss or a reduction in the amount of some product.

If alterations in DNA are responsible for aging or if they are at least a contributory factor, then the alteration must be beyond reach of rapid repair. For this condition to be met, both strands of DNA must be affected, a base change must occur, or repair systems must be deficient or slow to act.

The likelihood of parallel damage to both strands of DNA is very small. Bjorksten (1968) proposed that aging is a result of the cross-linking of various biomolecules. Insofar as DNA is concerned, such an event would join two DNA strands irreversibly. The theory lacks experimental support, although Massie *et al.* (1975) provide suggestive evidence for the presence of cross-linked DNA in old *Drosophila*. However, no age-related effect was found. One must consider that mitotic cells with both strands of DNA broken or cross linked, will not divide. Non mitotic cells with cross-linked DNA might survive, but they would have to become numerous for the tissues to show senescent properties in the presence of a mass of normal cells. Therefore, if nonmitotic tissues show senescent properties, one should find a considerable amount of DNA showing an increased molecular weight. No age-related accumulation of such DNA has been observed.

Mutations that affect single strands may have different ramifications, depending upon whether or not the cells involved are mitotic or postmitotic. In the former case, the error would be incorporated into the genetic material of the daughter cells. These cells might suffer a biological disadvantage and perhaps would not survive many replications. On the other hand, they might create a neoplasm. In nonmitotic or slowly dividing cells, the effect of the mutation or accumulation of mutations would be expressed only by the affected cells. To provide the manifestations of senescence, a substantial part of the tissues would have to consist of mutated cells. Otherwise, it is difficult to see how mutation could play more than a secondary role in the normal aging process, exclusive of initiating the development of pathological conditions. Such conditions are not present in the typically smooth transition from healthy mature to healthy old organisms.

In the case of damage to the DNA, unrepaired breaks in the nucleotide chain would presumably prevent replication of the damaged strand. For nonmitotic cells, such breaks occurring in a gene could result in the loss of information by preventing transcription of that gene. Breaks in the area of control genes could have the same effect. The ultimate consequence of such damage to DNA would, however, depend upon the efficiency of the repair mechanisms.

What is the evidence for the existence of damaged DNA or chromatin in old organisms? As seems unfortunately to be the case in many aspects of aging research, the available information is somewhat sketchy and contradictory. A number of investigators have provided evidence for strand breaks in old animals. Modak and Price (1971) showed that the nuclei from fixed brain sections of old mice (3–4 versus 30 months) incorporated more labeled dATP than did sections from young animals (approximately 2.5-fold increase) after treatment with calf thymus DNA polymerase. These results indicated to the authors that more 3'-OH groups were available for the polymerase. That is, there were more strand breaks in the "old" DNA. Similar results were obtained by Price et al. (1971) with nerve, Kupffer, and cardiac muscle cells from young and old mice. Massie et al. (1972) reported that in old rat liver, a 10-fold decrease of molecular weight in both single- and double-stranded DNA occurred, mostly by 400 days of age. No other investigators have reported changes of this magnitude. Wheeler and Lett (1974) obtained indirect evidence for an age-related decrease in size of DNA molecules extractable from neurons of the cerebellum of old beagles. Chetsanga et al. (1975) reported an increased digestion by S_1 nuclease (specific for single-stranded DNA) of DNA from old mouse liver. Subsequently, Chetsanga et al. (1976) found that myocardial DNA from aging CBF mice (20, 25, 30 months) showed an increasingly broad band in CsCl gradients, suggesting the existence of a variety of molecular weights. DNA from young organisms (6, 15 months) gave a sharper profile. Centrifugation in alkaline sucrose gradients (showing single-stranded DNA) gave three peaks for "old" DNA (30 months) versus a single peak for "young" (6 months) material, indicating a variety of sizes for single chains in the former. S_1 nuclease digested "old" DNA more effectively than young DNA (15 versus 2.6%, respectively). In mouse brain tissue, "old" DNA showed more strand breaks both by analysis on alkaline sucrose gradients and by digestion with S_1 nuclease (Chetsanga et al., 1977).

The above evidence would seem to provide a clear indication of rather substantial changes in DNA structure with age in terms of nicks or single-stranded gaps. However, two recent reports throw doubt on the validity of this conclusion. Finch (1979) found no increase in digestion of mouse liver DNA by S_1 nuclease at 10 versus 30 months of age. The author added neonatal liver prelabeled with [3H]thymidine to act as a control for the preparative procedure. Experimental conditions were otherwise similar to those of the earlier work. Dean and Cutler (1978) also could find no age-related increase in the sensitivity of old mouse liver DNA to S_1 nuclease. The authors used young (4.5–7.5 months), mature (20 months), and old animals (28–31 months). Additional evidence that single-strandedness was not increased was indicated by an unchanged thermal melting profile and CsCl density gradient pattern. The authors took precautions against selective loss of single-stranded regions. Thus, we have three papers demonstrating an age-related increase in S_1 digestion of (single-stranded) DNA in mouse liver, heart, muscle, and brain, and two papers showing no such change in mouse liver. One cannot easily rationalize the conflicting evidence in terms of faulty experimental techniques. The interested investigator must, therefore, either await additional results or perform a new set of experiments for himself. Although the weight of evidence suggests that there is no change in the single-strandeness of DNA with age, the available literature cannot provide unequivocal proof at this time.

VII. OTHER TYPES OF DNA DAMAGE

Smith (1976a) points out that radiation can result in cross-linking of DNA to protein and to amino acids and suggests that the lowered percentage of DNA that can be readily isolated from older tissues (von Hahn, 1963) may be due to this fact. However, Gaubatz *et al.* (1976) obtained yields of DNA better than 90% from young and old mouse brain and liver with no reported differences in the extraction characteristics due to age. Although indirect evidence has been presented that UV radiation can cause DNA–protein cross-linking in human fibroblasts (Fornace and Kohn, 1976), such a process does not seem likely to occur in intact animals much below skin level. In fact, there is no evidence at this time that cross-linking of DNA with other molecules plays a significant role in the intact animal under normal aging conditions. An overview of nucleic acid adducts, both natural and radiation induced, can be found in a volume edited by Smith (1976b). Recently, Lesko *et al.* (1980) have assessed the role of superoxide and the formation of hydroxyl radical in DNA strand scission.

VIII. DNA REPAIR

The desirability of maintaining DNA in an error-free state throughout the lifetime of cells hardly needs exposition. In the face of continued chemical or physical challenge, the integrity of the gene and consequently, of gene expression depends upon the ability to repair damage as it occurs. The question, therefore, arises as to whether a reduction in repair capability occurs with age, and if so, could this reduced function be responsible for the deleterious effects of senescence.

If unrepaired but nonlethal errors occurred in mitotic cells, they might be passed on to progeny and presumably would contribute to a reduced function of the tissue involved. However, the numbers of cells with altered function would have to be considerable or one would expect them to be diluted out by more effective "normal" cells. Of course, a steady net production of DNA damage (that is, above the repair capacity) would provide a substantial accumulation of altered cells over the years. If the errors gave rise to transformed cells, neoplasms might be generated that would sweep all before them. Thus, at a minimum, repair systems must function to prevent types of damage that may lead to neoplasms or to modestly altered cells, which could gradually replace undamaged cells. On the other hand, an accumulation of unrepaired damage could itself lead to depreciation of the repair system. In this case, one should observe an age-related decline in this function unless of course, cells so affected quickly die from an excess of unrepaired damage. The involvement of DNA repair systems in the aging process is thus an important consideration, especially since it seems clear that damage to DNA does occur *in vivo*. Physical and chemical agents can bring about a number of changes that range from free radical attack to alkylation of bases. Briefly, the types of damage to be expected include formation of pyrimidine dimers (UV irradiation), attack on single-stranded regions (γ irradiation, free radicals, endonuclease nicking of damaged DNA), and apurinic or apyrimidinic sites (chemical reactions, γ irradiation). Damage is not necessarily random, as the structure of chromatin might make certain areas more or less susceptible to attack by outside agents. The same reasoning applies to susceptibility of different areas of chromatin to repair processes.

From a practical point of view, DNA repair can be put into four main categories: excision, strand break, postreplication, and photoreactivation repair. There is little information available on the effect of these processes on aging or vice versa. Hart and Trosko (1976) reviewed this area, and a further discussion may be found in Hart *et al.* (1979). The possibility of nicks and gaps in DNA of old animals has already been discussed (see Section VI). If this type of damage does indeed occur, is it a result of declining ability for repair? Unfortunately, as mentioned above, little is known about changes with age in the capacity of DNA repair mechanisms. It is extremely unlikely that a slowed rate of excision repair by itself, can be responsible for aging. For example, cells of individuals with Xeroderma pigmentosum are defective in excision repair (Cleaver, 1969), but the subjects do not age accordingly. The disease is a genetic one in which the skin is highly sensitive to UV light, and there is a high incidence of skin cancers. Repair replication and rejoining of strand breaks in skin fibroblasts from patients with the disease appears normal (Kleijer *et al.*, 1970), suggesting that the fault lies in dimer excision. There are variants in which dimer excision and subsequent repair are normal, but there is a defect in rejoining the repaired section to the DNA strand (Fornace *et al.*, 1976).

Little work has been done on the effect of age on the ability for DNA repair. Gensler (1981) found little change in the ability of lung and kidney tissue to perform UV-induced unscheduled DNA repair during two-thirds of the life-span of hamsters. Wheeler and Lett (1974) selected nonmitotic tissues (neurons from cerebellum) from dogs 7 weeks to 13 years of age. After irradiation with X rays, the resulting strand breaks were repaired equally rapidly at all ages. Most of the other studies of this nature have been carried out with cells in tissue culture (Section XII) in which early- and late-passage cells are compared. We do not know if the findings are applicable to intact organisms.

A thorough study of all types of DNA repair would be necessary before a proper assessment can be made of the importance of these systems to the aging process per se. Moreover, even if the ability for DNA repair decreases with age, it may well remain above the level needed to handle routine problems.

IX. EXCISION REPAIR AND LIFE-SPAN

Although changes in repair systems with age have not been studied in detail in higher organisms, a curious relationship has been observed between the life-spans of a number of mammalian species and the rate at which cells from the animals can perform dimer excision repair. For this repair, endonucleases recognize the type of damage and excise 10–200 bases. The gap is then repaired by the appropriate DNA polymerase, presumably using the intact strand as the template. Finally, a ligase joins the two ends. The experimental protocol for determining excision repair involves the culture of cells from skin biopsies taken from the various animals. At confluence, the cultures are irradiated with UV light to create thymine dimers. The cells are then placed in medium containing [^{3}H]dT and hydroxyurea, the latter agent to prevent scheduled DNA synthesis. Thus, incorporation of labeled thymidine represents excision repair. It is most typically determined by counting grains after radio-autography.

Hart and Setlow (1974) first reported that the ability for excision repair showed a linear relationship to life-span for shrew, mouse, rat, hamster, elephant, and human beings. The authors reasoned that their results are not due to differences in the numbers of photoproducts per unit of UV fluence (i.e., the sensitivity of the various DNAs to irradiation), the amount of DNA per cell, internal thymidylate pools, or the average size of repair regions. They concluded that the number of repair sites per unit length of DNA are indeed being measured. Subsequently, Hart *et al.* (1979) showed that the same life-span versus repair relationship held for two species of mice, one long-lived (*Peromyscus leucopus*, the white-footed mouse) and the other short-lived (*Mus musculus*, the house mouse). The rate of excision repair was 2.5-fold greater for cells of the former, about the same as the ratio of life-spans, *Peromyscus* to *Mus*.

These results and the idea engendered of a relationship between excision repair and life-span is fascinating. Most recently, Hall *et al.* (1982) in a blind study, examined excision repair in fibroblasts cultured from punch biopsies (fifth passage) and in lymphocytes from the blood of various primates maintained at the San Diego Zoo, the Los Angeles Zoo, and the UCLA Medical Center. Both test systems (fibroblasts and lymphocytes) showed a linear relationship between maximal life-span and excision repair. Unfortunately, recent work has come to an opposite conclusion about the correlation between repair ability and life-span of a species. Kato *et al.* (1980) examined excision repair in 34 species of mammals and corrected the values for genome size to human DNA content. Results comparable to those obtained by Hart and Setlow (1974) were observed for shrew, mouse, rat, golden hamster, and human beings. However, the primates were found to be particularly off scale. For example, the crab-eating macaque showed the same degree of repair as humans, but lives only 15.5 years. The experimental system differed somewhat from that of Hart and Setlow (1974) and Hall *et al.* (1981). Mostly, lung tissue was used and the authors point out that the donor animals were not at similar stages of aging.

Whatever the final outcome of the conflicting evidence for and against this interesting relationship, all species clearly do not fit the pattern. Woodhead *et al.* (1980) recently performed similar experiments on cells from three cold-blooded vertebrates—trout, turtle, and Amazon molly. The respective life-spans are up to 8 years, 118 + years, and about 3 years. The amount of repair in each case was small or nonexistent, being about 5% for the Amazon molly and turtle and nondetectable in trout.

Moreover, H. S. Targov and P. V. Hariharan at the Roswell Park Memorial Institute (personal communication) found that the free-living nematode *Turbatrix aceti* exhibits manyfold more rapid excision repair than for any of the mammalian cells reported, yet its maximal life-span is only about 40 days.

It should be stressed that other forms of repair have not shown a relationship to maximal life-span. For example, repair of single strand breaks produced by γ rays was the same in *M. musculus* and *P. leucopus* (Hart *et al.*, 1979). Hall *et al.* (1981) further support this view.

One should be aware that differences in ability for repair may vary with the tissue. Peleg *et al.* (1976) reported that cells from mouse kidney (mostly epithelial) show considerably more repair than skin fibroblasts. Interesting too, is the report of the authors that cells from 13- to 15-day-old fetuses show considerable dimer excision repair, whereas at 17–19 days or after birth, the level is small. Whether or not these early developmental results can be applied to mature versus older animals of the same species has not been determined. That is, differences in repair capability with respect to age rather than to life-span have not been demonstrated.

It is interesting to note that there seems to be an inverse relationship between species life-span and ability to activate 7,12-dimethylbenz(a)-anthracene to forms that can bind to DNA. The activation of the compound results from the action of mixed function oxidases. Thus, cells cultured from humans and elephants showed little DNA binding and those from cows, rabbits, guinea pigs, and rats showed increased binding in the order of their life-spans (Schwartz and Moore, 1979). Curiously aging increases the susceptibility of mouse skin to the drug (Ebbensen, 1974).

X. GENE EXPRESSION

Damage to chromatin that does not result in a mutation may be reflected in changes in the expression of individual genes. Such changes may vary from complete loss of function to an altered rate of function or to an altered sensitivity to signals for production of certain molecules. It seems reasonable to assume that the first of these alternatives would be lethal unless the gene in question were reiterated. Therefore, there have been a number of attempts to measure directly, age-related changes in the transcriptional ability of genes. Unfortunately, as is the case for other work on chromatin structure, the results are contradictory. Pelc (1970) seems first to have suggested that the amount of reiterated DNA could be related to aging. Subsequently, Medvedev (1972a) suggested that one of the functions of repetition in DNA sequences is to prevent an accumulation, with time, of miscoding from molecular accidents. He reasoned that random mutations would be more damaging to organisms with less reiterated genes. Medvedev thus theorized that "genomes with a relatively large number of repetitious vital genes will be associated with a longer life-span in differentiated cells." An expanded discussion of the theory was published giving greater emphasis to the role of nonrepeated genes on life-span (Medvedev, 1972b).

Experimental efforts to measure gene expression in aging have been made by a number of investigators, mostly involving the idea that there may be age-related changes in the amount of reiterated DNA or in gene dosage. The studies depend on rather crude systems, and the validity of the conclusions should be viewed with caution. Johnson *et al.* (1972), pursuing the idea that there might be a loss of genes during aging, utilized DNA–RNA hybridization techniques to study ribosomal RNA genes in selected mitotic and postmitotic tissues of beagles of 1–10 years of age. They obtained DNA from various tissues at different ages and performed hybridization studies with purified ^{3}H-labeled rRNA from yeast. The findings indicated no significant change with age in liver, spleen, and kidney, whereas DNA from heart, brain, and muscle of old animals show about a 25% decrease in ability to bind the labeled yeast rRNA. This loss is taken to represent a substantial decline in the number of RNA cistrons in the DNA from these organs. Johnson *et al.* (1975) subsequently examined human heart tissue from young and old human subjects and reported a significant decline in rRNA genes in myocardium from the older individuals (44, 73, and 75 years versus 18 and 20 years).

Using more or less the same basic approach, Cutler (1973–1974) considered the possible relationship of redundant DNA to the aging process by attempting to determine if there was a correlation between the amount of redundant DNA and the aging rate of several mammalian species. Brain tissue from young, old, and embryonic C57BL/6J mice and young and old humans was used for the experiments. Among the data compared were the percentage of reassociating classes of DNA (fast, intermediate, low-repetitive, and slow) for a number of mammalian species. No change was observed in the average percentage of extremely slowly reassociating DNA (unique DNA) between brain tissue of young and old mice and young and old humans, respectively. The percentage reiterated DNA sequences did not show a correlation with life-span. For example, mouse and man show the same values. The data tabulated by Cutler (1973–1974), including other information from the literature, showed rather substantial variation even for the same animal species, depending presumably upon the laboratory in which the work was performed. In subsequent work, Cutler (1975) measured the number of gene types being transcribed in mouse liver and whole brain at different ages (48–745 days). Using the basic techniques described above (DNA–DNA reassociation, DNA–RNA hybridization), a steady and substantial (up to twofold) decrease was found in the percentage of reiterated DNA sequences transcribed in the mouse liver with increasing age. Similar results were obtained for unique DNA. The hybridization studies showed no real trend. The author noted that the T_m values for the DNA–RNA hybrids were 6° below that of the native double-stranded DNA, indicating considerable mismatch.

Gaubatz and Cutler (1978) investigated the effect of age on the number or rRNA genes in liver, brain, kidney, and spleen of the young and old mice (C57BL/6J). Hybridization experiments were carried out using rRNA which was labeled with ^3H by methylation. The authors found no age-related change in the thermal stability of the DNA–RNA hybrids. At ages greater than 800 days, a decrease in rRNA dosage was observed in brain, kidney, spleen, and liver. The liver tissue started with a lower number of genes in young animals, but caught up to the values for the other tissues between 200 and 350 days. The declines with age were reported to be from 100 to 92 rRNA genes in liver at 790 versus 922 days (60 genes at 1200 days) and 100–77 genes in brain (61 genes at 1200 days); 91–57 genes in kidney (100 versus 922 days); and 91 versus 56 genes in spleen (100 versus 905 days). No statistical data are provided. Moreover, there are obvious anomalies. For example, kidney at 905 days shows 83 genes, but at 922 days, 17 days later, shows only 57, a loss of 26 genes. Brain drops sharply from 101 to 72 genes at 788 versus 850 days. The above experiments are an extension of an earlier paper (Gaubatz *et al.*, 1976) in which mouse liver and brain were examined, but only up to 788 days of age. These earlier experiments reported no change, presumably because the reported losses in genes appear to start only after this age. The earlier paper had also reported that samples of human liver (mean age 1.7–67 years) and brain (0.5–62 years) showed no significant changes in gene dosage with age. The trends suggest an initial high level, a drop, and then a recovery to original levels of rRNA genes. The individual variation in humans was exceedingly large (up to twofold), whereas there was little difference found between individual mice of the same age.

Strehler *et al.* (1979) extended the work performed earlier on postmortem human heart ribosomal genes (Johnson *et al.*, 1975). They eliminated protein from their DNA samples by digestion with pronase, RNA by alkali treatment, and short DNA segments by chromatography on

Sephadex G-100. Labeled rRNA for the experiments was obtained from liver after injection of [^{3}H]orotic acid into a 13-week-old rat. From a 23-year-old donor, the percentage of hybridization was greater than from a 63-year-old at saturating levels of rRNA (0.045 versus 0.059%). Hybridization took place slightly more slowly in old preparations. The authors tested human myocardial DNA from 29 individuals and claimed an average rate of loss of rRNA genes of 0.5% per year. Thus, these studies were taken to confirm the initial conclusions of Johnson *et al.* (1975) based on the original five cases. In a further study, Strehler and Chang (1979) ran comparable experiments on ribosomal DNA from the hippocampus and reported a similar loss of rRNA genes with age, the magnitude being 1.5 times that for heart.

Based upon the above experiments, there seems to be agreement that there is a loss of gene dosage with age although Gaubatz *et al.* (1976) reported no change in human brain or liver, whereas Strehler *et al.* (1979) claimed that there is a decline in the former tissue as well as in heart. Strehler *et al.* (1979) pointed out that their hybridization procedure is performed in liquid phase rather than on filters and that they obtain very little scatter between individual experiments, whereas the results of Gaubatz *et al.* (1976) with human tissue showed a considerable range.

In spite of the apparent agreement, there are enough anomalies in the results to remind one that these experiments may be subject to experimental artifacts. Moreover, statistical analysis of important results is lacking. Therefore, it would be best to await further studies before coming to firm conclusions. Perhaps "suggestive" would be a fair way of classing the present status of the work.

Ono and Cutler (1978) approached the problem of gene expression from the opposite direction. Rather than genes becoming "turned off," perhaps they become "turned on" during aging, resulting in products no longer normal for a differentiated tissue. Thus, the authors sought to determine if there were a greater degree of expression for globin synthesis in old versus young brain DNA. Labeled DNA complementary to purified globin mRNA was synthesized and used to titrate nuclear RNA from young, middle-aged, and old mouse brain (6–7, 19–20, and 26–27 months). The ratio of globin RNA to total RNA in the cytoplasm was reported to increase with age for each group. In the nucleus, the ratio increased only between young and middle-aged mice. The number of globin RNA molecules per cell was calculated to increase from 282 to 515 in the cytoplasm (15 versus 32 in the nucleus), but the calculation requires that RNA content remain constant with age. In liver, the number of globulin molecules was reported to increase with age for each group in both nucleus and cytoplasm.

In addition to the work involving DNA with respect to gene expression certain fractions of DNA have been checked for quantitative changes with age. Prashad and Cutler (1976) examined the amount of satellite DNA in liver, spleen, kidney, and brain of C57BL/6J mice as a function of age. They found that the yield of satellite DNA varied with the procedure used. Their own method (CsCl:chloroform extraction of the DNA) gave excellent yields (82–98%). Satellite DNA comprised 10% of the total in spleen, kidney, and brain tissue. Differences between individual animals of the same age were small. In these tissues, there was no significant change in satellite DNA with the age of the animals (10 to about 780 days). In liver, there appear to be an age-related increase in satellite DNA from 8 to 13% of the total DNA. However, this increase was already present at 455 days and may, therefore, not be related to aging.

Herrmann et al. (1975) reported that a fraction of spontaneously reassociating DNA increased from 0.6 to 1.8% in liver and brain of old (27-month) versus young (2- and 12-month) C57BL/6 mice. The difference was not observed if the DNA was treated with nuclease-free pronase or with alkali. The authors, therefore, concluded that the increased percentage of this fraction in old animals is due to a protein or peptide component that is necessary for reassociation.

It is clear that much more exploration must be performed in the area of gene expression and aging. If at least the studies of DNase I action on nucleosomes and the DNA hybrid studies were in agreement, one could assume that a firm groundwork for future efforts had been established. However, the results are still tentative and in some cases contradictory. One must wait patiently and hope that improvements in technical procedures can be applied along with a steady attack on the problem.

XI. DNA POLYMERASE

Reduced function of DNA in senescent animals could depend on several factors besides alterations of chromatin structure which mediate the degree of gene activity, and changes in chromatin proteins, areas which have already been discussed. Reduced function may result from a lessened ability to synthesize new DNA or more particularly, from a lowered fidelity during replication or repair. Of course, why the enzymes responsible for these processes would change their properties unless the DNA structure changed first is reminiscent of the "which comes first" principle of the chicken and the egg. Still, changes in the fidelity of a DNA polymerase could come about by changes in the cellular milieu, such as ion concentration or localized changes in pH. Once again, one would have to consider whether such changes are gene-mediated, are random, or are perhaps due to time-related changes in the structure of certain long-lived molecules or membrane components. Eichhorn (1979) discusses the effect of metal ions in bringing about misincorporation. He points out that not only the "wrong" metal (e.g., Mn^{2+} instead of Mg^{2+}), but the wrong concentration, can cause such difficulties. Metal ions including Cu^{2+}, Zn^{2+}, Co^{2+}, and Mn^{2+}, which bind to bases are also capable of cross-linking between strands, not only of DNA, but also of RNA. Furthermore, the specificity of DNase I can be changed by metal ions complexing with DNA. These and other reactions could be important in the aging process if the metal concentration of given tissues changed substantially with age. In this respect, it is interesting that the aluminum concentration of brain cells increases with age. Moreover, it has been reported that the metal is concentrated in chromatin (DeBoni *et al.*, 1974). Recent studies of the fidelity of DNA polymerase from human placenta (Krauss and Linn, 1980; Seale *et al.*, 1979) show differences in the accuracy of eukaryotic α, β, and γ forms of the enzyme and again point up the errors resulting from the presence of Mn^{2+}. Parenthetically, DNA polymerase α is believed to be responsible for replicative DNA synthesis, polymerase β for repair synthesis and extension of DNA chains, and polymerase γ for strand displacement synthesis and mitochondrial DNA (Weissbach, 1979.)

A. Enzyme Activity

There is little evidence that DNA polymerase activity changes in old animals. Barton and Wang (1975) examined DNA polymerase activity in crude preparations from the spleens of young (3–8 months) and old (24–30 months) mice (BALB/c). A high molecular weight polymerase (presumably polymerase α) showed no changes in activity with age. A lower molecular weight polymerase (presumably polymerase β) showed substantially reduced activity in old animals. However, a similar decline in activity was not observed in liver or kidney or in the spleen of a different strain of mice (BC3F$_1$). Müller *et al.* (1980) found a decrease with age in polymerase activity as determined in homogenates of bone marrow cells of "old" mice that were 15 months of age. Since a short-lived strain of CBA(J) mice was used (mean life-span, 7 months), the authors consider these animals "senescent." Deoxynucleotidyltransferase activity was lower in cells from senescent versus mature animals (13 and 34%, respectively, of the values for immature animals).

Ove and Coetzee (1978) examined incorporation of labeled TTP into DNA in liver nuclei from "mature" and "old" rats, aged 4 and 19 months, respectively. No differences were observed. Bleomycin-stimulated repair synthesis was greater in the younger animals. The amounts of DNA polymerase β in the nuclei, however, appeared to be the same.

B. Enzyme Fidelity

In a recent paper, Fry *et al.* (1981) compared both the activity and fidelity of DNA polymerase β in chromatin from livers of young and old short-lived mice (*Mus musculus*) and long-lived mice (*Peromyscus leucopus*). The authors found that neither the activity or the accuracy of DNA synthesis changed with age, and these characteristics were not related to the longevity of the species. The fidelity of the polymerase in the chromatin was determined using poly [d(A-T)] as an external template and [^{3}H]dGTP to determine erroneous incorporation. [^{32}P]dTTP was used as an identifying component for the synthesized nucleotide. The amount of chromatin-directed synthesis of DNA using exogenous templates was the same for *M. musculus* at 12 and 30 months and for *P. leucopus* at 12 and 72 months, respectively. Intermediate ages were reported to yield similar results. The activity of the polymerase in the two species was not greatly different. Inhibitor studies showed that polymerase β was the enzyme responsible for copying poly [d(A-T)]. For measurements of fidelity, endogenous DNA was removed by gel filtration after treatment of the chromatin with restriction endonucleases, which do not attack the poly [d(A-T)]. The average rate of error for *M. musculus* was calculated to be about 1/800, and this figure was constant from 12 months on. In *P. leucopus*, there was a considerable range in the small number of samples, but it seems clear that there is no loss of fidelity between 24 and 72 months. Overall fidelity is about the same in the two species. Use of partially purified DNA polymerase β gave similar results (errors, 1/800 and 1/1200 for *Mus* and *Peromyscus*, respectively) and again there were no age-related changes. Agarwal *et al.* (1978) had previously reported that DNA polymerase in lymphocytes from old human subjects was more heat-sensitive but fidelity using poly (dC) as template was unchanged with age.

It seems safe to conclude that aging is not related to an excess of errors caused by altered DNA polymerase. The work of Fry *et al.* (1981) is particularly convincing in this respect, because the fidelity of the polymerase was measured in chromatin, and thus no losses of "functionally altered enzymes" occurred during the isolation procedure. The use of long-lived and shorter-lived species of mice adds an interesting sidelight to speculation about the relationship of excision repair to life-span (Section IX).

The continued inability of experimenters to demonstrate age-related increases in errors, whether in proteins (Chapter 8) or nucleic acids makes it more and more likely that aging is a programed process.

XII. TISSUE CULTURE

Studies of chromatin and other aspects of DNA metabolism have been carried out in early- and late-passage cells, more or less in the same pattern as with intact animals. Unfortunately, the data obtained are often equally inconclusive. It is apparent that the results of work with tissue culture are not at a stage in which they can be interpreted with any degree of certainty, nor can a meaningful relationship of tissue culture work to aging in animals be drawn.

A. DNA Repair

The degree to which cells from animals can carry out excision repair has been related to the life-span of the donor species, although this relationship is not unanimously agreed upon (Section XIII). Study of DNA repair has also been undertaken in cells "aged" in tissue culture. By and large, the results of the various studies are at variance, although the differences can perhaps be explained by the use of different types of cells sometimes studied under different conditions. The conclusions drawn by various investigators are summarized briefly in Table 6.II. Earlier work in the field has been reviewed by Little (1976).

1. Single-Stranded Breaks

Single-stranded breaks in DNA are produced by irradiation with X rays or γ rays. Typically, DNA is first labeled in growing cells in the presence of isotopic thymidine. After irradiation, repair is allowed to proceed for various lengths of time, and the size of the DNA strands is then determined after centrifugation in alkaline sucrose gradients. A newer procedure involves alkaline elution of DNA from cells lysed on polyvinyl chloride filters. The rate of elution of labeled DNA molecules is a measure of the strand sizes. According to Clarkson and Painter (1974) and Bradley et al. (1976), single-stranded breaks resulting from X irradiation were repaired equally well in early- and late-passage WI-38

Table 6.II. Effect of Passage Number on DNA Repair

Cell type	Type of repair[a]	Passage number[b]	Effect of passage no.[c]	Remarks	Reference
WI-38	A	15/25	NC	Phase II versus terminal passage	Clarkson and Painter, 1974
WI-38	A	19/61	NC	Early versus middle-passage	Bradley et al., 1976
Human skin	A	5/15-25	−	Isolated nuclei used	Epstein et al., 1974
WI-38	A	23/39/47	−		Mattern and Cerutti, 1975
Human embryonic	A	12 → 60	NC	Slower repair of terminal passage	Suzuki et al., 1980
WI-38	A	18/48/60	−	Early versus late-passage	Hart and Setlow, 1976
Human diploid	A		NC	Slower only at late-passage	Suzuki et al., 1980
WI-38	B	20/41/44	NC	Late-passage	Painter et al., 1973
WI-38	B	18 → 60	NC	Late Phase III	Milo and Hart, 1976
WI-38	B	18 → 60	−	Continuous decline with passage number	Hart and Setlow, 1976
WI-38	B	27/45	−	Arrested cells	Bowman et al., 1976
Mouse embryo	B	1/9	−	After fourth transfer	Peleg et al., 1976
Human foreskin	B	12/59	NC	Confluent cells	Dell'Orco and Whittle, 1978
Human foreskin	B	12/59	+	Arrested for 9 days	Regan and Setlow, 1974
Progeric	A		NC		Epstein et al., 1974
	A		−		Rainbow and Howe, 1977
	A		−		

[a] A, Single strand breaks after X or γ irradiation; B, excision repair after UV irradiation.
[b] Arrows signify intermediate passage levels.
[c] NC, no change; +, increases; −, decreases.

cells. In fact, the latter group used "terminally senescent" cells, i.e., cells at or close to their last doubling. On the other hand, Epstein *et al.* (1974) reported that middle-passage human skin fibroblasts took longer to repair X-ray-induced strand breaks than cells in very early passage (five doublings). Total doubling potential for these cells is 50–55 doublings. Hart and Setlow (1976) subsequently studied the repair of DNA damage caused in WI-38 cells by UV irradiation, the chemical *N*-acetoxy-2-acetylamino-fluorene, and X irradiation. They found that after X irradiation, on average, late-passage cells do less unscheduled synthesis (repair) than early-passage cells. However, except for very late passage cells, almost all cells carry out some unscheduled synthesis. These results agree with those obtained by Mattern and Cerutti (1975) who found that isolated nuclei from late-passage WI-38 cells excise damaged bases after γ irradiation less well than nuclei from early- or mid-passage cells. Recently, Suzuki *et al.* (1980) studied rejoining of X-ray-induced single-stranded breaks in human diploid fibroblasts with the usual alkaline elution procedure, but using both labeled DNA and a more sensitive fluorometric method. In general, differences in the patterns after radiation and subsequent repair were small for cells up to the terminal passage. Only at the last passage was repair slower, but even in this case, the DNA eventually recovered to the size of unirradiated material. The authors concluded (as had Hart and Setlow, 1976) that a deficiency in DNA repair of single-stranded breaks is not the cause of *in vitro* aging in diploid cells particularly as repair took place at the same rate even in cells that had begun to show a declining growth rate. It seems safe to conclude that loss of ability to repair single-stranded breaks is not a necessary concomitant of aging *in vitro*.

2.   Excision Repair

Painter *et al.* (1973) observed no decline in the ability of WI-38 human fibroblasts to repair UV-induced damage with increasing passage. Milo and Hart (1976) also found repair to be affected only in late phase III. Only at the last doubling was there a deficiency in the repair system. These negative results are in contrast to a series of papers reporting a reduction of repair capability in late-passage cells. Hart and Setlow (1976) determined the average amount of repair in early- and late-passage cultures of WI-38 cells and also the distribution of repair rates among the cells by use of double labeling radioautographic procedures in which scheduled and unscheduled synthesis could be ascertained in the same cell. The authors observed a decrease in both processes with increasing passage number. They also found evidence that those cells that were not synthesizing DNA were also undergoing a sharply reduced degree of excision repair. The authors pointed out that interpretation of results can depend on the number of grains above background, which one selects to represent unscheduled DNA synthesis. Bowman *et al.* (1976) examined excision repair in WI-38 cells prevented from dividing by removal of fetal calf serum and arginine from the medium. Using radioautographic procedures, they observed that the UV-irradiated "young" cells incorporated more labeled thymidine into repaired DNA than did old cells. Reminiscent of the results of Hart and Setlow (1976), they observed that the percentage of cells capable of incorporating thymidine (i.e., the number of dividing cells, a factor which decreased with age) correlated with the percentage that underwent unscheduled synthesis. The authors suggested that since Painter *et al.* (1973) (see above) had prelabeled their cells with BudR and removed the "heavy" DNA, which represented material formed by semiconservative synthesis, many young, actively dividing cells could have been eliminated, thus affecting the results. Other work suggesting that excision repair is related to the ability of cells to divide is that of Peleg *et al.* (1976) who found that mouse embryo cells (13–15 days gestation), concomitantly with a reduced growth rate, lose their ability for excision repair with increased passage.

Dell'Orco and Whittle (1978) using human diploid fibroblasts (foreskin tissue) obtained rather different results. They found that confluent cultures showed no differences in excision repair regardless of passage number. However, in populations of cells arrested for 9 days, that is, cells which were prevented from dividing by reducing the amount of fetal calf serum in the medium—there was a substantial *increase* (30–50%) in DNA repair in the passages representing the latter two-thirds of the life-span. The results were not artifacts of the radiation dose, isotope concentration, or pool sizes. The authors suggest the use of preconfluent cultures by other investigators may emphasize the fact that the proportion of cells in S phase decreases in later passages. If repair is most active in S phase, then dividing cells at late-passage would show decreased repair. In confluent populations, 90–95% of the cells are in the G phase, regardless of age. One other possibility to be considered is that the foreskin fibroblasts behave differently than the fibroblasts used by the other investigators.

What the work described above means in terms of *in vivo* aging or even *in vitro* aging is, as yet, obscure. For late-passage cells, we have reports of loss of repair capacity, no change of capacity, and increased capacity (for arrested cells). Understanding must await further insights into the enzymatic mechanisms involved. Painter *et al.* (1973), Hart and Setlow (1976), Dell'Orco and Whittle (1978), and Suzuki *et al.* (1980) all agreed that the loss of repair ability is not the cause of loss of growth potential. This seems a reasonable conclusion, especially since Goldstein (1971) reports that cells from subjects with Xeroderma pigmentosum, which are deficient in excision repair, survive the same number of passage as cells that are normal in this respect.

3. Progeric Cells

A number of investigators have studied repair of single-stranded breaks in cells from subjects with progeria because the disease seemingly mimics accelerated aging. Epstein et al. (1973) found that cells from progeric subjects have a sharply reduced ability to repair strand breaks resulting from ionizing radiation (γ rays from a cobalt-60 source). Subsequently, Epstein et al. (1974) compared early- and middle-passage, normal and progeric human fibroblasts and again found that rejoining of γ-ray-induced strand breaks was impaired. In an accompanying article, Regan and Setlow (1974) reported normal repair in progeric cells under similar, but not identical, experimental conditions. In fact, some of the same progeric cell strains were included in both studies. Bradley et al. (1976) also found no evidence for deficient strand rejoining in progeroid cells after X irradiation. Later, Rainbow and Howes (1977), by an independent procedure involving repair of irradiated virus, again concluded that progeric cells are defective in repair of DNA damaged by γ rays. Although obviously of considerable intrinsic value to understanding defects in progeric cells and to basic knowledge of cellular behavior, it is doubtful that these conflicting results, even if resolved, will be greatly helpful in assessing the possible role of strand break repair in aging. There is little consistent evidence that progeric cells behave in the same manner as cells aging in vivo, or for that matter, like late-passage cells.

B. DNA Damage

Chemical or physical damage to DNA is not well studied in tissue culture. There is no evidence for, or are there any detailed studies on, such changes in DNA in early- versus late-passage cells except in the context of work on chromatin. Fornace and Kohn (1976) obtained indirect evidence for protein–DNA cross-linking resulting from UV irradiation of human fibroblasts. Such cross-links have been postulated by Smith (1976a) as a possible cause of aging. However, Bradley et al. (1976) could find no evidence for normally occurring DNA cross-linking in early- or late-passage cells. The methods of detection involve X irradiation of the chilled cells to produce single-stranded breaks, after which, alkaline solutions are used to elute DNA from cells lysed on filters. The rate of elution is related to the molecular weight of the DNA. Cross-linked DNA elutes very slowly. Cross-linking of DNA nuclei under normal conditions of cell culture, if it exists, has yet to be demonstrated.

C. Changes in Chromatin Properties

It is well established that DNA synthesis is reduced, i.e., the number of cells dividing decreases during late-passage (Chapter 2, Section III). Thus, a search for changes in chromatin that could be associated with this process is a logical approach to the study of *in vitro* aging. A number of investigators have explored various aspects of chromatin composition and behavior in early versus late passage cells. In general, the experiments have been exploratory rather than sharply focused. Perhaps for this reason, no clear direction seems to have emerged which gives promise that studies of chromatin will lead to an explanation of cellular senescence. Moreover, much of the work deals with whole cells, involving incorporation of an isotopic precursor into various molecules. The results are surely subject to unrecognized experimental artifacts and, therefore, will require a set of more detailed follow-up experiments before firm conclusions can be made.

As to actual experiments, several investigators have measured the protein and histone content of cells or chromatin at different passage numbers using quiescent and proliferating cells. Ryan and Quinn (1971) and Ryan and Cristofalo (1972) agree that there is no change in the histone:DNA ratio in chick embryo and WI-38 cells, respectively, in Phase II versus Phase III. However, Srivastava (1973) reported an increase in this ratio in chromatin from WI-38 cells with increasing passage number. Courtois (1974) found no change in the protein:DNA ratio of chromatin from chick embryo cells. He also reported that there was no change in the CD spectra of chromatin and no qualitative change in chromatin proteins on gels in early- or late-passage chick embryo cells. By contrast, Maizel *et al.* (1975) observe CD differences in chromatin from confluent Phase II versus Phase III WI-38 cells. These differences disappeared after salt extraction. The extracted proteins showed qualitative differences after gel electrophoresis.

The T_m of chromatin was reported to be unchanged in early- and late-passage WI-38 cells (Comings and Vance, 1971) and in chick fibroblasts (Courtois, 1974). Weisman-Shomer *et al.* (1979) found that intact chromatin from "senescent" cells (mouse embryo) copies exogenous DNA more slowly than does chromatin from "young" cells. Curiously, DNA polymerase extracted from the chromatin of senescent cells showed increased activity. The authors point out that factors besides DNA polymerase must be involved in control of DNA synthesis. Petes *et al.* (1974) had earlier reported that chain elongation of DNA took place more slowly in late-passage human diploid cells.

D. Histones and Non-Histone Proteins

Ryan and Cristofalo (1972) found that incorporation of labeled acetate into histones is reduced in Phase III compared to Phase II in WI-38 cells. No change was found in methylation or phosphorylation. Pochran *et al.* (1978) also investigated acetylation and deacetylation of histones in WI-38 cells, both quiescent and serum-stimulated. Little difference was observed between "young" and "presenescent" cells. Some reduction in acetylation was found between early- and late-passage cells when pool sizes were taken into account. Deacetylation was slow in senescent cells, either resting or stimulated, but at early-passage, deacetylation was more rapid in resting cells.

Courtois (1974) observed changes in $^{14}C:^{3}H$ in the main histone band after "young" and "old" chick fibroblasts were prelabeled with the respective isotopic leucines. These results indicate differences in the rate of synthesis. More specifically, Mitsui *et al.* (1980) reported that the relative amount of histone H1 decreased with passage number in human fetal lung cells, reaching a 36% reduction by late-passage. The other histones showed no change except perhaps for a modest increase (11%) in H4. The loss appears to be related to a lowered rate of synthesis.

Non-histone proteins also showed changes, according to several investigators. After administration of [^{14}C] and [^{3}H]leucine, Courtois (1974) observed different isotopic ratios in phase II versus phase III chick fibroblasts, indicating differing rates of synthesis. Mitsui *et al.* (1980) observed a substantial increase in the amount of nuclear proteins in late-passage human fetal lung cells and suggested that the increase was due to accumulation of acidic proteins. Stein and Burtner (1975) found changes in gel scans of proteins extracted from chromatin from early- and late-passage WI-38 cells. Using human foreskin fibroblasts Dell'Orco *et al.* (1978) found that labeled amino acids were incorporated into various histone and non-histone proteins in chromatin in amounts depending upon the cell conditions (log phase, confluence, arrested). There were quantitative, but not qualitative, differences in the labeling of chromosomal proteins at low and high population doublings for each state. Stein (1975) also observed quantitative changes in a DNA binding protein with passage number. From actively dividing chick fibroblasts, Kaftory *et al.* (1978) detected three major DNA binding proteins by use of a Sepharose–DNA column followed by SDS-gel electrophoresis. Exponentially growing "young" and "senescent" fibroblasts showed identical distributions of two of these proteins, the third being absent. Stationary young cells showed an altered pattern of the three proteins but resting senescent cells were unchanged from dividing young or old cells.

E. DNA Polymerase

The activity of DNA polymerase has been examined by several workers. Srivastava (1973) found little change with passage number of soluble or chromatin-bound DNA polymerase activity in WI-38 cells. No difference in specific activity of DNA α-polymerase was noted in old chick embryo fibroblasts, although other properties of the enzyme changed (Fry and Weisman-Shomer, 1976) (Chapter 9, Section IX). Weisman-Shomer *et al.* (1979) found that endogenous DNA synthesis by chromatin preparations was greater (2.5-fold) in late-passage mouse embryo cells than in either quiescent (serum starved) or logarithmically growing "young" cells. However, copying of exogenous templates by chromatin was lower for senescent than for growing early-passage cells. Quiescent cells showed the least activity. Extraction of DNA polymerase showed a greater level of activity in senescent cells for both α and β forms. Thus, the reduction of exogenous DNA synthesis in late-passage cells is not reflected by a loss of DNA polymerase activity. Moreover, as seen by the increased endogenous synthesis of DNA, isolated chromatin does not reflect the reduction of DNA synthesis characteristic of aging cells.

F. Loss of DNA Sequence

Recently, Reis and Goldstein (1980) reported that a specific group of repeated DNA sequences was reduced in late-passage (75–94% of the life-span) in three strains of human diploid fibroblasts. The data was obtained both from use of restriction enzymes and from hybridization experiments. Kinetic analysis of DNA–DNA reassociation further supported the idea of depletion of total highly reiterated sequences. The authors conclude that at late-passage, there are numerous small DNA deletions. Divergence, base modification, cell cycle changes, and chromosomal loss cannot account for the experimental results.

XIII. COMMENT

Although there are a number of interesting findings, it takes no great insight to recognize that the totality of the results reported for cells in culture presents a fragmentary picture of DNA metabolism in "aging" cells. The work available cannot be pieced into a unified concept of cellular aging. The task of relating the work to aging *in vivo* is even more difficult since the studies in whole animals are equally incomplete and more contradictory. A steady flow of high quality research will be required to fill out the details and to make plain the effects of aging on DNA metabolism and vice versa, both in intact animals and in tissue culture.

REFERENCES

Agarwal, S. S., Tuffner, M., and Loeb, L. A. (1978). *J. Cell Physiol.* **96,** 235–244.
Barton, R. W., and Wang, W. K. (1975). *Mech. Aging Dev.* **4,** 123–126.
Berdyshev, G. D. (1976). *Interdiscip. Top. Gerontol.* **10,** 70–82.
Berdyshev, G. D., and Zhelabovskaya, S. M. (1972). *Exp. Gerontol.* **7,** 321–330.
Bjorksten, J. (1968). *J. Am. Geriatr. Soc.* **16,** 408–427.
Bonner, J., and Huang, R. C. (1963). *J. Mol. Biol.* **6,** 169–174.
Bowman, P. D., Meek, R. L., and Daniel, C. W. (1976). *Exp. Cell Res.* **101,** 434–437.
Bradley, M. O., Erickson, L. C., and Kohn, K. W. (1976). *Mutat. Res.* **37,** 279–292.
Carter, D. B. (1979). *Exp. Gerontol.* **14,** 101–107.
Carter, D. B., and Chae, C. B. (1975). *J. Gerontol.* **30,** 28–32.
Castle, T., Katz, A., and Richardson, A. (1978). *Mech. Aging Dev.* **8,** 383–395.
Chetsanga, C. J., Boyd, V., Peterson, L., and Rushlow, K. (1975). *Nature (London)* **253,** 130–131.
Chetsanga, C. J., Tuttle, M., and Jacoboni, A. (1976). *Life Sci.* **18,** 1405–1412.
Chetsanga, C. J., Tuttle, M., Jacoboni, A., and Johnson, C. (1977). *Biochim. Biophys. Acta* **474,** 180–187.
Clarkson, J. M., and Painter, R. B. (1974). *Mutat. Res.* **23,** 107–112.
Cleaver, J. E. (1969). *Proc. Natl. Acad. Sci. U.S.A.* **63,** 428–435.
Collins, J. M. (1978). *J. Biol. Chem.* **253,** 5769–5773.
Comings, D. E., and Vance, C. K. (1971). *Gerontologia* **17,** 116–121.
Courtois, Y. G. C. (1974). *Mech. Aging Dev.* **3,** 51–63.
Curtis, H. J. (1963). *Science* **141,** 686–694.
Cutler, R. G. (1973–74). *Mech. Aging Dev.* **2,** 381–408.
Cutler, R. G. (1975). *Exp. Gerontol.* **10,** 37–60.
Das, R., and Kanungo, M. S. (1979). *Biochem. Biophys. Res. Commun.* **90,** 708–714.
Dean, R. G., and Cutler, R. G. (1978). *Exp. Gerontol.* **13,** 287–292.
DeBoni, U., Scott, J. W., and Crapper, D. R. (1974). *Histochemie* **40,** 31–37.
Dell'Orco, R. T., and Whittle, W. L. (1978). *Mech. Aging Dev.* **8,** 269–279.
Dell'Orco, R. T., Guthrie, P. L., and Simpson, D. L. (1978). *Mech. Aging Dev.* **8,** 435–444.
Ebbensen, P. (1974). *Science* **184,** 217–218.
Eichhorn, G. L. (1979). *Mech. Aging Dev.* **9,** 291–301.
Epstein, J., Williams, J. R., and Little, J. B. (1973). *Proc. Natl. Acad. Sci. U.S.A.* **70,** 977–981.
Epstein, J., Williams, J. R., and Little, J. B. (1974). *Biochem. Biophys. Res. Commun.* **59,** 850–857.
Finch, C. E. (1979). *Age* **2,** 45–46.

Fornace, A. J. Jr., and Kohn, K. W. (1976). *Biochim. Biophys. Acta* **435,** 95–103.
Fornace, A. J. Jr., Kohn, K. W., and Kann, H. E. Jr. (1976). *Proc. Natl. Acad. Sci. U.S.A.* **73,** 39–43.
Fry, M., and Weisman-Shomer, P. (1976). *Biochemistry* **19,** 4319–4329.
Fry, M., Loeb, L. A., and Martin, G. M. (1981). *J. Cell Physiol.* **106,** 435–444.
Gaubatz, J. W., and Cutler, R. G. (1978). *Gerontology* **24,** 179–207.
Gaubatz, J., Prashad, N., and Cutler, R. G. (1976). *Biochim. Biophys. Acta* **418,** 358–375.
Gaubatz, J., Ellis, M., and Chalkley, R. (1979a). *J. Gerontol.* **34,** 672–679.
Gaubatz, J., Ellis, M., and Chalkley, R. (1979b). *Fed. Proc. Fed. Am. Soc. Exp. Biol.* **38,** 1973–1978.
Gensler, H. L. (1981). *Exp. Gerontol.* **16,** 59–68.
Goldstein, S. (1971). *Proc. Soc. Exp. Biol. Med.* **137,** 730–734.
Hall, K. Y., Hart, R. W., Benirschke, A. K., and Walford, R. L. (1982), submitted.
Hart, R. W., and Setlow, R. B. (1974). *Proc. Natl. Acad. Sci. U.S.A.* **71,** 2169–2173.
Hart, R. W., and Setlow, R. B. (1976). *Mech. Aging Dev.* **5,** 67–77.
Hart, R. W., and Trosko, J. E. (1976). *Interdiscip. Top. Gerontol.* **9,** 134–167.
Hart, R. W., Sacher, G. A., and Hoskins, T. L. (1979). *J. Gerontol.* **34,** 808–817.
Herrmann, R. L., Dowling, L., Russell, A. P., and Bick, M. D. (1975). *Mech. Aging Dev.* **4,** 101–109.
Hewisch, D. R., and Burgoyne, L. S. (1973). *Biochem. Biophys. Res. Commun.* **52,** 504–510.
Hill, B. T. (1976). *Gerontology* **22,** 111–123.
Hirsch, G. P. (1978). *In* "The Genetics of Aging" (E. L. Schneider, ed.), pp. 91–135. Plenum, New York.
Johnson, L. K., Johnson, R. W., and Strehler, B. L. (1975). *J. Mol. Cell Cardiol.* **7,** 125–133.
Johnson, R., Chrisp, C., and Strehler, B. L. (1972). *Mech. Aging Dev.* **1,** 183–198.
Kaftory, A., Weisman Shomer, P., and Fry, M. (1978). *Mech. Aging Dev.* **8,** 75–84.
Kanungo, M. S., and Thakur, M. K. (1977). *Biochem. Biophys. Res. Commun.* **79,** 1031–1036.
Kanungo, M. S., and Thakur, M. K. (1979). *Biochem. Biophys. Res. Commun.* **87,** 266–271.
Kato, H., Harada, M., Tsuchiya, K., and Moriwaki, K. (1980). *Jpn. J. Genet.* **55,** 99–108.
Kleijer, W. J., Lohman, P. H., Mulder, M. P., and Bootsma, D. (1970). *Mutat. Res.* **9,** 517–523.
Krauss, S. W., and Linn, S. (1980). *Biochemistry* **19,** 220–228.
Kurz, D. I., and Sinex, F. M. (1967). *Biochim. Biophys. Acta* **145,** 840–842.
Kurz, D. I., Russell, A. R., and Sinex, F. M. (1974). *Mech. Aging Dev.* **3,** 37–49.
Lesko, S. A., Lorentzen, R. J., and T'So, P.O.P. (1980). *Biochemistry* **19,** 3023–3028.
Liew, C. C., and Gornall, A. G. (1975). *Fed. Proc. Fed. Am. Soc. Exp. Biol.* **34,** 186–187.
Little, J. B. (1976). *Gerontology* **22,** 28–55.
Liu, D. S. H., Ekstrom, R., Spicer, J. W., and Richardson, A. (1978). *Exp. Gerontol.* **13,** 197–205.
Mainwaring, W. I. P. (1968). *Biochem. J.* **110,** 79–86.
Maizel, A., Nicolini, C., and Baserga, R. (1975). *Exp. Cell Res.* **96,** 351–359.
Massie, H. R., Baird, M. B., Nicolosi, R. J., and Samis, H. V. (1972). *Arch. Biochem. Biophys.* **153,** 736–741.
Massie, H. R., Baird, M. B., and Williams, T. R. (1975). *Gerontologia* **21,** 73–80.
Mattern, M. R., and Cerutti, P. A. (1975). *Nature (London)* **254,** 450–452.
McGhee, J. D., and Felsenfeld, G. (1980). *Ann. Rev. Biochem.* **49,** 1115–1156.
Medvedev, Zh. A. (1972a). *Nature (London)* **237,** 453–454.
Medvedev, Zh. A. (1972b). *Exp. Gerontol.* **7,** 227–238.
Medvedev, Zh. A., Medvedeva, M. N., and Huschtscha, L. I. (1977). *Gerontology* **23,** 334–341.

Medvedev, Z. A., Medvedeva, M. N., and Robson, L. (1978). *Gerontology* **24**, 286–292.
Medvedev, Z. A., Medvedeva, M. N., and Robson, L. (1979). *Gerontology* **25**, 219–227.
Medvedev, Z. A., Medvedeva, M. N., and Robson, L. (1980). *Age* **3**, 74–77.
Medvedeva, M. N., Huschtscha, L. I., and Medvedev, Z. A. (1975). *FEBS Lett.* **53**, 253–257.
Milo, G. E., and Hart, R. W. (1976). *Arch. Biochem. Biophys.* **176**, 324–333.
Mitsui, Y., Sakagami, H., Murota, S. I., and Yamada, M. A. (1980). *Exp. Cell Res.* **126**, 289–298.
Modak, S. P., and Price, G. B. (1971). *Exp. Cell Res.* **65**, 289–296.
Müller, W. E. B., Zahn, R. K., and Geursten, W. (1980). *Mech. Aging Dev.* **13**, 119–126.
Oh, Y. H., and Conard, R. A. (1972). *Life Sci.* **11**, 1207–1214.
O'Meara, A. R., and Herrmann, R. L. (1972). *Biochim. Biophys. Acta* **269**, 419–427.
O'Meara, A. R., and Pochron, S. F. (1979). *Biochim. Biophys. Acta* **586**, 391–401.
Ono, T., and Cutler, R. G. (1978). *Proc. Natl. Acad. Sci. U.S.A.* **75**, 4431–4435.
Orgel, L. E. (1963). *Proc. Natl. Acad. Sci. U.S.A.* **49**, 517–521.
Orgel, L. E. (1970). *Proc. Natl. Acad. Sci. U.S.A.* **67**, 1476.
Ove, P., and Coetzee, M. L. (1978). *Mech. Aging Dev.* **8**, 363–375.
Painter, R. B., Clarkson, J. M., and Young, B. R. (1973). *Radiat. Res.* **56**, 560–564.
Panyim, S., and Chalkley, R. (1969). *Arch. Biochem. Biophys.* **130**, 337–346.
Pelc, S. R. (1970). *Exp. Gerontol.* **5**, 217–226.
Peleg, L., Raz, E., and Ben-Iskae, R. (1976). *Exp. Cell Res.* **104**, 301–307.
Petes, T. D., Farber, R. A., Tarrant, G. M., and Holliday, R. (1974). *Nature (London)* **251**, 434–436.
Pieri, C., Zs.Nagy, I., Giuli, C., Zs.Nagy, V., and Bertoni-Freddari, C. (1976). *Experentia* **32**, 891–892.
Pochran, S. F., O'Meara, A. R., and Kurz, M. J. (1978). *Exp. Cell Res.* **116**, 63–74.
Prashad, N., and Cutler, R. G. (1976). *Biochim. Biophys. Acta* **418**, 1–23.
Price, G. B., Modak, S. P., and Makinodan, T. (1971). *Science* **171**, 917–920.
Pyhtila, M. J., and Sherman, F. G. (1968). *Biochem. Biophys. Res. Commun.* **31**, 340–344.
Pyhtila, M. J., and Sherman, F. G. (1969). *Gerontologia* **15**, 321–327.
Rainbow, A. J., and Howes, M. (1977). *Biochem. Biophys. Res. Commun.* **74**, 714–719.
Regan, J. D., and Setlow, R. B. (1974). *Biochem. Biophys. Res. Commun.* **59**, 858–864.
Reis, R. J. S., and Goldstein, S. (1980). *Cell* **21**, 739–749.
Ryan, J. M., and Cristofalo, V. J. (1972). *Biochem. Biophys. Res. Commun.* **48**, 735–742.
Ryan, J. M., and Quinn, L. Y. (1971). *In vitro* **6**, 269–273.
Salser, J. S., and Balis, M. E. (1972). *J. Gerontol.* **27**, 1–9.
Samis, H. V. Jr., and Wulff, V. J. (1969). *Exp. Gerontol.* **4**, 111–117.
Samis, H. V. Jr., Poccia, D. L., and Wulff, V. J. (1968). *Biochim. Biophys. Acta* **166**, 410–418.
Sarkander, H. I., and Knoll-Köhler, E. (1978). *FEBS Lett.* **85**, 301–304.
Schwartz, A. G., and Moore, C. J. (1979). *Fed. Proc. Fed. Am. Soc. Exp. Biol.* **38**, 1989–1992.
Seale, G., Shearman, C. W., and Loeb, L. (1979). *J. Biol. Chem.* **254**, 5229–5237.
Shirey, T. L., and Sobel, H. (1972). *Exp. Gerontol.* **7**, 15–29.
Smith, K. C. (1976a). *Interdisc. Top. Gerontol.* **9**, 16–24.
Smith, K. C. (1976b). *In* "Aging, Carcinogenesis and Radiation Biology" (K. Smith, ed.), pp. 67–82. Plenum, New York.
Srivastava, B. I. S. (1973). *Exp. Cell Res.* **80**, 305–312.
Stein, G. (1975). *Exp. Cell Res.* **90**, 237–248.
Stein, G. S., and Burtner, D. L. (1975). *Biochim. Biophys. Acta* **390**, 56–68.
Stein, G. S., Wang, P. L., and Adelman, R. C. (1973). *Exp. Gerontol.* **8**, 123–133.
Steinhagen-Thiessen, E., and Hilz, H. (1976). *Mech. Aging Dev.* **5**, 447–457.

Strehler, B. L., and Chang, M. P. (1979). *Mech. Aging Dev.* **11**, 379–382.
Strehler, B. L., Chang, M. P., and Johnson, L. K. (1979). *Mech. Aging Dev.* **11**, 371–378.
Suzuki, F., Watanabe, E., and Horikawa, M. (1980). *Exp. Cell Res.* **127**, 299–307.
Szilard, L. (1959). *Proc. Natl. Acad. Sci. U.S.A.* **45**, 30–45.
Tas, S. (1976). *Exp. Gerontol.* **11**, 17–24.
Tas, S., Tam, C. F., and Walford, R. L. (1980). *Mech. Aging Dev.* **12**, 65–80.
Thakur, M. K., Ratna, D., and Kanungo, M. S. (1978). *Biochem. Biophys. Res. Commun.* **81**, 828–831.
Von Hahn, H. P. (1963). *Gerontologia* **8**, 123–131.
Von Hahn, H. P. (1964/65). *Gerontologia* **10**, 174–182.
Von Hahn, H. P., and Fritz, E. (1966). *Gerontologia* **12**, 237–250.
Von Hahn, H. P., and Verzar, F. (1963). *Gerontologia* **7**, 104–107.
Von Hahn, H. P., Miller, J., and Eichorn, G. L. (1969). *Gerontologia* **15**, 293–301.
Weisman-Shomer, P., Kattory, A., and Fry, M. (1979). *J. Cell Physiol.* **101**, 219–220.
Weissbach, A. (1979). *Arch. Biochem. Biophys.* **198**, 386–396.
Wheeler, K. T., and Lett, J. T. (1974). *Proc. Natl. Acad. Sci. U.S.A.* **71**, 1862–1865.
Woodhead, A. D., Setlow, R. B., and Grist, E. (1980). *Exp. Gerontol.* **15**, 301–304.
Zasloff, M., and Felsenfeld, G. (1977). *Biochemistry* **16**, 5135–5145.
Zongza, V., and Mathias, A. P. (1979). *Biochem. J.* **179**, 291–298.
Zubay, G., and Doty, P. (1959). *J. Mol. Biol.* **1**, 1–20.

Chapter *7*

RNA

I. OVERVIEW

It is easy to speculate on possible roles for RNA in the aging process. Most obviously, reduced synthesis of mRNA, tRNA, or rRNA would reduce production of proteins. Such an effect could be brought about by a reduction in gene expression or by reduction in the activity of various RNA polymerases. Although there are reports that RNA synthesis is decreased in old animals, there is no experimental evidence that it slows to the extent that a damaging lack of protein synthesis occurs nor is there undisputed evidence for loss of RNA polymerase activity with age, particularly a serious loss. Not only the amount of polymerase, but its accuracy could have far reaching effects on an aging organism. However, based on an inability to find evidence for proteins with sequence changes in old animals (Chapter 9, Section IV), it appears that fidelity of transcription is not affected by age.

As to the general status of research, a substantial amount of work has been performed on various aspects of RNA metabolism with respect to age. Earlier work in intact animals dealing with RNA content and RNA synthesis is contradictory. For this reason, the information provided throughout the 1960s and early 1970s cannot be accepted with any degree of security. There does seem to be agreement that there is a decrease in RNA synthesis in old animals, at least in rat liver. Unfortunately, disagreement continues to plague most other areas of research involving RNA. Most investigators agree that RNA polymerase activity declines in aged animals, but this conclusion is not unanimous. Transfer RNA has been proposed as a controlling element and there has been a proposal that undermethylation is involved in aging, but much of the available evidence does not support these ideas. The loss of rRNA genes (Chapter 6, Section X) has been suggested but remains speculative. In short, there is little unequivocal evidence upon which to base a role for changes in RNA metabolism as a causative factor in aging. In work using cells in culture, there is better accord on certain aspects of RNA metabolism. It is agreed that RNA content increases sharply in late-passage cells and that template activity and RNA polymerase activity in nuclei decrease. Partly because of contradictory results, direct comparisons of *in vivo* versus *in vitro* aging are difficult to make. Furthermore, some of the changes in cells occur abruptly at very late passage, making the events difficult to interpret with respect to an *in vivo* system. As with studies in other areas, one can produce agreement or disagreement by selective choice of appropriate reports. It is interesting to note that tRNA methylase activity is reported to drop sharply in senescent cells, but does not change in fibroblasts from young versus old donors.

In general, research on RNA with respect to aging appears to be increasing in sophistication and should now begin to provide more consistent and insightful results.

II. RNA CONTENT

A number of studies, many of them carried out in the 1960s, deal with the levels of RNA in various animal tissues at different ages. There are problems inherent in these measurements since they are often related to DNA content. However, the DNA content of some tissues may vary because cells may change in number. In liver cells, there are increases in ploidy with age (Chapter 6, Section II) resulting in an increase in DNA content. Thus, calculation of the amount of RNA in liver based on micrograms of DNA may contain an age-related error. Moreover, the RNA content itself is normally quite variable, depending on the physiological or developmental state of the organism. Indeed, RNA content might be expected to increase when animals are growing and synthesizing more protein. Therefore, changes in the amount of specific RNA molecules may well reflect metabolic conditions in old animals and not aging per se.

The early work dealing with RNA content with respect to age is inconclusive and contradictory. For the record, the results are summarized in Table 7.I. The reasons for the differences probably lie in the experimental procedures, which did not distinguish between various types of RNA, and cognizance was not always given to the possible action of RNase. In addition to the data in Table 7.I, a number of determinations have been made on the RNA content of aging brain. These reports are fragmentary and at present, lead to no conclusions as to effect or meaning, except that there are species differences and different brain regions

Table 7.I Early Reports of Changes in RNA Content with Age[a]

Reference[b]	Tissue	Age (months)	Measured	Age effect	Comment
1	Rat liver	10–12 versus 29–31	RNA	−	
			Nuclear RNA	+	
1	Rat muscle	10–12 versus 29–31	RNA	−	
2	Rat liver	4 versus 27	Nuclear RNA	−	Old nuclei showed a higher specific activity
3	Mouse muscle, liver Kidney	Up to 18	RNA	+	Radioautography
3	CNB neurons		RNA	NC	
4	Rat nerve and muscle	0.5–32	RNA	NC	Initial drop, then no change
5	Brain	10–13 versus 28–30	DNA/RNA	NC	Hypothalmus, cerebellum, hippocampus, septum. There was a 10% reduction in striatum

[a] (−), Decreases with age; (+), increases with age; NC, no change with age.
[b] Key to references: (1) Detwiler and Draper, 1962; (2) Samis *et al.*, 1964; (3) Wulff *et al.*, 1967; (4) Wulff *et al.*, 1963; (5) Chaconas and Finch, 1973.

give different results (Chaconas and Finch, 1973; Shaskan, 1977; Naber and Dahnke, 1979). In a recent study, Colman *et al.* (1980) reported that the sequence complexity of total poly(A) RNA and polysomal poly(A) RNA is unchanged with age in brain tissue of both Sprague-Dawley and Fischer male rats. The limits of detection (by hybridization studies) indicate that age-related differences, if they exist, are smaller than interstrain differences.

Liu *et al.* (1978) reported an increase in the RNA content of rat testes between 20 and 30 months of age. However, this result may arise from the presence of interstitial cell tumors. Buetow *et al.* (1977) provided a brief review of macromolecule content of various tissues at different ages.

In other organisms, the RNA content of adult insects has been shown not to change with age by Lang *et al.* (1965) (mosquito) and Samis *et al.* (1971) (*Drosophila*). Ring (1973) observed a decline in the blowfly. Klass and Smith-Sonneborn (1976) determined that RNA synthesis decreases with fission age in *Paramecium aurelia.*

For a review of histochemical and morphological attempts to determine changes in the RNA content of cells both in invertebrate and mammalian species, see Miquel and Johnson (1979).

III. RNA SYNTHESIS

One conclusion upon which there is some agreement is that there is a reduced level of RNA synthesis in old animals. Earlier work is summarized in Table 7.II. The differences observed by different investigators may be due to the effect of pool sizes or to the time between injection of the isotope and the sacrifice of the animals. In fact, Collins (1978) noted that the amount of injected (tail vein) [³H]uridine incorporated into liver RNA varied from rat to rat by 30% even within the same age-group. Castle *et al.* (1978) stated that the β-mercaptoethanol, used by many investigators, creates unstable nuclei and chromatin preparations. Moreover, the maintenance conditions, diets, and experimental protocols varied in the different laboratories. For example, Samis *et al.* (1964) fasted their rats 18 hours before their experimental procedure, whereas Kanungo *et al.* (1970) fasted their rats for 3 hours prior to injection of labeled precursors.

As mentioned, the idea that there is an age-related reduction in the synthesis of RNA is supported by recent work, although the degree to which the decrease occurs varies in the different reports. Castle *et al.* (1978) studied RNA synthesis in rat liver nuclei and observed a large

Table 7.II Early Reports of Age-Related Changes in Synthesis of RNA[a]

Reference[b]	Tissue	Age (months)	Measured	Age effect	Comment
1	Rat marrow cells	12 versus 24	RNA	NC	Precursor was orotic acid
2	Rat liver nuclei	1 versus 30	Nuclear RNA	−	
3	Mouse liver nuclei	6 versus 30	Nuclear RNA	−	
4	Rat liver nuclei	5 versus 22	Nuclear RNA	NC	
5	Rat liver nuclei	3 versus 25	Nuclear RNA	−	50% Decrease
6	Various rat tissues	3, 10, 22	RNA		Incorporation of injected label varied differently with age for each tissue
7	Rat liver slices	12–14 versus 24–26	RNA	−	

[a] (−), Decreases with age; NC, no change with age.
[b] Key to references: Menzies et al., 1967; (2) Devi et al., 1966; (3) Mainwaring, 1968; (4) Gibas and Harman, 1970; (5) Britton et al., 1972; (6) Kanungo et al., 1970; (7) Chen et al., 1973.

decrease (about twofold) in the ability of the old nuclei (12 versus 30 months) to incorporate labeled UTP, expressed on the basis of milligrams of DNA or per nucleus. The rates varied with the two media which were selected for measuring rRNA and mRNA synthesis, respectively. RNA synthesis was low in young animals and increased sharply (1.6- to 2.7-fold) between 3 and 6 months so that the net difference, if one was to measure only the ages of 3 versus 31 months, would appear to be very small. There were no age-related effects in UTP uptake or degradation, and no RNAse activity was present in the nuclei. Bolla and Denckla (1979) also observed a decrease in RNA synthesis in old rat liver nuclei although to a lesser degree. Their results showed a drop of 75% between animals of 6 and 24 months of age as determined by incorporation of [³H]UMP.

Further support for the existence of an age-related decrease in RNA synthesis in liver was provided by Kreamer *et al.* (1979), who compared the amount of incorporation of [³H]orotic acid into the RNA of hepatocytes of rats of 6, 12, 18, and 30 months of age. There was a 30% decrease between 12 and 18 months and 37% between 12 and 30 months. The decrease between 18 and 30 months was, therefore, very small (7%) and may reflect the fact that old hepatocytes have been reported to synthesize increased amounts of protein (Van Bezooijen *et al.* (1977) (Chapter 8, Section III). Uptake and ability to incorporate orotic acid into UTP were not affected by age.

In an interesting study, Collins (1978) investigated RNA synthesis in rat liver cells with increased ploidy brought about by increased age. The number of 4 *n* and 8 *n* cells increased between 13 and 24 months. On the other hand, 2 *n* cells dropped to a very small percentage of the total by 14 months. Surprisingly, RNA synthesis increased very little in the 4 *n* and 8 *n* cells, although the amount of DNA increased dramatically. With regard to the effect of age, after an early small drop, RNA synthesis showed a gradual decrease in 8 *n*, 4 *n*, and 2 *n* cells. After about 18 months, the decrease was minimal. The results were obtained using [³H]uridine and a pulse time which was claimed to achieve maximum incorporation into RNA. The inappreciable decrease in RNA synthesis at older ages observed by Collins (1978) does not differ greatly from the results obtained by Kreamer *et al.* (1979), who also used hepatocytes. It does differ from the conclusion of Castle *et al.* (1978) and Bolla and Denckla (1979) obtained using nuclei.

From the above reports, there appears to be general accord that RNA synthesis declines with age in liver cells, at least to late maturity. The quantitative aspects are in reasonable agreement. Castle *et al.* (1978) showed a 1.6- to 2.7-fold drop (6 versus 31 months) after an initial rise

(3 versus 6 months), whereas Bolla and Denckla (1979) showed a 75% drop (6 versus 24 months), both groups utilizing nuclei. The varied experimental conditions, the difference in strain of rats (Fischer 344 versus Sprague-Dawley), and the use of older animals by Castle *et al.* (1978) could account for the disparity. In the case of hepatocytes, Kreamer *et al.* (1979) observed a 30% loss of activity between 6 and 18 months, which is larger than the decrease found by Collins (1978) but not greatly so. Both investigations showed a small decrement thereafter.

The reported reduction in RNA synthesis can be due only to a limited number of causes. Experimentally derived problems such as pool size, transport, metabolism of substrate, or presence of nucleases have been considered by the various investigators. It is interesting that Bolla and Miller (1980) reported that in rat liver nuclei, the total nucleotide pool decreases linearly after 6 months of age, putting an important stress on measuring pool sizes during studies of RNA synthesis in intact nuclei. The remaining options would seem to be a lowered effectiveness of RNA polymerase (less, altered, or inhibited enzyme) or a change in chromatin structure.

There have been few attempts to show age-related differences in the types of RNA produced. Castle *et al.* (1978) showed small differences in size distribution of labeled RNA produced in rat liver nuclei. When media were selected for preferential synthesis of mRNA and rRNA, respectively, a similar decline was noted in both cases. Mori *et al.* (1978) examined gel electrophoretic patterns of cytoplasmic mouse kidney, liver, and brain and found no age-related differences. They did, however, observe an increase with age in the ratio of 18 to 28 S RNA in the latter two tissues. Moudgil *et al.* (1979) found no change with age in the proportion of ribosomes active in protein synthesis in adult versus senescent rat liver. The amount of poly(A)-containing RNA in the ribosomes was also the same. However, the data showed a large degree of scatter so that even considerable differences would not be detectable.

IV. RNA POLYMERASE

The role of RNA metabolism in aging may depend not only on the structure of chromatin, but also on the enzymes intimately concerned with its synthesis and breakdown. Therefore, age-related deficiencies in RNA polymerase would be of primary importance to studies of aging. At present, it is believed that polymerase I (or A) is responsible for preribosomal RNA and is located in the nucleolus. Polymerase II (or B) and polymerase III (or C) synthesize premessenger RNA and pretransfer

RNA, respectively, and are located in the nucleoplasm. α-Amanitin at low levels inhibits RNA Polymerase II, so that measurements of this activity and that of Polymerase I and III can readily be distinguished.

At this time, there is no indisputable information available that indicates that there are age-related changes in the properties of RNA polymerase. There are, however, a few reports that deal with changes in enzyme activity. Such reports are about evenly divided between those showing decreases and those showing no change with age. Mainwaring (1968) using purified nuclei found decreases in liver and prostate RNA polymerase activity in old mice (6 versus 30 months). Activity in lung tissue was too low and variable to determine age-related effects. Britton *et al.* (1972) reported a substantial decrease with age in mouse liver and muscle nuclear RNA polymerase activity (I, II, and total), although when the liver enzymes were solubilized, they showed only a small decrease. Benson and Harker (1978) also observed no change in the levels of solubilized RNA polymerase in total homogenates of mouse liver and brain between 18 and 31 months of age. The lack of difference reported in polymerase level in brain agrees with the lack of age-related change in RNA turnover in this tissue in rats (Menzies and Gold, 1972) and with earlier reports that indicate that there are no dramatic changes in brain RNA content (Table 7 I). In agreement with the above work, Bolla and Denckla (1979) found no age-related change in total RNA polymerase in nuclei from rat liver between 6 and 24 months of age. The authors did note a decrease in the amount of chromatin-bound enzyme, but only through 18 months, along with a concomitant increase in free enzyme. There was no effect of age on the total (chromatin-bound + free) RNA polymerase I and III and polymerase II. Castle *et al.* (1978) also found no changes in polymerase activity with age in rat liver nuclei, although there was a substantial age-related drop in RNA synthesis as measured by UTP incorporation.

On the other hand, in partially purified preparations of mouse heart (between 21 and 30 months of age), Benson (1978) observed that the activities of RNA polymerase I and II decline with age (57 and 15%, respectively). Polymerase III activity was unchanged. The three polymerases were separated on DEAE-Sephadex by elution with increasing concentrations of ammonium sulfate. Since the content of RNA and the total rate of RNA synthesis in aging heart has not been established *in vivo*, it cannot be determined if this lowered activity of RNA polymerase has any effect in the intact animal.

Chen *et al.* (1980) obtained a rather different picture of RNA polymerase using skin fibroblasts from human donors of various ages. They prepared cells permeable to nucleotides by use of a nonionic detergent and measured both free and bound RNA polymerase. Total bound RNA polymerase activity was found to be unchanged from fetal to 11 years of age. It then slowly declined to 65% of the initial activity through 94 years of age because of the loss of activity in RNA polymerase II.

In crude homogenates of the free-living nematode *Turbatrix aceti* Bolla and Brot (1975) noted a sharp early rise and then a decline in the specific activity of RNA polymerase with increasing age.

A brief summary of the reported effects of age on RNA polymerase is given in Table 7.III.

Lacking evidence for a decrease in available RNA polymerase or for increased action of RNase, Miller, *et al.* (1980) pursued the idea that perhaps age-related changes in RNA synthesis might be mediated by altered levels of initiation. Using rat liver nuclei from groups of rats of ages from 0.75 to 30 months, the authors reported a significant age-related decrease in the number of initiations of RNA chains as measured by incorporation of [^{32}P]GTP and ATP.

One must conclude that the reported reduction in RNA synthesis with age has not been satisfactorily connected to a deficiency of RNA polymerase, although several investigators have reported an age-related diminution of the enzyme. In this regard, since reduced synthesis of RNA has been observed both in isolated nuclei and intact cells, cytoplasmic factors do not seem to be involved. The idea of an age-related deficiency in enzymes of generalized function (e.g., RNA polymerase) is a scattergun approach which is intellectually unsatisfying, especially compared to the idea of a scrupulous regulation of the synthesis of specific RNA molecules. The most reasonable assumption would be that the lowered level of RNA in old tissues is due to differences in gene expression. On the other hand, unequivocal proof as demonstrated by measurable differences in gene expression or chromatin structure between young and old animals is yet to be obtained (Chapter 6).

V. RNA TURNOVER

Very little work has been performed on the effect of age on RNA turnover. If RNA synthesis in old organisms is really decreases *in vivo*, especially to the degree reported by some authors (Section III), then degradation must also be slowed or the cellular RNA will fall to very low levels. Although measurements of the RNA content of cells provide conflicting results (Table 7.I and Section II), even in those cases in which reductions are reported, they are not of great magnitude. Hence, one might expect RNA turnover to slow with age to maintain a normal RNA

Table 7.III Effect of Age on RNA Polymerase Activity

Tissue	Age[a]	Age effect[b]	Comment	Reference
Mouse liver nuclei	6 versus 30	−		Mainwaring, 1968
Mouse prostate nuclei	6 versus 30	−		Britton et al., 1972
Mouse liver nuclei		−	Solubilized	
Mouse muscle nuclei		−		
T. aceti	15 → 40 days		First 15 days, +	Bolla and Brot, 1975
Mouse liver	18 versus 31	NC		Benson and Harker, 1978
Mouse brain	18 versus 31	−	I and II	Benson, 1978
Mouse heart	21, 27, 30	NC	III	
	21, 27, 30	−		
Rat liver nuclei	6 → 24	NC	Chromatin bound	Bolla and Denckla, 1979
	6 → 24		Total polymerase (free + bound)	
Human skin fibroblasts	4, 6, 26	NC		Castle et al., 1978
	Fetal → 11 year	NC		Chen et al., 1980
	11 year → 94 year	−	Polymerase II	

[a] Months, unless otherwise indicated. Arrows indicate that several intermediate age groups were included.
[b] (+), Increases; (−), decreases, NC, no change.

level. The few experimental results so far available do not bear out this postulate. Menzies *et al.* (1972) studied RNA turnover in Wistar rats of 12 versus 24 months of age. For the first 2 hours, [^{14}C]orotic acid was incorporated faster into old rat liver and kidney RNA. Subsequently, in the period between 2 and 24 hours, the old animals showed a slower rate of incorporation than young animals. The authors draw attention to the effect of pool sizes and transport on this type of study. To circumvent these problems, they measured turnover of ribosomes and soluble RNA at the steady state. The term "soluble RNA" was used in case the tRNA preparations contained other low molecular weight species. For ribosomes, no age-related differences were found in liver, kidney, lung, spleen, and intestinal mucosa. For soluble RNA, no age-related difference was found in lung or intestinal mucosa. Spleen, however, showed a significant difference (6.46 versus 8.29 days for young and old rats, respectively). In liver and kidney, the soluble RNA from old animals contained a fast degrading component (half-life, 1.1–1.2 days), which was absent in the young preparations. In addition, the second "old" kidney component differed from the single "young" component, having a value of 9.51 versus 4.99 days for the latter.

Similar studies with brain tissue (Menzies and Gold, 1972) showed no significant age-related differences in half-lives of mitochondrial RNA or soluble RNA. Ribosomes, however, based upon turnover of RNA and protein, appeared to have a much longer half-life in old compared to young animals (18.2 versus 7.4 days).

It must be concluded that more definitive studies are needed before any assessment can be made regarding the effect of age on RNA turnover. Particularly, the work should also be extended to older animals to avoid the possibility that changes or lack of changes occur only between 12 and 24 months.

VI.   tRNA

Transfer RNA could be involved in aging in a number of ways. One idea that has been put forth is that certain tRNA isoaccepting species disappear as an organism ages, thus limiting the rate of protein synthesis by slowing the insertion of particular amino acids into growing protein chains. As can be seen below, what little experimental evidence is available does not support this idea. Another suggestion is that the formation of altered tRNA would result in insertion of the wrong amino acid into protein sequences. This idea would require a change in gene structure or perhaps a faulty RNA polymerase. Neither alternative has good experimental support. If anything, the contrary seems true. A more promising possibility is that there is a modification or lack of modification of certain tRNAs in old organisms. Undoubtedly, a number of other plausible scenarios can be invented that could involve tRNA in the aging process. In fact, Strehler *et al.* (1971) proposed a theory of aging involving the relative abundance of certain tRNAs as a mechanism that, together with the specificity of aminoacyl-tRNA synthetases, could control translation.

Although a number of studies of changes in tRNA during development have been carried out, little research has been performed with regard to aging. Some earlier work dealing with senescing plants has been reported and is summarized in a review of the role of tRNA and tRNA acylases in cellular control mechanisms (Andron and Strehler, 1973). Wust and Rosen (1972) investigated aminoacylation and methylation of tRNA in rat spleen and liver, but as their young and old animals were 2–3 months and 11 months of age, respectively, the results cannot be considered relevant to aging. Reitz and Sanadi (1972) examined tRNAs obtained from the free-living nematode *T. aceti*. The only significant change in old organisms was found in the quantitative relationships of the isoaccepting forms of tRNA for arginine and tyrosine, respectively. Frazer and Yang (1972) found no differences in the elution profiles of nine tRNAs from young (5–9 months) and old (26–36 months) mouse liver and six tRNAs from brain.

Mays *et al.* (1973, 1979) proposed that reduced methylation of tRNA could play a role in the aging process perhaps by altering fidelity or rate of translation. Indeed, a continued decline in methyltransferase activity was observed in mouse liver at 12, 18, and 30 months. The authors suggest that an age-related increase in glycine *N*-methyltransferase activity deprives the older tissue of enough methylating ability to cope with tRNA requirements. Hoffman and McCoy (1974) subsequently examined the base composition of nucleosides of tRNA from mature and old (12 and 30 months) C57BL/6J mouse liver and in young and old mosquitoes and found no age-related differences. Since these authors found no differences in methylated bases, they concluded that tRNA modifying enzymes play no role in aging. Borek (1974), in a letter to the editor, points out that changes in single isoaccpeting species of tRNA can have far-reaching physiological effects and that the methods used by Hoffman and McCoy and Frazer and Yang (see above) would not detect such small changes. Hoffman (1974) responded with the argument that an enzyme that modifies any tRNA species will act on many tRNAs so that the modification would be detectable. Furthermore, two recent papers negate the idea that changes in methylation enzymes affect the role of tRNA. Weber *et al.* (1979) recently reported that there are no differences in the activity of tRNA methyltransferases in crude preparations from whole brain or certain specific brain regions of aging mice (12, 18, and 30 months of age). Lin and Chang (1979) found that S-adenosylmethionine:tRNA transferase is present in a low but practically unchanged level in cells cultured from human donors from 3 months to 94 years of age. Fetal cells had a much higher activity.

Lawrence *et al.* (1979) (see also Mays *et al.*, 1979) observed an anomaly in that tRNAs obtained from old whole rat liver had a reduced acylation capacity compared to tRNAs from young preparations (3 versus 24 months). However, if the tRNAs were obtained from high speed supernatants, the differences tended to disappear. Old rats also had a higher proportion of the isoaccepting tRNA$_5^{Lys}$. Frazer and Yang (1972) had earlier noted that tRNAs from "young" liver and "old" brain showed greater acylation capacities than their respective old and young counterparts. However, the authors attributed these results to an inability to control the purity of the respective "young" and "old" tRNA preparations. Casting further doubt on an age-related role for aminoacylation of tRNA, Foote and Stulberg (1980) recently prepared a tRNA-dependent protein synthesizing system from ascites tumor cells. Ribosomes and a ribosome-free supernatant depleted of tRNA were utilized. RNA was obtained from encephalomyocarditis virus. Transfer RNAs were prepared separately from liver, kidney, heart, and spleen of C57BL/6 mice at 10–12 and 29 months of age. The authors found no age-related differences in the level of aminoacylation of tRNA from any of the tissues. As to the ability to translate mRNA, there were no age-related differences as a function of tRNA concentration in any of the tissues, although each tissue yielded a different rate of incorporation of labeled phenylalanine. The tRNA concentrations used in the experiments were based on capacities for amino acid acceptance rather than A_{260} units. No age-related differences in fidelity of the synthesized proteins could be found by SDS electrophoresis coupled with fluorography to enhance detection of labeled products.

Recently, three papers have appeared dealing with tRNA metabolism in aging *Drosophila*. Hosbach and Kubli (1979a) observed a drop (an average of 16%) in the ability of tRNA to become aminoacylated. Transfer RNA^{Leu} was particularly deficient, showing a 53% reduction in old flies (35 days) as compared to young organisms (5 days old). The charging capability of the synthetases was also reduced by more than 50% for several, but not all, of the amino acids. In a subsequent paper, the authors (Hosbach and Kubli, 1979b) found no change with age of isoacceptor patterns in tRNAs for alanine, leucine, methionine, or serine, but substantial differences for aspartate, asparagine, histidine, and tyrosine. The latter four tRNAs contain the Q base, a hypermodified form of guanosine. Owenby *et al.* (1979) observed that isoaccepting species of tRNA for histidine, tyrosine, and aspartate increased their content of Q with age in *Drosophila*. Since the total amount of the respective tRNAs remained the same, it appears that the isoacceptor tRNAs are partly converted to their respective Q forms. Aminoacylation and protein synthetic rates were not altered by the amount of Q in the tRNA. The authors found that diet has a considerable effect on the ratio of Q to non-Q isoacceptors. For example, Q is increased when flies are raised on 2–4% yeast. The rate of Q formation was also related to the strain of *Drosophila* tested. Thus, considerations other than aging are involved. Of course, the degree to which aging in insects parallels aging in higher animals is itself an unanswered question. One must keep an open mind until enough work has been done to establish which age related biochemical changes are general and which are species specific.

From the above results, it seems clear that changes in methylation or lack of it have not been shown to be a necessary adjunct of aging. Moreover, it has not been shown that changes in isoaccepting species play a significant role in the process. As to reduced function of tRNA (i.e., acylation capacity), the conclusions are contradictory. On balance, the evidence supports the idea that there are no consistently observed changes. For that matter, even if the reduced aminoacylation capacity reported for tRNA in old animals is real, it may simply be a reflection of a reduced requirement for protein synthesis in older tissues and thus a normal consequence of a given metabolic situation. In short, changes in tRNA are not necessarily a reflection of a basic inability to synthesize proteins. As in other fields of aging, we will simply have to wait for more data to become avialable on the relation of aminoacyl synthetases, tRNAs, and protein synthesis to the aging process. Meanwhile, the evidence seems to favor normal function and makes tRNA an unlikely candidate for a role as a regulator in the aging process. Furthermore, if the idea that protein synthesis is regulated by tRNA has any validity, then the large changes reported should be reflected in much slower rates of protein synthesis. Such an effect has not been shown *in vivo*. Indeed, lack of tRNA is not the reason for slowed protein synthesis observed in cell-free preparations (Chapter 8, Section III). In fact, in many cases, old animals are as capable of producing induced enzymes as young ones (Chapter 5, Section III and Chapter 10, Section XIII). Of course, without quantitative data, we do not know what percentage of the capability of the cell for protein synthesis is being utilized. If tRNAs are present in excess, even a substantial decrease in these molecules may not have noticeable effects *in vivo*.

VII. TISSUE CULTURE

The effect of passage number on RNA in cells in culture has been the subject of considerable attention. The results seem to be less contradictory than those obtained from intact animals and even provide a few undisputed conclusions. For example, there is general agreement that the RNA content of cells increases with age, particularly at late-passage. There is some disagreement over whether or not with this increase, the relative amount of various RNA species changes. Studies of the rate of incorporation of uridine into RNA has led to variable results, perhaps reflecting the effect of pool sizes, transport, and cell type. Studies of the template activity of chromatin using isolated nuclei show a reduction in the amount of transcription. Most papers agree that the reduction sets in only at late-passage. Thus, it occurs after loss of division potential has begun and is, therefore, not the cause of the inability of late-passage cells to divide.

A. RNA Content

There is general agreement that the RNA content increases as cells become senescent. Srivastava (1973) reported an increased RNA/DNA ratio in WI-38 cells approaching quiescence (Phase III). Schneider *et al.* (1975) and Schneider and Shorr (1975), among others, showed that RNA content increases sharply in WI-38 cells in the senescent phase. The latter authors noted that the increase applied equally to rRNA, mRNA, and tRNA and was not due to the accumulation of any particular RNA species. Hill *et al.* (1978) found increases in RNA with passage number in WI-38, HE-125, and HE-288 human lung fibroblasts. Whatley and Hill (1979) observed a gradual increase in the RNA content of human embryonic mesenchymal cells with passage number. The amount of RNA increased sharply at the senescent phase. As noted by others, the increased RNA content matched closely the increase in cell volume which occurs in senescent cells. In agreement with the conclusions of Schneider and Shorr (1975) Whatley and Hill (1979) found no change in the proportion of rRNA to total RNA or 28 to 18 S RNA. On the other hand, Johnson *et al.* (1977) also noted an increased RNA content, but they reported substantial changes in the proportion of various RNAs in senescent as compared to early-passage WI-38 cells. It is, therefore, interesting that Mori *et al.* (1978) reported age-related changes in the cytoplasmic RNA composition of various tissues in aged mice. However, these data are as yet too fragmentary for one to draw meaningful conclusions.

B.   RNA Synthesis

RNA synthesis in the context of this section refers to incorporation of a labeled precursor into RNA. The term "senescent" or late-passage refers to late phase III. One should bear in mind that measurements of total incorporation of labeled nucleotides yield average values that encompass the fact that many cells in late passage culture are nondividing and may be relatively inactive in transcription. Thus, a lack of RNA synthesis does not necessarily mean an incapacity, but perhaps may reflect the metabolic state of the cells. In dealing with intact cells, care must be taken to consider not only pool sizes and ability to take up the precursor, but also cell number. Differences in interpretation resulting from these factors could be substantial. For example, Schneider *et al.* (1975) found that the specific activity of RNA after labeling with [³H]uridine was decreased in senescent WI-38 cells. Since precursor pools and labeling kinetics were unchanged, the results could be interpreted as a decreased rate of RNA synthesis in senescent cells. However, the increase in the content of RNA distorts the figures. Calculated on a per cell or per microgram of DNA basis, the authors concluded that there is actually a slight increase in RNA synthesis in senescent cells. The increase applies equally to rRNA and tRNA. The authors point out that the slightly increased rate of synthesis is not enough to account for the substantially increased quantity of RNA found in senescent cells (Section VII,A). In a similar vein, Hill *et al.* (1978) observed a slightly elevated incorporation of labeled uridine into RNA in senescent WI-38 cells as well as in two other strains of human embryo lung fibroblasts. However, in these experiments, the results could simply be due to a more rapid uptake of the nucleotide rather than an increase in overall RNA synthesis.

By contrast with the above results, several authors reported reductions in RNA synthesis at late-passage. Macieira-Coelho *et al.* (1966) showed that at phase III versus phase II, a reduced synthesis of RNA occurred in postlogarithmic WI-38 cells, although all phase III cells carried out active RNA synthesis. Razin *et al.* (1977) also noted a sharp reduction of incorporation of labeled uridine into growing late phase III WI-38 cells. Both groups used autoradiographic methods, a technique that Schneider *et al.* (1975) point out may be responsible for the differing conclusions. Macieira-Coelho and Lima (1973) had earlier reported a decreasing incorporation of labeled uridine into the RNA of chick embryo cells with increased passage, which coincided with slower growth of the cells. Pool sizes were not determined. Ryan (1975) also noted a decrease in uridine incorporation into RNA in late-passage WI-38 cells. In this case, confluent cells were stimulated to divide by addition of 30% serum. There was no change for 9 hours, after which, the rate of synthesis increased, but did so much faster for early-passage cells. Stein and Burtner (1975), however, observed a 2.6-fold increase in incorporation of labeled uridine within 1 hour of serum stimulation of quiescent WI-38 cells. In contrast to the results of Ryan (1975), the magnitude of the increase was similar regardless of whether the cells were early- or late-passage. Incorporation was determined on a per milligram of RNA basis—a parameter that is variable. Pool sizes were not truly accounted for in either study. From autoradiographic studies, Bowman *et al.* (1976) attributed the decline reported for RNA synthesis to decreased synthesis of nucleolar RNA with no change in nucleoplasmic synthesis.

From the above work, it appears that a simple increase in the rate of synthesis is not the explanation for the increased RNA content of senescent cells. One possibility is that there is a slowed rate of degradation, although nuclease activity, where measured, is seemingly unchanged. Obviously, more work must be performed before the problem can be resolved.

The results obtained by various authors with respect to RNA synthesis in cells in culture are listed in Table 7.IV.

C. Template Activity

A reduction in template activity has been postulated as being a controlling factor in cellular aging. Therefore, several investigators have examined transcription both in isolated chromatin and in intact nuclei. Typically, RNA polymerase from *E. coli* has been used in conjunction with chromatin. Most commonly, incorporation of a labeled nucleotide into RNA is measured.

The results reported by various investigators are contradictory. No change in template activity of chromatin (*E. coli* polymerase) with passage number was observed by Stein and Burtner (1975), either in quiescent (confluent) or serum-stimulated WI-38 cells, although transcription increased under the latter conditions. Courtois (1974) found no passage-related change of activity in chromatin from quiescent chick embryo cells. Hill (1976b) found no change in human embryo lung fibroblasts (HE-104) between early- and middle-passage, either in the logarithmic

Table 7.IV Effect of Passage Number on RNA Synthesis in Intact Cells[a,b]

Cell type	Passage number[c]	Result	Comment	Reference
WI-38	21–25 versus 41–45(50)	+	Per cell	Schneider et al., 1975
WI-38	20 versus 43(44)	+	Per cell	Hill et al., 1978
HE-125	10 versus 20(25)	+	Per cell	Hill et al., 1978
HE-388	11 versus 29(22)	+	Per cell	Hill et al., 1978
WI-38	18 versus 46(55)	−	Radioautography, resting	Macieira-Coelho et al., 1966
WI-38	22/50 versus 69(69)	−	Radioautography	Razin et al., 1977
Chick embryo	5–30(34)	−	Per cell	Macieira-Coelho and Lima, 1973
WI-38	"young" versus "old"	−	Stimulated	Ryan, 1975
WI-38	17 versus 49–52	NC	Stimulated	Stein and Burtner, 1975
WI-38	33 versus 53	−	Stimulated	Bowman et al., 1976

[a] (+), Increases with passage number; (−), decreases, NC, no change.
[b] Measured by uridine incorporation.
[c] Numbers in parentheses refer to terminal passage.

or plateau phase of growth. Both endogenous and *E. coli* RNA polymerase were utilized with nuclear monolayers serving as the template. [The two types of polymerase yielded rather different results with chromatin from young and old mice (Hill, 1976a).]

In contrast to the above results, several investigators observed a decreased template activity in chromatin from senescent cells. Ryan and Cristofalo (1975) found that chromatin isolated from mid-passage WI-38 cells (65% of life-span completed) incorporated one-third as much labeled AMP into RNA as did early-passage cells (45% of life-span completed) using RNA polymerase from *E. coli*. These cells might be considered as "middle aged" and "young," respectively. These results are in agreement with the reduction in template activity noted by Srivastava (1973), but the decline in the latter case, came later in cell life, after an initial increase. Ryan and Cristofalo (1975) obtained results similar to those from their chromatin experiments by measuring template activity in nuclear monolayers so that endogenous RNA polymerase was utilized. They observed a twofold reduction in senescent cells. As the authors point out, measurements of this nature do not distinguish between the template activity of individual "old" cells and of the culture as whole in which some cells may be metabolically quiet. RNA synthesis is discussed by Ryan (1975) in the context of histone acylation.

At variance with the report of Ryan and Cristofalo (1975), several recent publications place the decrease of template activity at late phase III. In this respect, Evans (1976) noted a decline only in late-passage cells. He found no change in total RNA synthesizing capacity in early-versus mid-passages of mouse embryo fibroblasts, but the value dropped to 50% in late-passage cells under conditions of stimulation by serum. Pochran *et al.* (1978) studied RNA synthesis in nuclei from quiescent (confluent) WI-38 cells stimulated by serum to divide. Presenescent cells (late phase II) showed little difference from early-passage cells, but nuclei from late-passage cells (Phase III) showed a decrease of about 50% in RNA synthesis (labeled uridine incorporation) compared to nuclei from early-passage cells. Hill *et al.* (1978) examined the effect of passage number on transcription in isolated nuclei using both WI-38 and another human embryo lung fibroblast, HE-388. In both cases, template activity declined only close to the last passage in logarithmically growing and in plateau-phase cells. Subsequently, Whatley and Hill (1980a) studied nuclear template activity as a function of the ability of human diploid mesenchymal cells (HE-125) to divide in both "quiescent" (confluent) and logarithmically growing states. Nuclear template activity was significantly reduced in growing cultures only when the Cristofalo Index (ability of cells to divide) fell below 60% and in quiescent cells, when it fell below 50%, i.e., at very late passage (24–26 out of 27 passages). The loss of transcriptional activity in these experiments was not related to kinetics of substrate uptake or to changes in the levels of RNAse. It was manifested well after the cells began to lose their ability to divide. Therefore, the results are not consistent with the idea that cell aging (as exemplified by loss of division potential) results from an inability to transcribe chromatin.

A summary of template activities as reported by various authors is provided in Table 7.V. A convenient review of *in vitro* nuclear template function is provided by Whatley and Hill (1980b).

D. RNA Polymerase

Changes in nuclear template activity could be brought about by changes in conformation of the chromatin or they could reflect changes in the activity of RNA polymerase. Although there appears to be little or no change in the ratio of the types of RNA produced in late-passage cells, it is not unreasonable to assume that the amount of RNA synthesized could be limited, if not finely controlled by the levels of RNA polymerases I, II, and III. Evans (1976) found that total bound polymerase activity in nuclei declines about 50% at late-passage in serum-stimulated fetal mouse fibroblasts. In addition, almost all the free polymerase activity disappeared. In quiescent cells, there was no change in either bound or free polymerase, the level of the latter being very low. Whatley and Hill (1980a) also observed sharp decreases both in bound and free RNA polymerase in HE-125 cells (human embryo lung fibroblasts) just before the terminal passage. Little change occurred earlier. The decline occurred in both logarithmically growing and quiescent cells. The pattern of loss is similar to that for template activity (Section VII,C).

E. tRNA

Transfer RNA methylase in fetal lung fibroblasts (IMR-90) and fetal skin fibroblasts dropped with passage number, based on activity per milligram of protein (Lin and Chang, 1979). The possibility of a change in protein content in late-passage cells was not considered. In sharp contrast, cells from young versus old donors (3 months to 94 years) at low passage number showed no difference in enzyme level. Fetal cells had a higher activity.

Table 7.V Template Activity of Chromatin and Nuclei

Cell type	Preparation	Age (months)[a]	Effect[b]	Comment	Reference
WI-38	Chromatin	17 versus 49–52	NC	Quiescent or growing	Stein and Burtner, 1975
Chick embryo	Chromatin	3 versus 27(37)	NC	Quiescent	Courtois, 1974
WI-38	Chromatin	24 versus 34	−	Note mid-passage	Ryan and Cristofalo, 1975
WI-38	Chromatin	19 versus 45(53)	−	Only at late-passage	Srivastava, 1973
WI-38	Nuclei	20–21 versus 41–51	−		Ryan and Cristofalo, 1975
Mouse embryo	Nuclei	2 versus 10(11)	−	Only at late-passage	Evans, 1976
HE 104	Nuclei	5 versus 17(24)	NC	Young versus middle age	Hill, 1976b
WI-38	Nuclei	22–26/35–37/45(52–58)	−	Stimulated cells only late-passage	Pochran et al., 1978
WI-38	Nuclei	20 versus 43(44)	−	Quiescent or growing	Hill et al., 1978
HE 388	Nuclei	10 versus 20(22)	−	Quiescent or growing	Hill et al., 1978
HE 125	Nuclei	9–11 to 24–26(27)	−	Quiescent or growing	Whatley and Hill, 1980a

[a] Figures in parentheses represent the terminal passage.
[b] (+), Increase with age; (−), decrease, NC, no change.

F. Comment

In attempting to rationalize the differences in the results reported for RNA synthesis, important factors to be considered include pool sizes, transport, and increased content of RNA in senescent cells. In addition, measurements of template activity depend upon the integrity of the various chromatin preparations. As to nuclear transcription activity, all reports agree that there is a decline during aging, most dramatically toward the end of cell survival. How this finding can be correlated with the increased amount of RNA in senescent cells is problematical. One could argue that degradation slows so that RNA is accumulated. However, in those cases where it was measured, endogenous nuclease activity seemed unchanged, although this is a crude measure which may be meaningless *in situ*.

Since quiescent cells have a lower template activity than dividing cells (Hill, 1976b; Whatley and Hill, 1980a) senescent cultures would, by analogy, be expected to show reduced activity since they contain a substantial proportion of nondividing cells. Yet, one finding that emerges is that reduced template activity is not related to the reduced ability of late-passage cells to divide. Hill (1976b) observed that the reduction in growth potential of HE-104 human embryo lung fibroblasts was not accompanied by a reduction in chromatin template activity in either quiescent or growing cells between early- and middle-passage. At the latter time, a Cristofalo Index of about 70% was observed as compared to 90% for early-passage cells. Thus, the decline in cell division was not not in phase with altered levels of RNA synthesis. Evans (1976) and Pochran *et al.* (1978) also found no change in RNA synthesizing capacity at mid-passage in mouse fibroblasts and WI-38 cells, respectively. Hill *et al.* (1978) and Whatley and Hill (1980a) strongly reinforced the view that a decline of growth sets in before changes in transcription occur. In brief, there appears to be a rather abrupt decline in template activity and this occurs very late in the life-span of the cells. All of this is speculative, of course, and depends upon the accuracy with which the experimental systems reflect the situation in the intact cell.

REFERENCES

Andron, L., and Strehler, B. (1973). *Mech. Ageing Dev.* **2**, 97–116.
Benson, R. W. (1978). *Exp. Gerontol.* **13**, 305–310.
Benson, R. W., and Harker, C. W. (1978). *J. Gerontol.* **33**, 323–328.
Bolla, R., and Brot, N. (1975). *Arch. Biochem. Biophys.* **169**, 227–236.
Bolla, R., and Denckla, W. D. (1979). *Biochem. J.* **184**, 669–674.

Bolla, R. I., and Miller, J. K. (1980). *Mech. Ageing Dev.* **12,** 107–118.
Borek, E. (1974). *Nature (London)* **251,** 260.
Bowman, P. D., Meek, R. L., and Daniel, C. W. (1976). *Exp. Cell Res.* **101,** 434–437.
Britton, V. J., Sherman, F. G., and Florini, J. R. (1972). *J. Gerontol.* **27,** 188–192.
Buetow, D. C., Moudgil, P. G., Eichholz, R. L., and Cook, J. R. (1977). *In* "Liver and Ageing" (D. Platt, ed.), pp. 211–224. Schattauer Verlag, Stuttgart.
Castle, T., Katz, A., and Richardson, A. (1978). *Mech. Ageing Dev.* **8,** 383–395.
Chaconas, G., and Finch, C. E. (1973). *J. Neurochem.* **21,** 1469–1473.
Chen, J. C., Ove, P., and Lansing, A. I. (1973). *Biochem. Biophys. Acta* **312,** 598–607.
Chen, J., Brot, N., and Weissbach, H. (1980). *Mech. Ageing Dev.* **13,** 285–295.
Collins, J. M. (1978). *J. Biol. Chem.* **253,** 5769–5773.
Colman, P. D., Kaplan, B. B., Osterburg, H. H., and Finch, C. E. (1980). *J. Neurochem.* **34,** 335–345.
Courtois, Y. G. C. (1974). *Mech. Ageing Dev.* **3,** 51–63.
Detwiler, T. C., and Draper, H. H. (1962). *J. Gerontol.* **17,** 138–143.
Devi, A., Lindsey, P., Raina, P. L., and Sarkar, N. K. (1966). *Nature (London)* **212,** 474–475.
Evans, C. H. (1976). *Biochem. Soc. Trans.* **4,** 813–815.
Foote, R. S., and Stulberg, M. P. (1980). *Mech. Ageing Dev.* **13,** 93–104.
Frazer, J. M., and Yang, W. K. (1972). *Arch. Biochem. Biophys.* **153,** 610–618.
Gibas, M. A., and Harman, D. (1970). *J. Gerontol.* **25,** 105–107.
Hill, B. T. (1976a). *Gerontology* **22,** 111–123.
Hill, B. T. (1976b). *Mech. Ageing Dev.* **5,** 267–278.
Hill, B. T., Whelan, R. D. H., and Whatley, S. (1978). *Mech. Ageing Dev.* **8,** 85–95.
Hoffman, J. L. (1974). *Nature (London)* **251,** 260.
Hoffman, J. L., and McCoy, M. T. (1974). *Nature (London),* **249,** 558.
Hosbach, H. A., and Kubli, E. (1979a). *Mech. Ageing Dev.* **10,** 131–140.
Hosbach, H. A., and Kubli, E. (1979b). *Mech. Ageing Dev.* **10,** 141–149.
Johnson, L. F., Abelson, H. T., Penman, S., and Green, H. (1977). *J. Cell Physiol.* **90,** 465–470.
Kanungo, M. S., Kool, O., and Reddy, K. R. (1970). *Exp. Gerontol.* **5,** 261–269.
Klass, M. R., and Smith-Sonneborn, J. (1976). *Exp. Cell Res.* **98,** 63–72.
Kreamer, W., Zorich, N., Liu, D. S. H., and Richardson, A. (1979). *Exp. Gerontol.* **14,** 27–36.
Lang, C. A., Lau, H. Y., and Jefferson, D. J. (1965). *Biochem. J.* **95,** 372–377.
Lawrence, A. E., Readinger, J. Z., Ho, R. W., Ackley, S., Hollander, M., and Mays, L. L. (1979). *Age* **2,** 56–62.
Lin, F. K., and Chang, S. H. (1979). *Mech. Ageing Dev.* **11,** 383–392.
Liu, D. S., Ekstrom, R., Spicer, J. W., and Richardson, A. (1978). *Exp. Gerontol.* **13,** 197–205.
Macieira-Coelho, A., and Lima, L. (1973). *Mech. Ageing Dev.*
Macieira-Coelho, A., Ponten, J., and Philipson, L. (1966). *Exp. Cell Res.* **42,** 673–684.
Mainwaring, W. I. P. (1968). *Biochem. J.* **110,** 79–86.
Mays, L. L., Borek, E., and Finch. C. E. (1973). *Nature (London)* **243,** 411–413.
Mays, L. L., Lawrence, A. E., Ho, R. W., and Ackley, S. (1979). *Fed. Proc.* **38,** 1984–1988.
Menzies, R. A., and Gold, P. H. (1972). *J. Neurochem.* **19,** 1671–1683.
Menzies, R. A., Press, G. D., and Strehler, B. L. (1967). *Biochem. Biophys. Acta* **145,** 178–180.
Menzies, R. A., Miskra, R. K., and Gold, P. H. (1972). *Mech. Ageing Dev.* **1,** 117–132.
Miller, J. K., Bolla, R., and Denckla, D. (1980). *Biochem. J.* **188,** 55–60.
Miquel, J., and Johnson, J. E. Jr. (1979). *Mech. Ageing Dev.* **9,** 247–266.

Mori, N., Mizuno, D., and Goto, S. (1978). *Mech. Ageing Dev.* **8,** 285–297.
Moudgil, P. G., Cook, J. R., and Buetow, D. W. (1979). *Gerontology* **25,** 322–326.
Naber, D., and Dahnke, H. G. (1979). *Neuropathol. App. Neurobiol.* **5,** 17–24.
Owenby, R. K., Stulberg, M. P., and Jacobson, K. B. (1979). *Mech. Ageing Dev.* **11,** 91–103.
Pochran, S. F., O'Meara, A. R., and Kurtz, M. J. (1978). *Exp. Cell Res.* **116,** 63–74.
Razin, S., Pfendt, E. A., Matsamura, T., and Hayflick, L. (1977). *Mech. Ageing Dev.* **6,** 379–384.
Reitz, M. S., and Sanadi, D. R. (1972). *Exp. Gerontol.* **7,** 119–129.
Ring, R. A. (1973). *J. Insect Physiol.* **19,** 481–494.
Ryan, J. M. (1975). *Adv. Exp. Med. Biol.* **53,** 123–136.
Ryan, J. M., and Cristofalo, V. J. (1975). *Exp. Cell Res.* **90,** 456–458.
Samis, H. V. Jr., Wulff, V. J., and Falzone, J. A. Jr. (1964). *Biochim. Biophys. Acta* **91,** 223–232.
Samis, H. V. Jr., Erk, F. C., and Baird, M. B. (1971). *Exp. Gerontol.* **6,** 9–18.
Schneider, E. L., and Shorr, S. S. (1975). *Cell* **6,** 179–184.
Schneider, E. L., Mitsui, Y., Tice, R., Shorr, S. S., and Braunschweiger, K. (1975). *Mech. Ageing Dev.* **4,** 449–458.
Shaskan, E. G. (1977). *J. Neurochem.* **28,** 509–516.
Srivastava, B. I. S. (1973). *Exp. Cell Res.* **80,** 305–312.
Stein, G. S., and Burtner, D. L. (1975). *Biochim. Biophys. Acta* **390,** 56–68.
Strehler, B. L., Hirsch, G., Gusseck, D., Johnson, R., and Bick, M. (1971). *J. Theor. Biol.* **33,** 429–474.
Van Bezooijen, C. F. A., Grell, T., and Knook, D. L. (1977). *Mech. Ageing Dev.* **6,** 293–301.
Weber, G., Margetau, J., Finch. C. E., and Mays, L. L. (1979). *Exp. Gerontol.* **14,** 157–160.
Whatley, S. A., and Hill, B. T. (1979). *Cell Biol. Int. Reports* **3,** 671–683.
Whatley, S. A., and Hill, B. T. (1980a). *Gerontology* **26,** 129–137.
Whatley, S. A., and Hill, B. T. (1980b). *Gerontology* **26,** 138–154.
Wulff, V. J., Piekielniak, M., and Wayner, M. J. Jr. (1963). *J. Gerontol.* **18,** 322–325.
Wulff, V. J., Samis, H. V., and Falzone, J. A. (1967). *Adv. Gerontol. Res.* **2,** 37–76.
Wust, C. J., and Rosen, L. (1972). *Exp. Gerontol.* **7,** 331–343.

Chapter 8

Protein Metabolism

I. OVERVIEW

Age-related decrements in the ability to synthesize proteins would obviously have far-reaching effects on organisms. Inadequate or erroneous replacement of proteinaceous cell components, particularly under conditions of maximal demand, e.g., wound healing, severe stress, or even rapid adaptation to altered environmental conditions, could help explain the lowered physiological capabilities of senescent organisms. In spite of the obvious need to investigate protein metabolism in relation to aging, there is not enough information available to provide us with a coherent picture. We simply do not have enough data to draw meaningful relationships between protein metabolism and other aging parameters. One such relationship has been hypothesized by Rothstein and co-workers, who, to explain the formation of altered enzymes (Chapter 9, Section V) proposed the idea that they result from slowed protein turnover. This idea might appear to fly in the face of experimental evidence, which shows that altered proteins (denatured, incorporating amino acid analogs, or containing an error in sequence) are degraded more rapidly than normal proteins. However, none of this evidence relates to the kind of subtle changes in structure observed in "old" proteins. In fact, the data so far available make it clear that the faster rate of degradation attributed to altered proteins does not apply to the type of alteration found in proteins formed by old animals.

Another effect of age on protein synthesis might be a decrease in fidelity with age. This idea, too, has little experimental support in its favor and much against it. The topic is dealt with in Chapter 9, Section IV.

On balance, the findings to date indicate that protein turnover in

198

higher animals slows with age, although the evidence is not unequivocal or is it without contradiction. Only for the free-living nematode *T. aceti* is the case quite clear.

Study of cell-free preparations has provided reasonable, although not unanimous, agreement that there is a reduced ability, with age, to incorporate labeled amino acids into protein, although this effect may be related to particular types of tissues. The deficiency seems to lie in factors in the cytosol rather than in faulty ribosomes, although, again, the evidence is not clear cut. Work with intact rat liver cells shows an increased synthesis in very old animals that is not reflected in cell-free systems. The increase may be a liver response to compensate for protein losses in the urine of old rats.

The contradictory results in individual reports may stem from differences in the strain of animals used, the tissue examined, the particular age-groups selected for comparison or variations in technique, e.g., neglect of pool sizes, time of assay, or the concentration of substrates used. One cannot help but wish that there was a fundamental, systematic, and sustained attack on the problem by more investigators well versed in the technical problems involved in dealing with protein metabolism.

II. PROTEIN TURNOVER

In higher animals, the first measurements of protein turnover in young versus old rodents were reported by Menzies and Gold (1971). They found no age-related differences (12 versus 24 months) in mitochondrial proteins from liver, brain, heart, testes, kidney, lung, or intestinal mucosa, although the absolute half-lives varied from tissue to tissue. Comolli *et al.* (1972) using a double-labeling procedure, observed no significant change in isotope ratios in liver mitochondria, lysosomes, or microsomal fractions of adult versus old Wistar rats. Although the conclusions are in agreement, one can find various reasons to challenge their validity. For example, it could be argued that protein turnover in "old" and "young" mitochondria is indeed different, but that the "old" organelles responsible for the difference are more fragile and are lost during isolation (Chapter 5, Section II); that certain proteins in old animals do turn over at altered rates, but these changes are masked by measuring the average rate for total proteins in the various fractions; or even that the half-lives of mitochondrial proteins do not change, but cytoplasmic proteins do; that modest changes would not show up statistically in the limited number of samples typically available. Unfortunately, research in aging is plagued by many unsolved "buts" and "perhaps" such as these. Indeed, they may be legitimate objections. However, it is simplest, until proved otherwise, to accept the above results as being true for mitochondria and simply ask if perhaps slowed turnover in rats occurs in other parts of the cell, in other tissues, or later than 24 months of age. Generally, the old rodents used in the search for altered enzymes (which have been proposed to be related to slowed turnover) have been about 27–31 months of age. In fact, Barrows and Kokkonen (1980) claimed that altered liver aldolase appears in rats somewhere between 24 and 31 months of age. On the other hand, Sharma *et al.* (1980) provided immunological evidence for the presence of altered phosphoglycerate kinase by 18–20 months of age in homogenates of rat muscle and brain, although no change was noted in liver until 28–30 months.

Menzies and Gold (1972) subsequently investigated the turnover of mitochondrial proteins in rat brain and found that the half-life of total proteins did not vary much with age (26.8 versus 23.5 days for 12-month and 24-month-old animals, respectively). However, when they subfractionated the proteins by precipitation with perchloric acid and subsequent extraction with chloroform–methanol, the two resulting fractions showed substantial age-related differences in their half-lives. For insoluble and soluble "young" protein, the half-lives were 26.3 and 26.1 days, respectively. From old animals, the equivalent values were 17.4 and 30.4 days. The problem with this approach is that the results can be distorted by a slow or rapid flux of differing amounts of highly labeled proteins. In addition, the fractionation procedure itself may be subject to errors because of age-related differences in cell components.

Reznick *et al.* (1981) using a double-labeling procedure observed an increase in the half-life of mouse liver aldolase from 26 hours in young (4 months) to 37 hours in old (25 months) mice. Petell and Lebherz (1979) reported no age-related difference. Other papers dealing with protein turnover in intact "old" rodents include those of Ove *et al.* (1972) and Richter (1977). The "old" animals were 18 and 12 months old, respectively, and hence have little bearing on the effects of aging. Recently, Lavie *et al.* (1982) have provided evidence for decreased protein degradation in subcellular fractions of senescent mouse liver after labeling with $NaH^{14}CO_3$.

As to humans, the only data available are from whole body studies, typically based upon administering [^{15}N]glycine and measurement of urinary [^{15}N]urea, although [^{14}C]leucine was used in conjunction with [^{15}N]glycine for comparison of the two methods (Golden and Waterlow, 1977). Winterer *et al.* (1976) concluded, somewhat tentatively, that whole body protein synthesis declines with age. Subsequently, Aauy *et al.* (1978) concluded that rates of whole body and muscle protein breakdown were lower in elderly subjects than in young adults. Golden and Waterlow (1977) using four methods of calculation based on administration of [^{14}C]leucine and [^{15}N]glycine concluded that total body protein turnover was 20–30% lower in old (68–91 years) subjects compared with middle-aged subjects or young adults.

Although the results depend on a number of assumptions and conclusions cannot be firmly drawn, it does seem that in man, protein turnover, in a generalized sense, slows in old age. Whether or not this conclusion can be related directly to the enzyme or organelle level remains to be discovered. It is unfortunate that more extensive studies of protein turnover in aging organisms have not been performed. Insofar as higher animals are concerned, the evidence so far provided leaves us with a serious deficiency in a very important area.

Unlike the situation in higher animals, the results obtained with free-living nematodes are clear and consistent. Experiments with *T. aceti* showed that there is a dramatic slowing of protein turnover with age. Soluble proteins and the single protein enolase labeled with [^{3}H]leucine were studied, the latter being specifically precipitated by use of anti-enolase serum. Both synthetic and degradation rates slowed with age, the half-lives of soluble protein and enolase increasing dramatically (Sharma *et al.*, 1979). Prasanna and Lane (1979), using [^{35}S]methionine, also showed a large increase in the half-life of soluble proteins in *T. aceti.* The $t_{1/2}$ values increased about 10 fold, being 25 hours at 2 days of age and increasing to 269 hours by 20 days of age. Zeelon *et al.* (1973) reported the slowing of turnover with age of aldolase in *T. aceti* by blocking protein synthesis in the organisms with cycloheximide and determining the enzyme activity per milligram of protein at different times. The authors report a half-life of 40 hours at 7 days of age, increasing to 200 hours at 28 days. These figures are in reasonably good agreement with the more direct studies of Prasanna and Lane (1979) and Sharma *et al.* (1979). More recently, Reznick and Gershon (1979) obtained rather similar results in *T. aceti* using NaH^{14}CO$_3$ as the label. Thus, insofar as these nematodes are concerned, the results are unequivocal: protein turnover slows sharply with age. The data are summarized in Table 8.I.

III. PROTEIN SYNTHESIS

Studies of protein synthesis during aging have mostly been carried out using cell-free preparations from a variety of tissues. Slices and intact hepatocytes have also been utilized. Results from the various

Table 8.I   Protein Turnover in Young and Old *Turbatrix aceti*

Aging procedure[a]	Age	Enolase	Soluble proteins	Aldolase	Reference[b]
FudR	5 days	58	73		1
	22–30 days	161	163		
	III[c]	76	109		
	VIII[c]	228	226		
FudR[d]	7 days			40	2
	28 days			200	
FudR	2 days		25		3
	20 days		269		
FudR	7 days			40	4
	35 days			156	

[a] See Chapter 2, Section IV.

[b] Key to references: (1) Sharma *et al.*, 1979; (2) Zeelon *et al.*, 1973; (3) Prasanna and Lane, 1979; (4) Reznick and Gershon, 1979.

[c] Refers to number of screenings at 3- to 4-day intervals to remove newborn organisms from cultures. Estimated ages are 10–16 days and 25–31 days for III and VIII, respectively.

[d] Measured by blocking protein synthesis *in vivo* with cycloheximide.

systems are not always consistent, and there are clearly differences in tissues of different origin. In general, there is a strong trend indicating a reduced protein synthetic ability with age.

A.   Intact Animals

Using intact mice, Du *et al.* (1977) reported that between mature and senescent animals (10 versus 31 months) there was no significant change in the rate of incorporation of [^{14}C]leucine or δ-aminolevulinic acid into protein of heart tissue. In contrast to results with cell-free preparations, there was a 22% increase in leucine incorporation in the liver of old animals, perhaps reflecting a similar increase noted by Van Bezooijen *et al.* (1977) in intact parenchymal cells (see III,C). The old animals showed a 38% decrease in the incorporation of δ-aminolevulinic acid into the microsomal fraction. Dwyer *et al.* (1980) found a small (9%) decrease in incorporation of labeled lysine in postmitochondrial supernatants from rat brain (forebrain, cerebellum, and brain stem) between 16.5 and 22.5 months of age. This result is similar to the small decrease noted in cell-free preparations from brain (see Section B).

In the intact free-living nematode, *T. aceti*, protein synthesis decreases with age (Sharma *et al.*, 1979).

B.   Cell-Free Preparations

Many experiments have shown that in cell-free systems, "old" preparations are less able than "young" preparations to synthesize proteins.
Although these experiments have mostly been performed with crude
systems using endogenous mRNA or synthetic messenger, the results
are remarkably consistent. Unfortunately, they do not necessarily agree
with results obtained using tissue slices or intact liver cells isolated from
young and old animals.

Mainwaring (1969) showed that in mouse liver, [^{14}C]phenylalanine
was less well incorporated into protein by "old" (30 months) than
"young" (5 months) microsomal preparations. Hrachovec (1971) and
Buetow and Gandhi (1973) obtained similar results with rat liver microsomal preparations. Chen *et al.* (1973) found no change in protein synthesis in young (1 month) versus old (24 months) rat liver microsomes
(although the proportion of albumin synthesized was greater in the
latter). A summary of the results of experiments on protein synthesis
and turnover in aging liver has been provided by Buetow *et al.* (1977).

In other tissues, Liu *et al.* (1978) found an age-related reduction of 40%
in postmitochondrial supernatant preparations from rat testes between
12–13 and 28–30 months of age. However, the change between 18–20
versus 30 months was small (7%) and not statistically significant. Britton
and Sherman (1975) found that the activity of ribosomal preparations
from mouse muscle increased between 3 months and 20 months and
then decreased in old age (28 months). Recently, Hardwick *et al.* (1981)
reported a continuous and dramatic (73–87%) drop in cell-free protein
synthesis by old rat kidney (4.5 to 31 months). Recent work shows much
more limited effects in brain preparations. Ekstrom *et al.* (1980) observed
that in preparations from rat brain, the decrease in incorporation of
labeled valine was substantial between 6 months and older ages, but
was only 13% between 12–14 versus 25–32 months of age and was not
significant after 18–19 months. Hence, the results from cell-free preparations of rat brain show that at best, small changes occur in old age.

In an attempt to pinpoint the cause of the lowered incorporation of
amino acids into proteins by "old" cell-free preparations, Moldave *et al.*
(1979) carried out careful studies of incubation conditions utilizing
postmitochondrial supernatants and semipurified preparations of rat
liver ribosomes (no polysomes) from 3- versus 30-month-old animals.
With endogenous and globin mRNA, they noted that the "old" preparations showed a 30% lower level of incorporation of isotopic leucine into
protein. The authors found that the elongation factor EF-1 (but not EF-2)

was 30–40% lower in the old preparations, using a poly(U) translating system. There was no significant age-related difference in ribosome efficiency. That is, high salt washed ribosomes from either young or old animals showed equivalent activity in both liver and brain preparations. The authors report that the deficiency in EF-1 was not present at 24 months of age, but only later. The deficiency of EF-1 would seem to implicate this factor as the cause of the lowered rate of protein synthesis in the old preparations. However, the alteration in protein synthesis was not consistent from animal to animal (K. Moldave, private communication).

There have been a number of other experiments which attempted to determine whether the reduced rate of protein synthesis is caused by a deficiency in the supernatant or in the ribosomes, typically by mixing "old" ribosomes and "young" cytosol and vice versa (Hrachovec, 1971; Buetow and Gandhi, 1973; Britton and Sherman, 1975; Liu et al., 1978; Hardwick et al., 1981). However, these experiments involved crude preparations and the results have not been consistent. There does not seem to be a deficiency in tRNA or aminoacyl-tRNA synthetases. In fact, even when [4,5-^{3}H]leucyl-tRNA was used as substrate, there was lower incorporation of isotopic amino acid by old versus adult mouse muscle preparations (Britton and Sherman, 1975). Moldave et al. (1979) using semi-purified "young" and "old" ribosomes (rat brain and liver) stripped of endogenous mRNA and peptidyl-tRNA in a protein-synthesizing system containing poly(U) failed to show differences in their synthetic ability. Based on these results, the deficiency appears to lie in a component of the cytosol, not in the ribosomes. On the other hand, Ekstrom et al. (1980) using a similar system from rat brain, found that the "old" ribosomes had a lowered activity.

In spite of the general, but not unanimous, agreement that in cell-free systems protein synthetic ability is reduced with age, the situation as related to whole cells or intact organisms is far from clear. As Coniglio et al. (1979) point out, the rate of synthesis in cell-free systems is only 1% of that observed in vivo. Moldave et al. (1979) make a similar point in questioning whether the changes noted in EF-1 or mRNA translation in vitro are sufficient to affect the intact animal.

C. Slices and Intact Cells

Study of protein synthesis in tissue slices has provided less consistent results than cell-free systems. Kim et al. (1980) measured incorporation of labeled leucine into protein in slices from parotid glands of Sprague-Dawley rats aged from 2 to 30 months. Between 12 and 30 months, there was a continuous decline with age. Pool sizes were determined and uptake of label appeared similar at all ages. McMartin and Schedlbauer (1975) studied incorporation of [^{14}C]leucine into soluble proteins and tubulin in mouse brain slices and found no age-related differences between 5 and 25 months of age. Incubations were carried out for 40 minutes during which the incorporation remained linear. One cannot tell if the eventual maxima would be the same, nor are possible differences in pool sizes considered. The small quantities of isotopic leucine used would be affected substantially by the levels of endogenous leucine and this would in turn, be reflected in the specific activity of the proteins. Petell and Leblierz (1979) found no evidence for an age-related change in the synthesis of aldolase in mouse liver minces, using animals of 3 versus 30 months of age. On the other hand, Reznick *et al.* (1981) observed a slowing of turnover of this enzyme with age as measured *in vivo* by a double-labeling technique.

Measurement of protein synthesis in isolated liver cells has been undertaken by a number of investigators. Van Bezooijen *et al.* (1976) found that isolated parenchymal cells from 36-month-old rats (WAG/Rij) synthesized more albumin (measured immunologically) than cells from 12-month-old animals. The authors achieved synthetic rates equal to 70% of the values obtained *in vivo*. Cells from 3-month-old animals synthesized more albumin than the 12-month-old cells, but about the same amount as the 36-month-old cells, taking into account the increase in cellular protein that occurs with age. Van Bezooijen *et al.* (1977) subsequently reported that after a decrease between 3 and 12 months, there was no change in incorporation of labeled leucine into protein between 12 and 24 months of age, but a substantial increase between 24 and 36 months. Coniglio *et al.* (1979) also utilized intact parenchymal cells from young and old Fischer 344 rats that ranged in age from 2.5 to 30 months. They measured incorporation of radioactive valine into soluble proteins (pool sizes were accounted for) and found an age-related decline (44%) until 18 months. Between 18 and 30 months there was a small although statistically insignificant, increase calculated on a per milligram of protein basis. Since the protein content of the cells increases with age, the results show an apparently lower incorporation. On the basis used by Van Bezooijen *et al.* (1977) (incorporation per 10^6 cells), the figures from both laboratories become similar at around a 70% increase for the old animals. The reasons for the increased protein synthesis reported in intact old liver cells may be due to compensation for the reported loss of protein through the kidneys which occurs with increasing severity in senescence (Van Bezooijen *et al.*, 1977). Alt *et al.* (1980) reported that total protein excretion in rat urine increases 10-fold between 21 and 38 months of age. The fact that protein synthesis in liver cells appears to go down and then up with increasing age would make results taken at only two points, say 5 and 30 months, very hazardous to interpret.

Viskup *et al.* (1979) also studied protein synthesis in isolated rat (Sprague-Dawley) hepatocytes and obtained quite different results. They found that leucine incorporation into total protein or the protein of subcellular fractions decreased substantially between 13–14 and 26–28 months of age, but not between young and mature preparations. Results were expressed on the basis of counts per 10^6 cells so that differing protein content of the cells should not be a factor. Unfortunately, the standard deviations were very large.

The generalized finding of a reduced synthetic ability in cell-free liver preparations from old animals is anomalous as it seems that intact cells *increase* their protein synthetic rate in old age. Readers should bear this contradiction in mind when attempting to translate *in vitro* into *in vivo* effects.

D.   Ribosomes

If protein synthesis is affected adversely by age, there is a possibility that ribosomes are responsible for the phenomenon. If there is indeed a defect in the ribosomal part of the synthesizing machinery, it is not one that produces errors in sequence (Chapter 9, Section IV), but one affecting the rate of synthesis. A number of investigators have explored the effect of age on ribosomal function. As seems all too typical, the results are inconsistent and at times, contradictory. Comolli *et al.* (1977) reported that the dissociation of rat liver ribosomes into subunits was mediated by factors in the 0.5 M KCl ribosomal wash. The dissociating activities were greatest in 13-month-old animals and lower in young (6 months) and old (25 months) animals. However, the large fluctuations in experimental results make the work unconvincing.

Kurtz (1978) suggested that during aging, there was a decrease in the number of active ribosomes in mouse liver (11 versus 28 months of age). Various "old" ribosomal preparations (microsomes, extracted ribosomes) showed a small reduction in endogenous ability to incorporate labeled phenylalanine into protein. With poly(U), the stimulation of incorporation was greater for young preparations. The authors concluded that, based on formation of labeled peptidyl puromycin, there was probably a smaller proportion of active ribosomes in the old preparations rather than a reduced efficiency in individual ribosomes. However, Moudgil *et al.* (1979) found no age-related differences in the proportion of ribosomes engaged in protein synthesis in rat liver. Wallach and

Gershon (1974) had earlier found that in nematodes, although the total number of ribosomes was the same for all ages, the percentage of polysomes decreased from 63% of the ribosomes to 35% in 5- versus 53-day-old organisms.

Mori *et al.* (1979) prepared polysomes and ribosomes free from factors, mRNAs, and peptidyl tRNAs from mice of 2–5 months and 15–26 months of age. Translational factors were prepared from ribosomes of rabbit reticulocytes and tRNAs were obtained from mouse liver. The "old" puromycin-treated ribosomes and polysomes were reported to possess decreased synthetic ability (10–40%), but, as the authors note, the individual variation between animals was extremely large and the significance of the figures is in doubt. The "old" preparations exhibit particularly large variation—perhaps due to greater individual differences in "physiological" aging. The thermosensitivity of the ribosomes showed no age-related change. No increase had earlier been found in rRNA strand breaks (Mori *et al.*, 1978).

Britton and Sherman (1975) reported a small, but significant, decrease in the ability of "old" ribosomes to catalyze protein synthesis in preparations from mouse muscle. Ekstrom *et al.* (1980) also reported that ribosomes of old rats were less effective in translating poly(U), in this case, in brain tissue. If these results can be applied to conditions *in vivo*, then the ribosomes are responsible for the age-related decrease noted in protein synthesis in cell-free preparations. However, as noted above (Section III,B), Moldave *et al.* (1979) found the problem to lie elsewhere. It is of interest that, using old rat brain ribosomes, fidelity of poly(U) translation was not changed (Ekstrom *et al.*, 1980). A similar retention of fidelity had been reported by Mori *et al.* (1979) with mouse liver ribosomes and by Kurtz (1975) using mouse liver microsomes (Chapter 9, Section IV).

Baker and Schmidt (1976) reported changes with age in ribosomes from *Drosophila*. They found a fivefold increase in the amount of protein dissociated from salt-washed ribosomes from old insects after extraction with 2.0 *M* KCl. Ribosomes from old *Drosophila* (30 days) also have a lower T_m and reassociate poorly, compared to ribosomes from young (4 day) organisms (Baker *et al.*, 1979). Examination of the proteins removed by the high salt buffer showed no detectable differences between young and old samples after two-dimensional polyacrylamide gel electrophoresis. Thus, the reason for the change in thermolability and extractability presumably lies in the rRNA or in other unobserved factors.

Miquel and Johnson (1979) have reviewed morphological and histochemical aspects of the effects of aging on ribosomes.

One is forced to conclude that the studies dealing with the effect of age on ribosomes have not provided an intelligible data base. The variability noted by Mori *et al.* (1979), the crude preparations utilized in many cases, the danger inherent in the presence of intrinsic proteases and nucleases, all could lead to uncertain results. On balance, it does not appear the ribosomes change markedly with age or does it appear that they are responsible for changes in the level of protein synthesis *in vitro.*

IV. PROTEIN METABOLISM IN TISSUE CULTURE

The study of protein synthesis and degradation in cells in culture is complicated by the response to components of the medium, the problem of amino acid pools, and transport of metabolites. There have been a number of studies performed on the effect of medium composition on protein synthesis (Warburton and Poole, 1977; Hershko and Tomkins, 1971; Hassell and Engelhardt, 1976; Poole and Wibo, 1973; Epstein *et al.,* 1975; Amenta *et al.,* 1976). In brief, protein synthesis is slowed and degradation increases by serum deprivation. With regard to certain enzymes, turnover may be affected by serum factors even beyond the degree noted for proteins in general. For example, Mellman *et al.* (1972) noted that catalase turnover in human skin fibroblasts was affected by a factor in fetal cell serum. Components such as glucocorticoids may have quite an effect, as the hormone induces formation of tyrosine aminotransferase. Insulin further increases the rate of synthesis and affects degradation (Spencer *et al.,* 1978) in rat hepatoma cells. In fact, there is growing evidence that various factors present in the medium greatly affect cell metabolism.

The effect of amino acid pools appears complex. Amino acids are claimed to be taken up from the medium and incorporated directly into proteins; they are also claimed to equilibrate with a pool before being utilized for synthesis (Hod and Hershko, 1976; Robertson and Wheatley, 1979). Scheibel *et al.* (1981) show clearly that in cultured muscle cells, leucine has different compartments. These considerations make difficult the interpretation of results of protein turnover studies obtained with early- versus late-passage cells. In fact, the work that has been reported on this subject has provided contradictory results. Bradley *et al.* (1976) concluded that short-lived proteins in early phase III WI-38 cells are degraded more rapidly than in phase II cells. Long-lived proteins showed no change. At the end of phase III, the latter are also degraded more rapidly than in early-passage cells. Goldstein *et al.* (1976), using human skin fibroblasts also found no change in long-lived proteins, but, unlike the above results, they found the degradation of short-lived proteins is slower at late-passage. It should be pointed out that the idea of two classes of proteins is not uniformly agreed upon. Shakespeare and Buchanan (1976) also found increased protein turnover rates for visibly senescent compared to early-passage MRC-5 cells. The authors pointed out the difficulties engendered by continued uptake of label from pools even after use of a "chase."

In other work, Macieira-Coelho *et al.* (1975) found a lower level of protein synthesis with "age" in chick embryo cells. Dell'Orco and Guthrie (1976) examined human diploid fibroblasts, which had been prevented from dividing by reducing the serum concentration in the medium to 0.5%. In the first 4 days of the arrested period, the cells lost about 20% of their protein and protein synthesis was slowed, but there were no significant age-related differences. In this respect, Hendil (1981) recently observed that protein is degraded faster in quiescent (density-inhibited) cells than in growing cells. After the 4 days, late-passage cells had a faster protein turnover than early-passage cells ($t_{1/2}$ = 5.4 versus 9.6 days), although both sets of cells maintained a constant protein content.

From the above results, it can be seen that we have the kind of conflicting results that are all too typical of aging research. The reasons for the differences may lie in technicalities—pool sizes, differential uptakes and effluxes by the different cell types, reutilization of isotope (although this is usually considered in the experimental protocol or calculations), and the role of compartments. In measuring degradative rates, both Goldstein *et al.* (1976) and Shakespeare and Buchanan (1976) checked actual precipitated counts rather than measuring radioactivity lost by the cells into the culture medium, but their conclusions are not in agreement. Perhaps the results reflect real differences that are typical of the different cell types, or perhaps they reflect experiments carried out under different laboratory conditions using different batches of calf serum.

One aspect of protein turnover with which many investigators agree is that proteins altered by incorporation of analogs are more rapidly degraded than normal proteins (Goldberg and St. John, 1976). Canavanine and 6-fluorotryptophan (Knowles *et al.*, 1975) in hepatoma cells; *p*-fluorophenylalanine (Shakespeare and Buchanan, 1976) in MRC-5 cells, *p*-fluorophenylalanine, canavanine and azetidine dicarboxylic acid (Bradley *et al.*, 1976) in WI-38 cells are examples. Although these are not age-related results, they are generally interpreted to mean that altered proteins are degraded faster than normal proteins. In fact, missense mutants of hypoxanthine–guanine phosphoribosyltransferase are degraded faster than the normal enzyme in mouse L cells (Capecchi *et al.*, 1974). Thus, the increase in degradation noted in some cases is taken by certain investigators as evidence that late-passage cells are synthesizing "faulty" proteins that are being rapidly degraded. As an example, Bradley *et al.* (1975) reported that the proteins from WI-38 cells at the last population doubling were degraded at increased rates by pronase, compared to proteins from cells from earlier passages. However, there is no firm evidence that the kind of modification found in the age-related alteration of enzymes results in rapid breakdown of these molecules. In fact, old free-living nematodes contain altered enolase, but the enzyme has a greatly *slowed* rate of turnover compared to the normal enolase in young organisms (Sharma *et al.*, 1979) (Section II). In accord with this view, Wheatley *et al.* (1977) argued that errors caused in proteins can vary from trivial to major, depending on the particular protein involved. The authors found that in HeLa cells, incorporation of *p*-fluorophenylalanine or 6-fluorotryptophan did not increase the rate of protein breakdown, but canavanine did. Pronase digestion of analog-containing proteins gave similar results (Giddings and Wheatley, 1977). Interestingly, Reznick and Gershon (1979) did not observe enhanced degradation of nematode proteins incorporating canavanine. A detailed overview of protein synthesis and use of amino acid analogs is provided by Wheatley (1978).

One must regretfully conclude that the relationship of protein turnover to the "aging" of cells is even less well understood than the situation in intact organisms. The fact that half-lives of cellular proteins have been reported with a wide range of values, from a few hours to several days, argues that there are many technical problems to be solved before the data becomes truly acceptable.

REFERENCES

Aauy, R., Winterer, J. C., Bilmazes, C., Haverberg, L. N., Scrimshaw, N. S., Munro, H. N., and Young, V. R. (1978). *J. Gerontol.* **33**, 663–671.

Alt, J. M., Hackbarth, H., Deerburg, F., and Stolte, H. (1980). *Lab. Anim.* **14**, 95–101.

Amenta, J. S., Baccino, F. M., and Sargus, M. J. (1976). *Biochim. Biophys. Acta* **451**, 511–516.

Baker, G. T., and Schmidt, T. (1976). *Experentia* **32**, 1505–1506.

Baker, G. T., Schunke, R. E. Z., and Podgorski, E. M. Jr. (1979). *Experentia* **35**, 1053–1054.

Barrows, C. H. Jr., and Kokkonen, G. C. (1980). *Age* **3**, 53–58.

Bradley, M. O., Dice, J. F., Hayflick, L., and Schimke, R. T. (1975). *Exp. Cell Res.* **96**, 103–112.

Bradley, M. O., Hayflick, L., and Schimka, R. T. (1976). *J. Biol. Chem.* **251**, 3521–3529.

Britton, G. W., and Sherman, F. G. (1975). *Exp. Gerontol.* **10**, 67–77.

Buetow, D. E., and Gandhi, P. S. (1973). *Exp. Gerontol.* **8**, 243–249.

Buetow, D. E., Moudgil, P. G., Eichholz, R. L., and Cook, J. R. (1977). *In* "Liver and Aging" (D. Platt, ed.), pp. 211–224. Schattauer Verlag, Stuttgart.

Capecchi, M. R., Capecchi, N. E., Hughes, S. H., and Wahl, G. H. (1974). *Proc. Natl. Acad. Sci. U.S.A.* **71**, 4732–4736.
Chen, J. C., Ove, P., and Lansing, A. I. (1973). *Biochim. Biophys. Acta* **312**, 598–607.
Comolli, R., Ferioli, M. E., and Azzola, S. (1972). *Exp. Gerontol* **7**, 369–376.
Comolli, R., Schubert, A. C., and Delpiano, C. (1977). *Exp. Gerontol.* **12**, 89–96.
Coniglio, J. J., Liu, D. S. H., and Richardson, A. (1979). *Mech. Ageing Dev.* **11**, 77–90.
Dell'Orco, R. T., and Guthrie, P. L. (1976). *Mech. Ageing Dev.* **5**, 399–407.
Du, J. T., Beyer, T. A., and Lang, C. A. (1977). *Exp. Gerontol.* **12**, 181–191.
Dwyer, B. E., Fando, J. L., and Wasterlain, C. G. (1980). *J. Neurochem.* **35**, 746–749.
Ekstrom, R., Liu, D. S. H., and Richardson, A. (1980). *Gerontology* **26**, 121–128.
Epstein, D., Elias-Bishko, S., and Hershko, A. (1975). *Biochemistry* **23**, 5199–5204.
Giddings, M. R., and Wheatley, D. N. (1977). *Microbios. Lett.* **8**, 31–46.
Goldberg, A. L. and St. John, A. C. (1976). *Ann. Rev. Biochem. Part* **2**, 747–803.
Golden, M. H. N., and Waterlow, J. C. (1977). *Clin. Sci. Mol. Med.* **53**, 277–288.
Goldstein, S., Stotland, D., and Cordeiro, R. A. J. (1976). *Mech. Ageing Dev.* **5**, 221–233.
Hardwick, J., Hsieh, W. H., Liu, D. S. H., and Richardson, A. (1981). *Biochim. Biophys. Acta* **652**, 204–217.
Hassell, J. A., and Engelhardt, D. L. (1976). *Biochemistry* **15**, 1375–1381.
Hendil, K. B. (1981). *FEBS Lett.* **129**, 77–79.
Hershko, A., and Tomkins, G. M. (1971). *J. Biol. Chem.* **246**, 710–714.
Hod, Y., and Hershko, A. (1976). *J. Biol. Chem.* **251**, 4458–4467.
Hrachovec, J. P. (1971). *Gerontologia* **17**, 75–86.
Kim, S. K., Weinhold, P. A., Harr, S. S., and Wagner, D. J. (1980). *Exp. Gerontol.* **15**, 77–85.
Knowles, S. E., Gunn, J. M., Hanson, R. W., and Ballard, F. J. (1975). *Biochem. J.* **146**, 595–600.
Kurtz, D. I. (1975). *Biochim. Biophys. Acta* **407**, 479–484.
Kurtz, D. I. (1978). *Exp. Gerontol.* **13**, 397–402.
Lavie, L., Resnick, A. L., and Garshon, D. (1982). *Biochem. J.* **202**, 47–51.
Liu, D. S. H., Ekstrom, R., Spicer, J. W., and Richardson, A. (1978). *Exp. Gerontol.* **13**, 197–205.
Macieira-Coelho, A., Lavia, E., and Berumen, L. (1975). *In* "Cell Impairment in Aging and Development" (V. J. Cristofalo and E. Holeckova, eds.), pp. 51–66. Plenum, New York.
Mainwaring, W. I. P. (1969). *Biochem. J.* **113**, 869–878.
McMartin, D. N., and Schedlbauer, L. M. (1975). *J. Gerontol.* **30**, 132–136.
Mellman, W. J., Schimke, R. T., and Hayflick, L. (1972). *Exp. Cell Res.* **73**, 399–409.
Menzies, R. A., and Gold, P. H. (1971). *J. Biol. Chem.* **246**, 2425–2429.
Menzies, R. A., and Gold, P. H. (1972). *J. Neurochem.* **19**, 1671–1683.
Miquel, J., and Johnson, J. E. Jr. (1979). *Mech. Ageing Dev.* **9**, 247–266.
Moldave, K., Harris, J., Sabo, W., and Sadnik, I. (1979). *Fed. Proc. Fed. Am. Soc. Exp. Biol.* **38**, 1979–1983.
Mori, N., Mizuno, D., and Goto, S. (1978). *Mech. Ageing Dev.* **8**, 285–297.
Mori, N., Mizuno, D., and Goto, S. (1979). *Mech. Ageing Dev.* **10**, 379–398.
Moudgil, P. G., Cook, J. R., and Buetow, D. E. (1979). *Gerontology* **25**, 322–326.
Ove, P., Obenrader, M., and Lansing, A. (1972). *Biochim. Biophys. Acta* **277**, 211–221.
Petell, J. K., and Lebherz, H. G. (1979). *J. Biol. Chem.* **254**, 8179–8184.
Poole, B., and Wibo, M. (1973). *J. Biol. Chem.* **248**, 6221–6226.
Prasanna, H. R., and Lane, R. S. (1979). *Biochem. Biophys. Res. Commun.* **3**, 552–559.
Reznick, A. Z., and Gershon, D. (1979). *Mech. Ageing Dev.* **11**, 403–415.
Reznick, A. Z., Lavie, L., Gershon, H. E., and Gershon, D. (1981). *FEBS Lett.* **128**, 221–224.

Richter, V. (1977). *Acta Biol. Med. Germ.* **36,** 1833–1836.

Robertson, J. H., and Wheatley, D. N. (1979). *Biochem. J.* **178,** 699–709.

Scheible, P. A., Airhart, J., and Low, R. B. (1981). *J. Biol. Chem.* **256,** 4888–4894.

Shakespeare, V., and Buchanan, J. H. (1976). *Exp. Cell Res.* **100,** 1–8.

Sharma, H. K., Prasanna, H. R., Lane, R. S., and Rothstein, M. (1979). *Arch. Biochem. Biophys.* **194,** 275–282.

Sharma, H. K., Prasanna, P. R., and Rothstein, M. (1980). *J. Biol. Chem.* **255,** 5043–5050.

Spencer, C. J., Heaton, J. H., Gelehrter, T. D., Richardson, K. I., and Garwin, J. L. (1978). *J. Biol. Chem.* **253,** 7677–7682.

Van Bezooijen, C. F. A., Grell, T., and Knook, D. L. (1976). *Biochem. Biophys. Res. Commun.* **71,** 513–519.

Van Bezooijen, C. F. A., Grell, T., and Knook, D. L. (1977). *Mech. Ageing Dev.* **6,** 293–304.

Viskup, R. W., Baker, M., Holbrook, J. P., and Penniall, R. (1979). *Exp. Ageing Res.* **5,** 487–496.

Wallach, Z., and Gershon, D. (1974). *Mech. Ageing Dev.* **3,** 225–234.

Warburton, M. J., and Poole, B. (1977). *Proc. Natl. Acad. Sci. U.S.A.* **74,** 2427–2431.

Wheatley, D. N. (1978). *Int. Rev. Cytol.* **55,** 109–169.

Wheatley, D. N., Giddings, M. R., Inglis, M. S., and Robertson, J. H. (1977). *Microbios. Lett.* **4,** 233–245.

Winterer, J. D., Stefee, W. P., Davy, W., Perera, A., Uauy, R., Scrimshaw, N. S., and Young, V. R. (1976). *Exp. Gerontol.* **11,** 79–87.

Zeelon, P., Gershon, H., and Gershon, D. (1973). *Biochemistry* **12,** 1743–1750.

Chapter 9

Enzymes and Altered Proteins

I. OVERVIEW

A simple explanation for the manifestations of senescence could lie in finding that the activity of certain enzymes is reduced with age. Under such conditions, cells would have less and less ability to respond to stress, cell components would be replaced more slowly than normal, and physiological processes would suffer accordingly. Beginning generally in the early 1960s, a number of studies were made to determine the level of various enzymes with respect to age. Although many changes were noted, the results from different laboratories were inconsistent or even contradictory. Technical considerations were no doubt partly responsible for this situation. In any case, no pattern of change in enzyme levels has been detected which can be applied to the observable effects of aging.

A more sophisticated approach to enzyme involvement in aging is the idea that perhaps as tissues age, they begin to make mistakes in the synthesis of macromolecules. In 1963, Orgel proposed an "error catastrophe" theory which he modified in 1970. In essence, the theory proposed that, if an error occurred in a protein that was involved in the protein-synthesizing system, it would generate further errors, which in turn would cause even more errors until an "error catastrophe" occurred. In short, above a certain level, errors would become amplified. The error hypothesis stimulated a search for altered proteins as well as studies on the fidelity of protein synthesis in old organisms. Indeed, it was such a search that led to the original discovery of an altered enzyme by Gershon and Gershon in 1970. Rothstein and co-workers (Section

V,B) subsequently provided firm evidence that the alteration of enzymes comes about by a postsynthetic process and not by errors in sequence. As more data accumulated, they proposed that the alteration was conformational in nature and that there were no covalent changes. Unequivocal proof for this thesis, at least with one enzyme, was obtained by unfolding–refolding experiments with unaltered and altered enolase from the free-living nematode *T. aceti*. Both forms of the enzyme yielded a new but identical form of the enzyme, proving that the original molecules had differed only in conformation. One should caution that similar studies with altered enzymes from higher animals remain to be performed.

The many studies, both on the nature of altered enzymes and on fidelity of protein synthesis make it virtually impossible that an error catastrophe takes place, and unlikely that small errors occur. Still, the idea of errors seems to have a particular fascination for gerontologists. Many recent papers on aging continue to have in their introduction or discussion sections, some reference to this hypothesis.

It has been shown by various investigators that a substantial number of enzymes do not become altered in old organisms. In fact, altered and unaltered enzymes may exist in the same tissue. For example, unaltered enolase and altered phosphoglycerate kinase are found in old rat muscle and unaltered enolase and altered superoxide dismutase and phosphoglycerate kinase are found in rat liver. A mechanism has been proposed by Rothstein and co-workers for the formation of altered enzymes, which takes into account both altered and unaltered enzymes. The basis of the hypothesis involves a slowing of protein turnover in old age. The effect would be to increase the "dwell time" of proteins in the cell so that changes, presumably kinetic rather than enzymatic, would occur in less stable enzymes. The slowed turnover would prevent altered enzyme molecules so formed from being removed and replaced. They would, therefore, accumulate. This idea is supported by the fact that in nematodes, protein turnover slows dramatically with age. However, for higher animals, such an effect is less certain (Chapter 8).

Early- and late-passage cells in culture have also been used to explore the idea of the error theory. As with intact animals, there is some evidence for enzyme alteration in senescent cells, and there is other evidence that enzymes are not altered. In some cases, the same enzyme is involved. Unfortunately, the enzyme most commonly studied, glucose 6-phosphate dehydrogenase, is subject to a number of modifications unrelated to aging. In any case, there is evidence that at least some of the changes observed in enzymes from late-passage cells are brought about by cytoplasmic factors and that they are posttranslational in nature. In short, work with cells in tissue culture has not generally supported the concept of errors.

Red blood cells have been used as a model for studying changes in enzymes as they cannot synthesize new protein. A number of enzymes have been shown to undergo changes in properties in old cells (not cells from old subjects), and these changes are therefore posttranslational. Finally, deamidation and racemization of amino acids in proteins have been considered as parts of the development of senescence. Although these processes undoubtedly exist, the importance of their role as a cause of aging would seem to be modest.

II. ENZYME LEVELS

Early research dealing with the effect of aging on enzymes was concerned with comparison of enzyme levels in various tissues of young versus old animals. This work, the bulk of which seems to have been carried out in the 1960s, was performed in a number of laboratories under a variety of conditions. The results have been tabled and summarized in reviews by Finch (1972) and Wilson (1973). Typically, enzyme assays were carried out with crude preparations and results were mostly related to weight, DNA, or protein content, a situation guaranteed to provide differing results, especially since protein content, water content, and weight can vary substantially with age. For example, according to Wilson (1972), the DNA/protein ratio decreased with age in livers of adult female mice, but increased in males. DNA/100 mg of tissue was relatively stable as total DNA increased more or less in concert with wet weight. In lungs, sex and age differences were less apparent after maturity, although wet weight and total DNA increased. Steinhagen-Thiessen and Hilz (1976) reported a substantial increase in DNA per gram net weight of old human muscle, probably due to loss of tissue water. However, Paterniti *et al.* (1980) did not observe a significant change in liver protein content per gram of tissue (wet weight) between 12 and 21–24 months in Sprague-Dawley rats. Schlenska and Kleine (1980) recently performed detailed studies on age-related changes in activity of enzymes involved in glycolysis, gluconeogenisis, fatty acid degradation, and energy transfer in human muscle and obtained substantially different results depending upon whether the data were expressed as activity per gram of single fiber weight or per gram of pure muscle weight. They concluded that there were presenile changes in cytoplasmic glycolysis and gluconeogenesis.

From these observations, it is obvious that reports of enzyme activity based on wet weight or milligrams of protein can be misleading. Further-more, the ages selected to represent young and old rats varied from laboratory to laboratory. One must take care to see that mature animals (say 6–12 months) are compared with old animals (say 24 months or above). Use of 1- to 2-month-old rats for comparison with 20-month-old animals may mask, for example, a rise to 12 months with a subsequent lack of change, thus presenting a fallacious gain in activity with age. Similarly, studies ending at 12 months or perhaps even 20 months may not reflect senescence. Strain and sex also vary in the various tabulated results. The changes in enzyme levels due to differences in strain may be of similar or greater magnitude than changes seen in aging animals. For example, Wang and Mays (1977) provided evidence that glucose 6-phos-phate dehydrogenase in rat liver decreases by one-half in aging Sprague-Dawley rats (3 versus 24 months), but doubles in Fischer 344 rats. Hosoi *et al.* (1978) found a 10-fold greater activity in L-glutamine D-fructose-6-phosphate aminotransferase activity in the submandibular gland of female compared to male mice. The sublingual and parotid glands showed no difference in this respect. Undoubtedly, diet and the routine of animal care are additional variables. Under these circumstances, it is perhaps not surprising that the data from different laboratories are more than occasionally conflicting. For example, hepatic alkaline phosphatase was reported to decrease by 30% (Ross, 1969) and remain unchanged (Barrows *et al.*, 1962) in old rats, yet increase by 30% in old mice (Zorzoli, 1955). Similarly, cathepsin was reported to increase by 25% with age in the liver of male Wistar rats (Beauchene *et al.*, 1967) and decrease by 20% (Ross and Ely, 1954).

Wilson and Franks (1971) determined the activities of nine enzymes in kidneys of male and female C57BL mice at four ages: infant (up to 10 days); young adult (6 months); mature (18 months); and old (30 months). Wilson (1972) subsequently carried out the same type of studies in liver and lungs, also using C3H mice at 6 and 12 months of age. Results were based upon DNA content of the tissues as providing the most constant baseline. Even under these standardized conditions, no consistent pat-tern emerged. With increasing age of the mice, some enzymes increased, some decreased, some were unchanged, some increased and then de-creased, and some decreased and then increased. There were substantial sex and tissue differences as well as species differences. The authors found no change in the localization of enzymes as observed by histo-chemical procedures. However, in a study of the distribution of several enzymes in lobules of mouse liver, Wilson (1978) observed a change with age in alkaline phosphatase, which appeared to be related to structural changes.

In another set of matched experiments, Kerr and Frankel (1976) used male and female rats (CFN) of 3, 12, 18, and 24 months of age. Liver and brain were examined for three dehydrogenases (glutamic, lactic, and malic dehydrogenase) as well as the cofactors NADH and NADP. Results were based on enzyme activity per gram of liver. Again, no consistent patterns of change were observed. In fact, the enzyme levels changed up or down, often by 30% or more, almost at random between age-groups. Paterniti *et al.* (1980) examined changes in the levels of heme-containing proteins and mitochondrial components in liver and kidney of young, mature, and old male Sprague-Dawley rats. Comparing mature (12 months) and old (21–24 months) animals, cytochrome *P*-450 showed no change. However, the changes reported for catalase activity in the old animals were dramatic: up by 43% in liver and by 70% in heart and down by 44% in kidney. The same authors (Paterniti *et al.*, 1978) had earlier studied levels of δ-aminolevulinic acid synthetase, the limiting enzyme of heme biosynthesis. In young (1 month), middle-aged (12 months), and old (24 months) Sprague-Dawley rats there were sharp decreases (generally 50–60%) in the basal levels of the enzyme between 12 and 24 months in liver, brain, and heart, but the age-related changes in kidney, although substantial, were not statistically significant.

It is clear that meaningful differences in the level of enzyme activities have not been established insofar as they can be related to the deleterious effects of the aging process. Of the large number of assays reported, few enzymes show more than 30% variation between young and old animals. Variations of this magnitude would seem of dubious importance, particularly given the vagaries of the experimental systems as witnessed by the almost random-appearing fluctuations. If old animals consumed more or less food than young ones or there were different degrees of physical activity (perhaps old animals are more sedentary than young ones), or if the increasing amount of fat in some strains of old rats (Chapter 2, Section II) affects intermediary metabolism and hence the relative enzyme activities in different pathways, one might expect to see a consistent tendency. However, even with animals maintained under the same regimen in the same facility, results are greatly variable. Granted that statistically significant numbers of animals are utilized, the results still cannot be readily interpreted in terms of aging. Lindena *et al.* (1980) followed the activities of ten enzymes in plasma, lung, spleen, and skeletal muscle of Han:Wistar rats from 35 until 1115 days of age, by which time, only 10% of the animals survived. In general, each enzyme followed its own behavior pattern, and this varied with the tissue. There were substantial sex differences. The authors conclude that the enzyme activity patterns are not greatly useful in understanding aging.

It would be nice if some enzyme had been found that showed a consistent and substantial age-related loss of activity in all species test-ed. Attention could then be focused on the functions related to that enzyme. Unfortunately, no such finding has been made. It must be concluded that the work described above, namely, the measurement of enzyme levels in crude tissue extracts, has provided few rewards. This is not to say that seeking for changes of enzyme levels in aging tissues is a useless approach. Quite the contrary. Enzymes, which are keys to spe-cific biochemical pathways or to the function of organelles, e.g., mito-chondrial enzymes, certainly must be part of physiological or biochemi-cal deficits known to be related to aging. The widely varying and often conflicting results described above suggest that experiments on a more profound level will be needed before the data acquired from changes of enzyme levels can be integrated into a complete picture of biological function.

III.   THE ERROR CATASTROPHE HYPOTHESIS

The idea of altered enzymes arose from the Orgel (1963) error catastro-phe hypothesis, which proposed that errors in proteins involved in the protein-synthesizing machinery would produce an increase in errors in subsequent proteins synthesized, and these in turn would produce even more errors until an "error catastrophe" was attained. This idea was modified (Orgel, 1970) to include the idea that catastrophe could be avoided if the frequency of errors were small. Then, instead of amplifica-tion, errors would be maintained at a steady state, i.e., reach a limiting value. The ideas expressed in these theories raised much speculation, and a substantial number of papers appeared in subsequent years, argu-ing for or against the theory. Many of the discussions were without direct experimental basis but drew upon results of experiments that showed errors or lack of errors in proteins and purported changes or lack of changes in the fidelity of translational systems. Such papers include that of Gershon and Gershon (1976), Dreyfus *et al.* (1977), Roth-stein (1977), and Kirkwood (1977). Except for the last, these discussions are against the error catastrophe hypothesis. Orgel himself (1973), in discussing various possibilities for aging processes, withdrew his sup-port for the idea of increasing errors and favored the idea that a small, stable error frequency could be maintained.

On the experimental side, Edelmann and Gallant (1977) showed that the large error frequency (up to 50-fold) induced in flagellin synthesis in *E. coli* by addition of streptomycin (as measured by the incorporation of radioactive cysteine which is normally not present) reached a limiting value and did not continue to increase. Removal of the streptomycin after two generations quickly restored the normal small rate of error. In short, the large error frequency did not create conditions for more errors, but tended toward a stable value. In a later paper, Gallant and Palmer (1979) pointed out that in *E. coli*, the error frequency in alkaline phosphatase can be artificially increased using streptomycin, by at least an order of magnitude without generating dead cells or creating clonal senescence. Errors stabilize at a high level (13-fold increase). Even this high level is not sufficient to set off an "error catastrophe." The authors pointed out that the frequency of errors in somatic cells, as far as is known, is very low and hence they concluded that error propagation is unrelated to aging. Arguing to the contrary, Rosenberger *et al.* (1980) reported that under their conditions of culture, generation of errors in β-galactosidase in *E. coli* was quite unlike that in alkaline phosphatase. The authors suggested that the case against error propagation should be reevaluated. Two recent papers have appeared that deal with the idea of error propagation on mathematical and theoretical grounds. Gallant and Prothero (1980) developed the thesis that errors are limited to a stable value. Kirkwood (1980) concluded that the available experimental evidence is not adequate to decide one way or the other. Both papers discussed mathematical models proposed by others. It should be pointed out that from the practical point of view, there has been no firm evidence for the existence of erroneous molecules that are a product of the aging process (Section IV,A).

Menninger (1977) has proposed a mechanism of "ribosomal editing," a process which dissociates inappropriate peptidyl-tRNA molecules from the ribosome in *E. coli*. He relates this possible mechanism of control of error levels to the error catastrophe model. Although other authors have proposed editing for aminoacyl-tRNA synthetases (Wright, 1980; Savageau and Freter, 1979), such mechanisms have not been related to aging.

IV. FIDELITY OF PROTEIN SYNTHESIS

Although there is little likelihood that an error catastrophe occurs, it seems possible that small errors might occur in proteins with increasing age. Thus, altered enzymes could perhaps result from the change of a single amino acid. Several sets of experiments have been reported that attempt to determine if there is a loss of ability with age to discriminate against inserting wrong amino acids into proteins. Evidence for such a loss of fidelity in protein translation is weak. On the other hand, evidence that no loss of fidelity occurs with age is quite substantial. Ogrodnik *et al.* (1975) measured incorporation of methionine versus ethionine into crude ribosomal proteins of C57CL/6J mice. The authors made a number of assumptions and complex calculations and concluded that there was a general decrease in the discrimination ratio as a function of age, even though their old animals were only 20 months of age. On the other hand, Kurtz (1975) found that microsomes from old mouse liver made fewer errors than preparations from young animals. The mice (C57BL/6) ranged from 1 to 31 months of age. Mariotti and Ruscitto (1977) used a cell-free protein synthesizing system from rat liver. They measured the incorporation of leucine versus phenylalanine using poly(U) as messenger. The authors found an increased leucine/phenylanine ratio with age. However, since the "old" rats were 1 year old, these results cannot be related to aging per se. That such changes were found in young animals throws suspicion on findings of this nature. Neither these results nor those of Kurtz (1975) are significant at the 95% confidence level, as pointed out by Buchanan *et al.* (1980).

Medvedev and Medvedeva (1978) looked for the increased incorporation with age of [^{35}S]methionine into Hl histones, which do not normally contain this amino acid. No age-related differences were found. Coniglio *et al.* (1979) studied the incorporation of labeled *p*-fluorophenylalanine into acid-soluble fractions by isolated liver parenchymal cells of the rat. The ratio of incorporated analog to incorporated valine was unchanged in 4- versus 18-month old animals. Unfortunately, older animals were not studied. In addition, one might argue that experiments such as these do not so much measure errors, as the ability to distinguish fluorophenylalanine from the natural amino acid.

Mori *et al.* (1979) treated ribosomes with puromycin and high ionic strength buffer to remove endogenous mRNA and peptidyl-tRNA. The order of infideltiy—incorporation of [^{3}H]leucine in poly(U)-directed systems—was about 10^{-3} with no age-related effects. The individual variation in animals of the same age-group was very large even though the experiments were carried out simultaneously. By way of confirmation of these results, neither ribosomes from rat brain (Ekstrom *et al.*, 1980) or kidneys (Hardwick *et al.*, 1981) showed increased infidelity with age, nor did tRNAs from a variety of tissues in old mice (Foote and Stulberg, 1980). Butzow *et al.* (1981) also found no age-related differences in the fidelity of rat liver ribosomes when magnesium levels were manipulated. However, the error frequency was greater in "old" ribosomes at concentrations of paromomycin between 20 and 60 μM.

It is fair to say that attempts thus far to show an increased tendency with age for errors to be made in protein synthesis have been unconvincing. On the contrary, the evidence for lack of errors is much stronger.

V. ALTERED ENZYMES

A more direct approach to the idea of error propagation has been made by investigators who have attempted to find altered proteins in old organisms thus observing directly, age-related changes in the fidelity of transcription or translation. The first evidence for the existence of altered enzymes in aged organisms was provided by Gershon and Gershon (1970). They utilized the free-living nematode *T. aceti*. The organism was aged by adding FudR to cultures of young organisms (Gershon, 1970). The FudR, a DNA inhibitor, prevents reproduction so that the cultures, as they age, do not become contaminated with newborn organisms (Chapter 2, Section IV). Homogenates of young and old organisms were prepared and the activity of isocitrate lyase was titrated with an antiserum prepared in rabbits to the "young" homogenate. Starting with equal amounts of enzyme activity, it required greater amounts of antiserum to remove the enzyme activity from the "old" preparation. That is, there were more "old" molecules per unit of activity. Put another way, "old" isocitrate lyase was catalytically less effective than "young" isocitrate lyase and was, therefore, altered in some way.

Subsequent to the report of Gershon and Gershon (1970), a number of altered enzymes have been reported to be present in old animals. These are listed in Table 9.I. It should be borne in mind that many of these results were obtained with crude homogenates and may be subject to experimental artifacts, as noted below.

Before discussing altered enzymes further, it would be useful to consider techniques, procedures, and precautions that should be observed. The search for altered enzymes in crude homogenates can provide misleading information. The typical criteria first sought are a lowered specific activity and an altered sensitivity to heat. Obviously, if there is more (or less) protein present in homogenates of old organisms, the specific activity of the enzyme (based on milligram of protein) will appear to change accordingly. Alternatively, there may be normal enzyme present in the "old" homogenate, but in lesser quantity. In these cases, the observed reduction in specific activity would not be due to a reduced catalytic ability. An example is triosephosphate isomerase from old *T. aceti*. The enzyme in crude "old" homogenates shows about one-half the specific activity of the enzyme obtained in "young" preparations (Gupta

Table 9.I Enzymes Reported to Be Altered with Age

Source	Enzyme	Ages	Preparation	Tests[a]	Specific activity[b]	Immuno.[c]	Heat sens.[c]	Remarks	Reference[d]
T. aceti	Isocitrate lyase	0–45 days	Crude homog.		29 versus 8	+	+	Original report of altered enzymes	1
T. aceti	Isocitrate lyase	6 versus 27 days	Pure	1,2	210 versus 83 (11) (3)	+	+	Enzyme consists of five isozymes. Results not due to change in isozyme composition	2
T. aceti	Phospho glycerate kinase	9 versus 27 days	Pure	1	700 versus 350 (3.9) (2.2)	•	NC		3
T. aceti	Enolase	5 versus 24 days	Pure	1,2,3	1200 versus 700 (16) (11)	+[e]	+	Spectral differences	4
T. aceti	Aldolase	9 versus 35 months	Pure	1,2	8.0 versus 4.2 (0.13) (0.072)	+	+		5,6
Mouse muscle	Aldolase	2.5 versus 31 months	Crude homog.		No change	+			7
Mouse liver	Aldolase	3 versus 31 months	Crude homog.		(0.058) versus (0.029)	+		Enzyme reported unaltered (see Table 9.II)	8
Mouse liver	Aldolase	5–6 versus 27–28 months	Crude homog.		(0.056) versus (0.029)	+			
Mouse liver	Basal tyrosine aminotransferase	3 versus 24 months	Crude homog.			+	+	Enzyme reported unaltered (see Table 9.II)	9
	Basal ornithine decarboxylase						+		

222

Source	Enzyme	Age	Preparation	Method[a]	Specific activity[b]		Change[c]	Comments	Reference[d]
Mouse heart	Phosphorylase	6 versus 24 months	Crude homog.		(150) versus (13)f	+	−	Changes thought to be associated with aggregation of enzyme	10
Mouse liver brain, spleen, kidney, lung	G6PD	6 versus 25 months	Crude homog.				+		11
Mouse heart	Aldolase	3 versus 30 months	Purified		0.44 versus 0.028		+		12
Mouse muscle	Aldolase	3 versus 30 months	Purified		0.49 versus 0.332		+		
Rat liver	Superoxide dismutase	6 versus 27 months	Pure	1,2	2460 (21) versus 1008 (8.5)	+	+	Enzyme reported unaltered (see Table 9.II)	6,13
Rat liver	Aldolase	12 versus 32 months	Purified		1048 versus 888	+	+		14
Rat lens	Aldolase	1 versus 24 months	Crude homog.		(43) versus (11)	+	+	Aldolase A and C	15
Rat muscle	Phosphoglycerate kinase	8–10 versus 30–34 months	Pure	1,2,3	No change	+	+	Spectral differences	16
Rat liver	Phosphoglycerate kinase	8–10 versus 32 months	Pure	1,2,3	No change	+	+	Spectral differences	18
Rat liver	Lactic dehydrogenase	4–5 versus 29 months	Crude homog.		(6.0) versus (2.8)	+	+	Activity is based on activity/antigen amount. Regenerating old liver yields more active enzyme	17

[a] (1) Gel electrophoresis; (2) isoelectric focusing; (3) C- and N-terminal analysis.

[b] Specific activities are usually given as units/mg of protein. Figures in parenthesis are values for crude homogenates.

[c] (+), Changes with age; NC, no change with age.

[d] Key to references: (1) Gershon and Gershon, 1970; (2) Reiss and Rothstein, 1975; (3) Gupta and Rothstein, 1976a; (4) Sharma et al., 1976; (5) Reznick and Gershon, 1977; (6) Goren et al., 1977; (7) Gershon and Gershon, 1973b; (8) Reznick et al., 1981; (9) Jacobus and Gershon, 1980; (10) Capasso and Zimmerman, 1980; (11) Wulf and Cutler, 1975; (12) Chetsanga and Liskiwiskyi, 1977; (13) Reiss and Gershon, 1976a; (14) Barrows and Kokkonen, 1980; (15) Ohrloff et al., 1980; (16) Sharma et al., 1980; (17) Schapira et al., 1978; (18) Hiremath and Rothstein, 1982, submitted.

[e] Antiserum prepared to "old" enzyme was less effective with "young" enolase (Sharma and Rothstein, 1978b).

[f] After stress.

and Rothstein, 1976a). However, after complete purification, both enzymes were shown to have identical properties. The apparent lowered specific activity in crude "old" preparations was a result of having about one-half as much enzyme present, and not from any alteration of its properties.

The difficulties with measuring specific activity in crude homogenates can be overcome by use of immunotitration with antisera specific for the enzyme in question. This technique, so far, appears to give dependable results. In fact, the amount of antiserum required to precipitate 50% of a given activity is identical for both pure and crude preparations of "young" and "old" nematode triosephosphate isomerase (Gupta and Rothstein, 1976a), pure and crude "young" rat muscle phosphoglycerate kinase, and pure and crude "old" rat muscle phosphoglycerate kinase (Sharma et al., 1980). Thus, there is no immunosensitive material in the original homogenates exclusive of the enzyme being measured.

Heat-sensitivity measurements obtained from enzymes in crude homogenates can also be deceiving. For example, the heat-sensitivity patterns of "young" and "old" nematode aldolase in crude preparations were found to be identical (Zeelon et al., 1973). Yet, when the enzyme was later purified, it was found that the patterns differed, the "old" aldolase actually being the more stable of the two (Reznick and Gershon, 1977). Other examples of the difficulties that may arise from using crude homogenates can be seen from the heat-sensitivity curves of glucose 6-phosphate dehydrogenase (G6PD) in extracts of a variety of tissues from young and old mice (Wulf and Cutler, 1975). Although there are differences reported between young and old preparations, the rate of loss of activity when the enzyme is heated varies considerably from tissue to tissue, although the enzyme should be the same in each case. Of importance in measuring this enzyme, Schofield and Hadfield (1978) showed that 6-phosphogluconate dehydrogenase contributes to the activity as measured in the above system.

Differing amounts of protein in the samples, the presence of proteolytic enzymes, or other unknown factors in the preparations would also affect the results. In fact, Kahn et al. (1977a) demonstrated that G6PD could be altered by components in cell homogenates. The authors showed that the enzyme in "young" (34 passages) human liver cells yielded a straight-line pattern of heat sensitivity and "old" (61 passages) cells yielded a biphasic curve. The G6PD was removed by addition of specific, antihuman G6PD and replaced with an equal amount of pure "young" G6PD. In the "young" cytosol, the heat sensitivity of the added enzyme was unchanged, but in the "old" cytosol, it became biphasic, similar to the G6PD which had been removed by the antiserum. Without question, the "old" cytosol contains some component that alters G6PD postsynthetically.

For enzymes with isozymic forms, use of crude homogenates could lead to equivocal results even if antisera are used for immunotitrations. The reason is that proportions of the isozymes may change with age so that measurements would not properly reflect specific activity. At the least, gel electrophoresis should be used to insure against such changes in enzyme composition. As examples, Reiss and Rothstein (1974, 1975) found that the ratios of isozymes of nematode isocitrate lyase change with age; Patnaik (1979) observed a shift from one form of alanine aminotransferase to another in immature rat liver in the period between 5 and 100 weeks; age-related changes have been reported in the amounts of hexameric, tetrameric, and dimeric forms of G6PD found in rat liver (Wang and Mays, 1978). Capasso and Zimmerman (1980) noted a similar difference in the amount of tetrameric myocardial phosphorylase in homogenates of old mice.

With the dangers of drawing conclusions from crude homogenates in mind, it is nonetheless of interest that Reiss and Gershon (1976b) performed immunotitrations of cytoplasmic superoxide dismutase in rat and mouse liver, brain, and heart. With respect to age, only the liver showed a loss of specific activity. Sharma *et al.* (1980), titrated homogenates of young, middle-aged, and senescent rat muscle, brain, liver, kidney, lung, and heart with antiphosphoglycerate kinase serum. Senescent muscle, liver, and brain showed substantial amounts of cross-reacting material. Of these, pure muscle and liver enzymes have been shown to be altered, confirming the results obtained with crude homogenates. The other tissues were unchanged or too slightly so to draw conclusions.

As "young–old" pairs of enzymes were identified, purified, and compared, a number of common properties emerged. In all cases studied, yields of the two forms were similar. That is, no evidence could be found that significant amounts of altered enzymes were being "purified away" in the "old" preparations. Moreover, the lack of additional cross-reacting material in crude homogenates clearly shows that multiforms of altered molecules, if they exist at all, can be present in only trace amounts. Essentially, the only altered enzyme present in the "old" homogenates is the enzyme which is subsequently purified. Superficial kinetic behavior and appearance on SDS gels and after polyacrylamide gel electrophoresis are the same for each respective "young–old" pair studied. The *N*-terminal and *C*-terminal groups are unchanged (nematode enolase; rat muscle phosphoglycerate kinase) proving that there are no cuts or losses of one or two amino acids from the ends of the "old" molecules. Amino acid analysis, although showing no changes between several old–young pairs, is too crude an indicator to distinguish small differences. Most important, isoelectric focusing was performed on rat liver superoxide dismutase (Goren *et al.*, 1977), nematode enolase (Sharma *et al.*, 1976), nematode aldolase (Goren *et al.*, 1977), and rat muscle phosphoglycerate kinase (Sharma *et al.*, 1980). In none of these experiments could any differences in charge be detected between the "young" and "old" forms of the respective enzymes. Immunodiffusion plates prepared to indicate purity of the isolated enzymes, also showed no differences between "young" and "old" enzymes (nematode enolase (Sharma *et al.*, 1976); rat liver superoxide dismutase (Reiss and Gershon, 1976a); nematode aldolase (Reznick and Gershon, 1977); nematode isocitrate lyase (Reiss and Rothstein, 1975); rat muscle phosphoglycerate kinase (Sharma *et al.*, 1980)). However, in all of the above cases, more "antiyoung" serum was required to precipitate a given amount of activity in an "old" enzyme compared to its "young" counterpart. Figure 9.1 shows a typical immunotitration of a "young" versus "old" enzyme. For this enzyme (nematode enolase), when the antiserum is prepared to the "old" enzyme, the curves are reversed, the "old" enzyme requiring less antiserum for a given loss of activity (Sharma and Rothstein, 1978b).

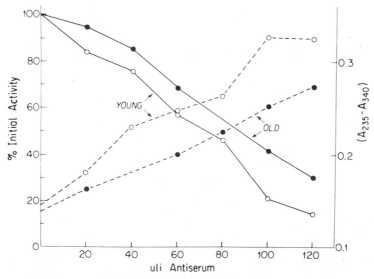

Fig. 9.1. Immunotitration of "young" and "old" enolase from *Turbatrix aceti*. Antiserum was prepared to pure "young" enzyme. The A_{235}–A_{340} (dotted lines) measures the protein in the immunoprecipitate. Reprinted by permission of Elsevier Sequoia S. A., Lausanne (see Sharma and Rothstein, 1978b).

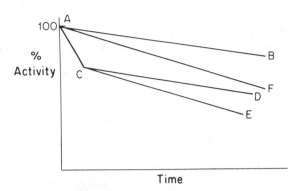

Fig. 9.2. Typical heat-sensitivity patterns of "young" and "old" enzymes.

The most apparent differences in altered enzymes are thus specific activity, response to immunotitration, and heat sensitivity. Measurement of the latter provided the three general patterns shown in Fig. 9.2. Discussing only the cases for pure enzymes, the "young" products typically are represented by the straight line relationship, AB. "Old" nematode isocitrate lyase is represented by ACD (Reiss and Rothstein, 1975) where the more stable component (CD) parallels the "young" pattern (AB). "Old" and "young" nematode phosphoglycerate kinase (Gupta and Rothstein, 1976b) are identical; "old" nematode enolase shows a pattern ACE, in which the second component (CE) falls more steeply than the "young" line (AB) (Sharma *et al.*, 1976). "Old" rat liver superoxide dismutase also follows this pattern (Reiss and Gershon, 1976a) but it is inverted, the first phase having a shallow slope and the second, a steeper one. Pure nematode aldolase is biphasic for both "young" and "old" enzyme (Reznick and Gershon, 1977), but the slopes of both components are steeper in the case of the "young" enzyme. That is, the "young" enzyme is more heat sensitive. "Old" rat muscle phosphoglycerate kinase is represented by AB and "young" by AF, although there is a small steeper component (Sharma *et al.*, 1980).

One question that arises and that has not been resolved is whether or not altered enzymes consist entirely of altered molecules or contain a mixture of altered and unaltered molecules. At first glance, the heat-sensitivity curves should provide this information. That is, a line such as AF or ACE should represent molecules, all of which are altered. ACD (e.g., "old" isocitrate lyase) would consist of a mixture of altered and unaltered molecules, the CD portion being the same as "young" enzyme. However, the possibility exists that the molecules represented by AC are in part converted to CD by the heating process. Moreover, "young" and "old" nematode phosphoglycerate kinase give identical patterns. One cannot be certain that the heat-sensitivity curves are valid for determining if one has a mixture of altered and unaltered molecules. Only where there is no similarity between "young" versus "old" patterns would the results seem to provide a valid indication of an entirely altered population of molecules.

A. Are Altered Enzymes a Consequence of Errors in Sequence?

Proof of the existence of altered enzymes in aged organisms brought forth arguments as to whether the causes were due to sequence changes (errors) or to postsynthetic modifications. Results obtained with pure enzymes (isocitrate lyase, superoxide dismutase, enolase, phosphoglycerate kinase) militated strongly against the idea of an error catastrophe, for if many errors were occurring in the molecules or small errors were occurring in many molecules, some of these error-laden proteins should be observed on immunodiffusion plates and would result in a lower specific activity based on immunotitration of enzyme in crude "old" homogenates versus pure "old" enzyme. Moreover, there should be charge differences after isoelectric focusing. None of these conditions have been observed. In fact, a single change of charge is detectable in aldolase (Gürtler and Leuthardt, 1970) and in a human phosphoglycerate kinase variant (Fujii et al., 1980). Thus, any changes in sequence must have replaced neutral amino acids with neutrals, basic amino acids with bases, and acidic amino acids with acids—an extremely unlikely restriction.

Further evidence against the error hypothesis is the fact that a number of enzymes have been shown to remain unchanged in old animals (Section VI). If errors are intrinsic to the protein-synthesizing machinery and involve transcription or translation, then all proteins must become error laden, a situation that is clearly not the case. Moreover, as will be seen below, altered proteins appear to result from conformational rather than covalent changes.

Even without direct proof, it is clear from the above information that an error catastrophe theory is untenable. The occurrence of a low level of errors—even a single sequence change in proteins—is also highly improbable as the arguments against a catastrophe apply equally.

Parenthetically, enzymes with single amino acid changes brought about by mutation show rather substantially altered properties. For example, many variants of human enzymes have been reported and characterized, particularly G6PD. (Yoshida, 1973; Kahn et al., 1978a). It is well to keep in mind that these types of altered enzymes were investigated because they were different and caused health problems. Alterations that had no effect would not be searched for. Therefore, the conclusion that changing a single amino acid always causes a profound effect on the enzyme may be misleading. The general comparison of the properties of "old" enzymes and mutationally changed enzymes is a highly speculative approach. The effect of a number of different single amino acid substitutions on the stability of tryptophan synthetase from *E. coli* has been reported by Yutani *et al.*, 1977.

B. The Nature of Altered Enzymes

The general properties of altered versus unaltered enzymes are listed in Table 9.II. In contrast to the idea of errors, Rothstein and co-workers (Reiss and Rothstein, 1974; Rothstein, 1975, 1977, 1979; Sharma and Rothstein, 1978a, 1978b; 1980) proposed that altered enzymes result from conformational changes without covalent modifications. They further postulated that the enzymes became altered because of a slowing protein turnover. Thus, the "dwell time" of proteins in the cells would

Table 9.II Properties of Altered Enzymes

Little or No Change
  $K_m{}^a$
  Molecular weight
  Terminal amino acid residues
  Charge (isoelectric focusing)
  Gel electrophoresis pattern
  Number of SH groups
  Absence of methionine sulfoxide
  Immunodiffusion pattern
Usually Altered
  Heat sensitivity[b]
  Specific activity[c]
  Spectral properties[d]
  Inactivation by proteases
  Immunotitration

[a] The K_m value sometimes shows small but consistent change which may be significant.

[b] No change for nematode phosphoglycerate kinase. For nematode aldolase and rat muscle phosphoglycerate kinase, the "old" enzyme is the more stable form.

[c] No change for rat muscle phosphoglycerate kinase.

[d] Differences disappear in guanidine solutions (nematode enolase; rat muscle phosphoglycerate kinase).

increase, and they would have ample opportunity to become subtly denatured, probably kinetically, but possibly by enzyme action. Moreover, the altered enzymes would tend to accumulate rather than be removed by the normal turnover process. As will be seen, this theory fits all of the experimental facts so far available.

The information, which proves that altered enzymes derive from postsynthetic modifications and probably result from conformational changes, was developed through study of a number of enzymes, the key being provided by intensive work on "young" and "old" nematode enolase (Sharma and Rothstein, 1978a,b) and rat muscle phosphoglycerate kinase (Sharma et al., 1980). For the respective enzymes, the young and old forms behaved identically on immunodiffusion plates and after isoelectric focusing. There were no young–old differences in C-terminal amino acids and the N-terminal residue was blocked in all cases, presumably by an acetyl group. Enolase possesses four cysteine residues, and these are all in the SH form for both young and old forms of the enzyme. Similarly, "young" phosphoglycerate kinase has one fast-reacting and three slow-reacting SH groups, and none of these becomes oxidized with age. No methionine sulfoxide could be detected in either form of either enzyme.

The possibility that covalent changes are responsible for altered enzymes is very small. From the information given above, it is obvious that sequence changes are highly improbable. Changes due to phosphorylation, acylation, esterification, and deamidation do not occur, or they would be seen after isoelectric focusing of the enzymes. Oxidation of SH groups, formation of methionine sulfoxide, and proteolytic cuts have all been negated. Methylation is not a factor as shown by amino acid analysis. The remaining possibility is that the change that creates altered enzymes is conformational in nature. There is considerable evidence to support this view. Both enolase and phosphoglycerate kinase show differences in structure between young and old forms which disappear when the molecules are unfolded. Thus, "old" enolase shows two extra tyrosine residues exposed to the medium but in 6 M guanidine, both forms of the enzyme show the same total number of residues. One extra tryptophan residue is exposed in the "old" enzyme but at concentrations of 0.8 M urea and higher, the values become the same. Circular dichroism confirms these results, showing differences between "young" and "old" enolase in the region of aromatic amino acids, which disappear in 6 M guanidine. The increased values for A_{280} per milligram noted for "old" enolase and phosphoglycerate kinase, respectively, also disappear in guanidine solutions.

Further support for the idea of conformational changes as the basic cause of altered enzymes comes from the isolation and identification of an inactive form of enolase and evidence for its relationship to "young" and "old" enzyme. In the last step of purification, enolase is chromatographed on the ion exchanger DE-52. In the case of the "old" enzyme, just before the activity emerges from the column, a protein appears, which although inactive, reacts with the antiserum prepared to "young" or "old" enolase. Some of this inactive product is formed from "old" enolase each time the pure enzyme is passed through the column (Sharma and Rothstein, 1978a). The material is obviously denatured enolase. An immunologically identical product is found in homogenates of old (but not young) *T. aceti* (Sharma and Rothstein, 1978b). Indeed, the amount of this material present in the homogenates increases with age.

If pure "young" enolase is chromatographed on the ion exchanger DE-52 3 times in succession, some of it is converted to the same inactive enolase as is formed from "old" enolase. Moreover, the remaining "young" enzyme changes its properties and becomes somewhat similar to "old" enzyme. It shows a biphasic heat-sensitivity curve, which does not, however, exactly match that of the "old" enolase. Antiserum prepared to this thrice-columned enzyme reacts more effectively with "old" than with "young" enzyme. It appears from these results, that *in vitro*, "young" enolase → "similar-to-old" enolase → denatured enolase. Since the last of these products is found in old worms *in vivo*, it is a fair assumption that this scheme represents the normal pathway for the formation of altered enzymes.

Unequivocal proof that altered enzymes, at least those from nematodes are simply conformational isomers has recently been obtained by Sharma and Rothstein (1980). They found that both "young" and "old" nematode enolase, respectively, could be unfolded in guanidine solutions with quantitative recovery of the enzyme activity after removal of the reagent. By rigorous criteria, the refolded enzyme was found to be identical whether formed from "young" or "old" enolase. Such criteria included UV and CD spectra, immunotitration, rate of inactivation by protease, and heat sensitivity curves. The refolded enzyme is similar but not quite identical to "old" enzyme, based on these criteria. The heat-sensitivity curve appears to be the same as that for thrice-columned "young" enolase (see above), suggesting that the "young" enzyme tends to refold this way even with the level of stress on molecular structure which is generated by passage through the ion exchanger DE-52. The above results prove beyond dispute that altered enolase is formed by modification of its conformation. It is of interest that Gockel and Lebherz

(1981) recently found that enolase is normally made up of conformational isozymes in *Ascaris,* although in this case, there is a charge difference in the respective forms of the enzyme. More to the point, A. Gafni (private communication) has recently obtained evidence that conformational changes also appear to be responsible for altered enzymes in higher organisms. He found that the altered properties of glyceraldehyde-3-phosphate dehydrogenase from old rat muscle (28 months) resemble those of "young" enzyme which has been modified by treatment with iodine and then reduced by mercaptoethanol. It seems probably that oxidation of an SH group causes conformational changes which are retained when the group is once again reduced.

With the caution that similar evidence for conformational changes has not yet been forthcoming for altered mammalian enzymes, the slowed turnover theory of Rothstein and co-workers (Section V,B) provides an appropriate mechanism for the formation of altered enzymes. Synthesis and degradation of proteins (A and B, Fig. 9.3) would slow with age and the "dwell time" of enzymes in the cell would increase accordingly. They would thus have time to become subtly denatured by kinetic action, and the altered molecules would gradually accumulate instead of being replaced with normal molecules. Depending upon their native stability, some enzymes would remain unchanged (e.g., rat muscle enolase) or some would denature beyond recognition, leaving behind a reduced amount of detectable normal enzyme (triosephosphate isomerase, human muscle creatine kinase, and aldolase). Other enzymes might convert to one or two altered forms (e.g., the two heat-sensitive compo-

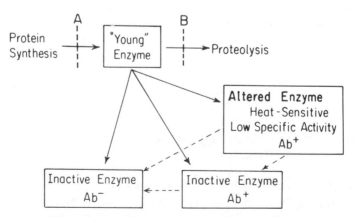

Fig. 9.3. Postulated effects of slowed protein turnover on the formation of altered enzymes. Ab^+ and AB^- indicate reaction and nonreaction, respectively, with antiserum prepared to the "young" enzyme.

nents of "old" enolase) and to other denatured products that could still react with antibodies to the enzyme (e.g., inactive enolase).

Does protein turnover indeed slow with age? The only conclusive data so far available are provided by work performed with free-living nematodes and the answer from these organisms is unequivocally positive. For higher animals we have little information, but work by Lavie *et al.* (1982) showed a decrease in the rate of degradation of mouse liver proteins. Thus, the information available supports the idea of slowed turnover in senescent animals (Chapter 8, Section II). It may be concluded that altered enzymes in nematodes are due to conformational changes in their structure resulting from a slowing of protein turnover. There is strong circumstantial evidence for a similar situation in higher animals, but the postulate remains to be proved.

VI. UNALTERED ENZYMES

A serious problem with the error theory is its inability to explain the fact that a substantial number of enzymes have been shown to remain unchanged in old animals (Table 9.III). For example, human parotid α-amylase from old subjects (60–100 years) and mature subjects (20–50 years) did not vary in specific activity, and heat-labile material was not present (Helfman and Price, 1974). Although purification or even gel electrophoresis of the enzyme was not carried out, the lack of differences is convincing.

Immunotitration of human creatine kinase and aldolase in crude homogenates of young and old muscle (Steinhagen-Thiessen and Hilz, 1976) showed no change in the amount of cross-reacting material. The heat-sensitivity curves for muscle aldolase (but not creatine kinase) showed differences that did not relate to the age of the samples. These anomalies may reflect the fact that the samples were taken postmortem and were determined in crude homogenates. In any case, the immunochemical evidence seems valid.

Other enzymes have been purified to homogeneity and shown to be unchanged in old animals. One of these is triosephosphate isomerase from young and old *T. aceti* (Gupta and Rothstein, 1976a). Although the "old" enzyme showed a lower specific activity in crude homogenates, this turned out to be the result of less enzyme being present. Subsequently, Rothstein *et al.* (1980) purified rat muscle and liver enolase to homogeniety and showed that the respective young–old enzyme pairs are identical. Both Fischer 344 and Sprague-Dawley rats were used for the

Table 9.III Enzymes Reported to Be Unaltered with Age

Source	Enzyme	Age	Preparation	Purity[a]	Test[b]	Remarks	Reference[c]
T. aceti	Triosephosphate isomerase	9 versus 26 days	Pure	1	I,H,S	Crude homogenate showed a 50% lower activity presumably less enzyme present	1
Human parotid	α-Amylase	21–49 versus 64–99	Crude		H,S		2
Human muscle	Creatine kinase Aldolase	27–47 versus 68–84	Crude		I,H I,H[d], S	Specific activities are decreased with age in crude homogenates—presumably due to less enzyme being present	3
Human lymphocyte	Aldolase A	18–41 versus 56–84 years	Crude		I,S		4
Human leukocytes	Pyruvate kinase G6PD GPI LDH α-Mannosidase β-Glycuronidase	20–30 versus 80 years	Crude		I		5
Human erythrocytes	Superoxide dismutase	2–69 years	Crude		I,H		6

Tissue	Enzyme	Age	Preparation	[a]	[b]	Comments	Reference[c]
Rat liver	Aldolase	2 versus 24 months	Crude[e]		I,H,S		7
Rat liver	Aldolase B	3–5 versus 27–30	Crude		I,H,S		8
Mouse liver	Aldolase B	3 versus 30	Pure	1	H,S		9
Rat muscle	Enolase	6–12 versus 28–32	Pure	1,2	H,S		10
Rat liver	Enolase	10–12 versus 30	Pure	1,2	H,S		10
Rat liver	Basal tyrosine aminotransferase	3–6 versus 27–30 months	Crude	2	I,H,S	No change with age of the three isozymes seen by isoelectric focusing	11
Rat liver	Ornithine decarboxylase						
Rat prostate	Ornithine decarboxylase	5 versus 24 months	Crude		I	Induced enzyme	12
Dog liver	Superoxide dismutase	1–2 versus 10–12 years	Pure	1	I,S		13
Rat liver	Superoxide dismutase	9–12 versus 32–35 months	Pure	1	I,H,S		14

[a] (1) Gel electrophoresis; (2) isoelectric focusing.

[b] I, Immunotitration; H, heat sensitivity; S, specific activity.

[c] Key to references: (1) Gupta and Rothstein, 1976a; (2) Helfman and Price, 1974; (3) Steinhagen-Thiessen and Hilz, 1976; (4) Steinhagen-Thiessen and Hilz, 1976; (5) Rubinson et al., 1976; (6) Joenje et al., 1978; (7) Anderson, 1976; (8) Weber et al., 1976; (9) Petell and Lebherz, 1979; (10) Rothstein et al., 1980; (11) Weber et al., 1980a; (12) Obenrader and Prouty, 1977; (13) Burrows and Davison, 1980; (14) Ghai and Lane, 1980.

[d] Heat sensitivity in the homogenates varied but not in an age-related manner.

[e] Based on activity per amount of peptide fragment from enzyme.

muscle enzyme and the latter for the liver enzyme. Young and old animals were 6–12 months and 28–30 months of age, respectively. Pure mouse aldolase A (Petell and Lebhertz, 1979) and aldolase A and B (Burrows and Davison, 1980) have been reported to be unchanged in old mice. The results with the former enzyme contribute to an ongoing controversy involving this enzyme (Section VII). The latter authors also found no change in aldolase A, aldolase B, or liver cytoplasmic superoxide dismutase in young versus old dogs. Old dog liver aldolase did show a reduced specific activity per unit of antiserum, but the results were not statistically significant. Schofield (1980) found that pure albumin from young and mature mice (8 versus 21 months) was essentially unchanged in terms of thermal denaturation, unless the protein was heated to about 50°C. In that case, the "old" albumin showed a greater degree of conformational change as measured by circular dichroism. Rubinson *et al.* (1976) showed by immunotitration that there were no differences in the specific activities of seven enzymes (pyruvate kinase, G6PD, glucose phosphate isomerase, 6-phosphogluconate dehydrogenase, LDH, α-mannosidase, β-glycuronidase) in human granulocytes from aged, young, and newborn human subjects. Similarly, aldolase A in lymphocyte extracts showed no change in cross-reacting material as determined by use of monospecific anti-aldolase A antibodies (Steinhagen-Thiessen and Hilz, 1979). Human erythrocyte superoxide dismutase gave similar results from human donors aged from 1 to 98 years (Joenje *et al.*, 1978). The white cell enzymes are of particular interest because they are being replaced quite rapidly and thus should reflect any defects in the protein-synthesizing system. Clearly, there are none. On the other hand, the enzymes should not show postsynthetic alterations because of rapid turnover.

From the above results, it is clear that altered enzymes are not a necessary consequence of aging. Even in the same tissue, enzymes may be altered or unaltered. For example, old rat muscle enolase and rat liver enolase are not altered, but old muscle and liver phosphoglycerate kinase are. Any explanation of the mechanism of formation of altered enzymes must take such facts into account.

VII. CONTRADICTORY RESULTS

Although some enzymes clearly become altered with age and others clearly do not, aldolase, superoxide dismutase, and tyrosine aminotransferase have been reported to be both altered and unaltered in aging organisms.

As noted above, Gershon and Gershon (1973a) found immunological evidence for the presence of altered aldolase in senescent mouse liver. Other investigators have obtained negative results. Anderson (1976), by analysis of the amount of a known peptide sequence, determined that there was no difference in the specific activity of liver aldolase in 2- and 24-month-old Fischer 344 rats. At about the same time, Weber *et al.* (1976) showed by immunotitration of homogenates of rat liver (Wistar) that there was no statistical difference in the specific activity of the aldolase B from 2- to 5-versus to 27- to 30-month-old animals or in the heat-sensitivity curves. Burrows and Davison (1980) found no significant differences in pure liver aldolase from young and old mice (3 versus 36 months of age). Although the specific activity of the enzyme was very low using one purification procedure and about sevenfold higher with another, there were no age-related differences. Immunotitration as carried out with antiserum prepared to dog liver aldolase B also showed no differences.

The differing results seemed to be explained by Petell and Lebherz (1979), who reported that if homogenates of C57BL/6 mouse liver were stored at 4°C, the "old" preparations yielded aldolase which had lost a C-terminal tyrosine residue. The "young" preparations were unchanged. The "old" aldolase behaved much as would be expected from an altered enzyme. There was a 60% reduction in specific activity, but behavior on gel electrophoresis, behavior on immunodiffusion plates, and the fructose diphosphate/fructose monophosphate activity ratio were not changed. Addition of proteolytic inhibitors (phenylmethylsulfonyl fluoride and leupeptin) prevented the "alteration." When fresh preparations were used, the aldolase isolated from 3- and 30-month-old mouse liver was unchanged. However, the situation is not so clear cut. Barrows and Kokkonen (1980) found by immunotitration, a small (13–22%) increase in cross-reacting material in crude preparations of old rat liver aldolase and a statistically valid 15% lower specific activity of the pure enzyme in Fischer 344 rats. More to the point, Reznick *et al.* (1981) recently reconfirmed the finding of altered aldolase in aging mouse liver. The authors attributed the use of hypotonic buffer and release of lysosomal enzymes by Petell and Lebherz (1979) as the probable cause of the proteolytic change in stored samples of "old" aldolase reported by the latter workers, and they suggested that altered molecules in freshly prepared "old" homogenates might have been lost during purification on a phosphocellulose column. In this respect, Pontremoli *et al.* (1979; 1980) reported that when rabbits are fasted, two modified forms of aldolase are present in livers. Although both are immunoreactive, one has a low specific activity and the second has no activity. The latter is not retained on the phosphocellulose column used for purification of the enzyme. The C-terminal residue is unchanged.

It should be borne in mind that proteolysis has not been a problem in the isolation of other altered enzymes. The N- and C-terminal analysis shows no change for "young" and "old" nematode enolase (Sharma and Rothstein, 1978a) and rat muscle phosphoglycerate kinase (Sharma *et al.*, 1980), and the use of proteolytic inhibitors during the isolation of isocitrate lyase (Reiss and Rothstein, 1975) caused no change in results. Moreover, the unaltered enzymes isolated from old organisms (Table 9.III) obviously were not attacked by proteolytic enzymes. Still, care should be taken during enzyme isolations, as Chappel *et al.* (1978) demonstrated that rabbit liver aldolase was subject to proteolytic attack unless lysosomes were first removed.

Although the result obtained by Petell and Lebherz may indeed come about for the reasons outlined by Reznick *et al.* (1981), the negative results of the other investigators are difficult to explain. According to Barrows and Kokkonen (1980), altered aldolase appears between 24 and 31 months of age. This result might explain why Anderson (1976) would have observed no change, as his old animals were "at least 24 months old." The results of Weber *et al.* (1976), who used older animals, would probably not show changes in the order of 10–15%. Their data, although the differences are not statistically significant, do suggest a slightly lowered specific activity and the presence of a component which is heat sensitive. However, these changes are at best, considerably smaller than the 35–40% lowering of specific activity noted by Gershon and Gershon (1973a) and Reznick *et al.* (1981). Thus, the reasons for the conflicting results obtained with aldolase B remain obscure. Perhaps the strain of animals or their condition at the time of sacrifice is involved.

As to other conflicting results, Weber *et al.* (1980a) found no cross-reacting material in basal liver tyrosine aminotransferase in old versus adult rats (27–31 versus 3–6 months) and they did not find any change in heat sensitivity or in the isoelectric focusing pattern. On the other hand, Jacobus and Gershon (1980) reported the presence of such cross-reacting material in homogenates of liver from old (20–26 months) mice. Heat sensitivity of the enzyme was also altered. There seems to be no simple explanation for these opposing results. Perhaps differences in the old liver created altered ratios of the multiple forms of the enzyme, which can occur during isolation (Hargrove *et al.*, 1980). Although Weber *et al.* (1980a) checked this point by isoelectric focusing, Jacobus and Gershon, (1980) did not. Alternatively, the situation may differ with respect to mice and rats. Curiously, Weber *et al.* (1980b) found evidence that induced, rather than basal, tyrosine aminotransferase becomes altered in old rats.

In this case, the formation of cross-reacting material is suppressed by serine protease inhibitors, suggesting that the altered protein arises as a result of proteolytic activity. All of the above work was performed using crude homogenates so that there is a possibility for undetermined artifacts.

Along similar lines, Obenrader and Prouty (1977) found no immunological evidence for increased cross-reacting material in the induced ornithine decarboxylase of liver or prostate tissue of old (24 months) versus young (5 months) rats. Jacobus and Gershon (1980) reported changes in heat sensitivity of the enzyme in mouse liver.

Pure cytoplasmic superoxide dismutase has been shown to be altered in the liver of senescent rats (Reiss and Gershon, 1976a). On the other hand, Burrows and Davison (1980) found no change in the purified enzyme from the liver of aged dogs. Ghai and Lane (1980) in a preliminary note, also found no age-related differences in the pure enzyme from the liver of Fischer rats.

VIII. MEMBRANE-BOUND ENZYMES

Little work has been performed on the possible alteration of membrane-bound enzymes. Grinna and Barber (1972) examined NADH-cytochrome c reductase and glucose 6-phosphatase in microsomes, and NADPH-cytochrome c reductase and β-hydroxybutyrate dehydrogenase in mitochondria from heart, liver, and kidney of 6-and 24-month-old rats. Although changes in specific activity were observed, the scatter of the results makes it difficult to determine their true magnitude. In older animals, NADPH-cytochrome c reductase and glucose 6-phosphatase from old kidney showed the biggest age-related loss of activity (41 and 47%, respectively). Since the phospholipid to protein ratios were apparently altered, the age-related differences could be due to changes in membrane conformation or perhaps to age-related changes affecting the isolation of the membrane fractions. Grinna and Barber (1975) subsequently reported that the loss of specific activity of glucose 6-phosphatase was due to the presence of less enzyme in the membranes of old rats. O'Bryan and Lowenstein (1974) compared renal membrane-bound enzymes in 3-, 6-, and 24-month-old rats and found that maltase was 50% lower in the latter; alkaline phosphatase was reduced 36% in one strain of rats and 18% in the other; phosphodiesterase was reduced 33%. The K_m value was unaltered in all cases. The authors concluded that the losses of activity were probably due to decreases in enzyme concentration. Gold and Widnell (1974) also found that the specific activity of glucose 6-

phosphatase and NADPH-cytochrome c reductase were lower in liver microsomes from old rats (3 versus 24 months). These results and others are considered in some detail in the discussion of microsomes. In fact, a substantial amount of work has been performed on membrane-bound microsomal enzymes, particularly with respect to their induction by drugs or hormones (Chapters 5, Section III and Chapter 10, Section XIII). Generally, reductions in activity appear to be a result of less enzyme molecules being present although there are indications that in some cases, changes in membrane composition are responsible. The latter situation may apply particularly to membrane-bound mitochondrial enzymes such as β-hydroxybutyrate dehydrogenase (Chapter 5, Section II). On the other hand, Schmucker and Wang (1980) suggest that the age-related decline of some of the microsomal enzymes is due to a reduction in the amount of smooth endothelial reticulum. The possibility of membrane changes makes it difficult to evaluate membrane-bound enzymes in terms of altered proteins.

IX. TISSUE CULTURE

The search for altered enzymes in late-passage cells began shortly after the report of Gershon and Gershon (1970) on the presence of altered isocitrate lyase in senescent free-living nematodes. As in intact animals, both altered and unaltered enzymes have been identified (although not purified) in cells in culture. In general, reports of altered enzymes are based upon comparison of heat-sensitivity patterns and sometimes on immunotitration.

A. Altered Enzymes

In an exploration of the validity of the error catastrophe hypothesis (Section III), Holliday and Tarrant (1972) compared the heat sensitivity of G6PD and 6-phosphogluconate dehydrogenase in early- and late-passage human fetal lung fibroblasts (MRC-5). In homogenates of the latter, they found an increased proportion of the enzymes was sensitive to heat. The authors interpreted this result as support for the error hypothesis. These first results, obtained from crude homogenates, provide an uncertain base from which to draw such conclusions. For example, protein concentrations of the various runs are not reported. If these vary, they might affect stability of the enzyme. If proteolytic enzymes are more prevalent in late-passage preparations, G6PD activity may be destroyed differentially, especially at the beginning of the period of elevated temperature used for the experiments, until the proteolytic enzymes themselves are destroyed. In fact, as was the case for Ogrodnik *et al.* (1975), who used crude homogenates of mouse tissues, the slope of the line representing the heat inactivation of the enzyme is not consistent from experiment to experiment. Moreover, G6PD is less than an ideal enzyme for this type of study as it exists in several molecular forms (see Section V).

In spite of these shortcomings, work with homogenates of cells in culture seems to show a real increase in the heat-labile fraction of certain enzymes in late-passage cells. Goldstein and Moerman (1975; 1976) studied enzyme lability in cells from patients with Werner's Syndrome and progeria. In examining skin fibroblasts from normal subjects used for control values, they found a statistically significant increase in thermolabile G6PD in early- versus late-passage cells (1.1 and 4.3%, respectively). 6-Phosphogluconate dehydrogenase also showed an increase in a thermolabile component (0.8 versus 4.9%). Hypoxanthine guanine phosphoribosyltransferase showed a large increase (7.7 versus 24.4%). Holliday *et al.* (1974) also examined skin fibroblasts from subjects with Werner's Syndrome. They used MRC-5 cells for control values and observed an increase in the thermolability of G6PD from 5.8 to 16.7% in early- versus late-passage cells. Fry and Weisman-Shomer (1976) partially purified DNA α-polymerase from the nuclei of chick fibroblasts and observed two distinct activities from phosphocellulose columns. No activity due to DNA polymerase β or γ was present. Neither polymerase in crude homogenates showed differences in specific or chromotographic behavior activity with increased passage number. However, both enzymes showed increased thermolability in phase III versus phase II cells and behavior toward inhibitors was altered.

Houben and Remacle (1978) examined the heat-lability patterns of three lysosomal enzymes (N-acetyl-β-D-glucosaminidase; α-D-galactosidase; N-acetyl-α-D-galactosidase), one mitochondrial enzyme (sulfate cytochrome *e* reductase), and one cytoplasmic enzyme (G6PD) in homogenates of WI-38 cells. Albumin was added to protect against effects of proteolysis. G6PD was the only enzyme that showed changes in late-passage (42–45) versus early-passage (22–25) cells. Kahn *et al.* (1977a) also showed that altered G6PD was present in late-passage human liver-derived cells and human fetal lung cells. More recently, Viceps-Madore and Cristofalo (1978) found that glutamine synthetase became more heat stable in late-passage WI-38 cells, indicating a change in the properties of the enzyme.

As shown by some of the investigators, the results do not seem to derive from differences in protein concentration, and chances of proteolytic attack have been reduced by addition of serum albumin. Therefore, the experiments reported above, although subject to errors inherent in work with crude homogenates, seem to demonstrate that altered enzymes exist in late-passage cells. However, their mechanism of formation is not clear. The creation of errors is subject to the same limitations as the whole animal studies discussed earlier in this chapter (Section V). Although there has been no work on pure enzymes, it is clear that many enzymes in tissue culture are unaltered (Section IX,B), and in fact as mentioned above, Houben and Remacle (1978) found four enzymes to be unaltered in the same late-passage cells in which G6PD showed an increase in heat lability. Therefore, errors in the protein-synthesizing machinery cannot exist as they would cause errors in all proteins.

The case for postsynthetic alterations is much stronger. In fact, in an ingenious set of experiments, Kahn *et al.* (1977a) demonstrated that changes in G6PD are brought about by the cytoplasm of late-passage cells. They found that G6PD from "young" cells gave a straight-line rate of loss on heating, whereas late-passage cells gave a biphasic pattern. Pure G6PD from human leukocytes also gave a straight line. The G6PD from early- and late-passage cells was removed by addition of an anti-G6PD serum and the removed enzyme was replaced with the same amount of pure G6PD. In the "young" cell supernatant, the original straight-line relationship was retained. However, when placed in the "old" cell supernatant, the pure "young" type enzyme yielded a biphasic pattern. The results prove that some component in the soluble fraction of the late-passage cells confers heat sensitivity on G6PD and that the changes must, therefore, be postsynthetic in nature. This conclusion is supported by Alekseav *et al.* (1979), who observed an increase in heat lability of about 8% for G6PD in late-passage human diploid cells. The purified enzyme was more stable than when measured in the homogenate, although the difference between "young" and "old" cells was preserved. Duncan *et al.* (1977) also suggested that the changes in heat lability of G6PD are due to the cell environment and not to errors. They investigated thermolability changes in the enzyme in human foreskin cells and observed an increase from 7 to 20% in heat-labile enzyme with increased passage number. Interestingly, enzyme from cells arrested for 14 days at any passage level showed a considerable decrease in heat lability over enzyme in proliferating control cells. When the G6PD was examined by gel electrophoresis, two forms of the enzyme were found. The evidence strongly indicates that one of these is a dimer and one a tetramer. There is a shift toward increasing amounts of the latter with increasing passage number. Since the tetramer is more heat sensitive than the dimer, the change in heat lability with increasing passage number is attributed to a shift in the enzyme from dimer to tetramer. Wang and Mays (1978) also found evidence for a shift in proportions of G6PD isozymes in old rat liver.

Still other possible problems with G6PD are enzymatic changes affecting the $NADP^+$ bound to the enzyme as postulated in red blood cells (Kahn *et al.* 1978b) or deamidation reactions as postulated in granulocytes (Kahn *et al.*, 1976). To make matters more complex, Hunter (1980) obtained isoelectric focusing patterns of the enzyme that vary with the cell type and the donor, yielding from four to six bands, although there seem to be some technical problems involving the type of appartus used. All of these complexities notwithstanding, it seems clear that whatever is happening to G6PD in late-passage cells, it is almost certainly a result of posttranslational effects and not due to errors in sequence.

The case for altered enzymes in tissue culture, although they undoubtedly occur, is difficult to interpret. All of the work has been performed in crude homogenates and unfortunately, most investigators seem to have chosen to study G6PD, an enzyme with complex behavior that exists as a monomer, a dimer, and a tetramer and requires bound NADP for structural integrity and stability. Whether or not cytoplasmic factors also cause the changes of other enzymes reported to be altered (e.g., 6-phosphogluconate dehydrogenase, and hypoxanthine guanine phosphoribosyltransferase) must await further investigation.

B. Unaltered Enzymes

As is the case with animal experiments, certain enzymes have been shown not to change in "senescent" cells in tissue culture. Strangely, one of these enzymes is G6PD. Pendergrass *et al.* (1976) could detect no increase in the heat lability of G6PD in late-passage human skin fibroblasts beyond the 5% figure obtained with early-passage cells. Immunotitration confirmed the results. One strain of their cells contained 40–60% heat-sensitive G6PD which in late-passage cells was reduced to less than 10%. Electrophoretic examination showed that these cells contained a mutant form of the enzyme, which presumably was being selected against with increasing passages. The authors pointed out the differences in experimental protocol between their work and that of Holliday and Tarrant (1972) (see Section IX,A). In the latter work, protein extracts were 100 times more concentrated; trypsin was used rather than scraping off the cells; cells were broken by sonication rather than by centrifugation; lung tissue was used rather than skin. In the light of later work, which confirms that altered or unaltered enzymes may both exist in late-passage cells, all but the last of these considerations have taken on less importance in explaining the differences in results. In fact, Evans (1977) also reported that altered G6PD was not formed in late-passage fibroblasts from mouse embryonic tissue (10–12 doublings) based on heat-sensitivity measurements. Little change in thermosensitivity was noted in RNA polymerase (Evans, 1976).

With regard to other enzymes, Danot *et al.* (1975) found no evidence for altered aldolase A in stationary (seven doublings) mouse embryo fibroblasts compared to early-passage cells, based upon immunotitration and heat-sensitivity procedures. Shakespeare and Buchanan (1978) found no changes in the thermostability or specific activity (as measured by specific antiserum) of phosphoglucose isomerase in early- versus late-passage MRC-5 cells. Kahn *et al.* (1977b), utilizing immunotitration and isoelectric focusing procedures, found no differences in four enzymes (phosphoglycerate kinase, M_2 type pyruvate kinase, glucosephosphate isomerase, and G6PD) in phase II versus phase III cells from human liver. Increased heat sensitivity observed for G6PD in intermediate and late-passage cells disappeared when the enzyme was partially purified. These results again emphasize the presence of factors in old cell cytoplasm, which affect G6PD. Houben and Remacle (1978), using WI-38 cells, examined three lysosomal enzymes (N-acetyl-β-D-glucosaminidase, D-glucosidase, and n-acetyl-D-galactosaminidase) and found no change in late-passage cells. Sulfite cytochrome *c* reductase, a mitochondrial enzyme, also showed no change. Only G6PD in the cytoplasm showed indications of the presence of heat-labile molecules. Similar findings were observed in fibroblasts (not "aged") from subjects with Werner's Syndrome (Houben *et al.*, 1980). Somville and Remacle (1980) in a brief report, note that in "old" WI-38 cells cytoplasmic superoxide dismutase is altered, but the mitochondrial form of the enzyme is not.

C. Fidelity of Protein Synthesis

The evidence strongly supports the idea that changes in enzymes in late-passage cells are due not to errors, but to posttranslational effects. In addition, there is other evidence that late-passage cells do not make error-laden proteins. For example, a number of investigators have shown that viral infection leads to the production of perfectly normal viral proteins in late-passage cells. Holland *et al.* (1973) reported that late-passage WI-38 cells supported normal production of three viruses (poliovirus type I, Herpes simplex virus type 1; vesicular stomatitis virus). Similar conclusions were drawn by Pitha *et al.* (1974; 1975) using a vesicular stomatitis virus and poliovirus, respectively, in WI-38 cells. Tomkins *et al.* (1974) used an RNA virus (poliovirus type 1) and a DNA virus (herpesvirus type

1) to infect WI-38 cells and found no changes in the electrophoretic pattern of viral proteins from early- versus late-passage cells. An interesting variation of these viral experiments was performed by Danner *et al.* (1978). Instead of using early- versus late-passage cells, they used skin fibroblasts from young and old subjects, the latter being 65–81 years of age. No significant difference in viral RNA synthesis or viral yield was obtained, on an average basis. Although individual variation was substantial the results indicate that in cells from old donors, viral proteins are made without error.

Because the search for altered proteins in late-passage cells was predicated upon the possibility of errors, direct attempts to study fidelity of protein synthesis have been made. The efforts are analogous to some of the work performed on whole animals. Linn *et al.* (1976) found that more errors were made when synthetic templates were acted upon by semi-purified DNA polymerase from late-passage compared to early-passage MRC-5 cells. As the authors point out, the levels of misincorporation are very high, being at an unrealistic level for *in vivo* synthesis of DNA. On the other hand, Buchanan and Stevens (1978) attempted to find evidence for misincorporation in "senescent" MRC-5 cells by measuring the presence of [^{35}S]methionine in histone Hl, a protein which normally does not contain that amino acid. The experiments did not detect any errors in late-passage cells. However, the authors note that the conclusion is not unequivocal. The upper limit for detection was less than seven methionine errors per 10^5 amino acids and less than two per 10^4 amino acids in early- and late-passage cells, respectively. Moreover, if cells that make errors do not divide, then only those cells which do not make errors would be analyzed. However, Buchanan *et al.* (1980) recently provided additional evidence against the idea of errors in protein synthesis. Using cell-free extracts of MRC-5 cells, they studied the level of misincorporation of leucine in poly(U)-directed protein synthesis and found it to be 1–2%, much higher than found *in vivo*. However, there were no differences between early- and late-passage cells and no age-related difference was observed in the increased level of errors caused by addition of paromomycin. The amounts of phenylalanine and leucine in the experimental medium affect the error rate. The authors point out that day-to-day variations in incorporation were of the order of 10%. Almost exactly the same type of experiment has been conducted by Wojtyk and Goldstein (1980), involving misincorporation of leucine into protein by poly(U)-directed cell-free systems. The authors found no differences between early- and late-passage human skin fibroblasts. No differences were observed in cells from subjects with progeria or Werner's Syndrome.

Interesting evidence that errors in protein synthesis do not increase in late-passage cells is provided by Harley *et al.* (1980). They utilized a technique designed to induce errors by starving cells of particular amino acids in the expectation that substitutions would be made and altered proteins would be synthesized (Parker *et al.*, 1978). Using this technique Harley *et al.* (1980) removed histidine from the medium and added histidinol to displace available histidine from cellular pools. Labeled methionine provided a marker. Proteins synthesized under these conditions showed "stutter" spots on two-dimensional gels—extra spots theoretically from proteins in which neutral glutamine is substituted for basic histidine. The error frequencies were unchanged (or in fact reduced) at late-passage in four types of cells or in cells from old donors. Under normal conditions, old cells produced labeling patterns of proteins indistinguishable from young cells. The error frequency was not related to the replicative span of the different cells. It rises in SV-40-transformed cells (immortal cells), contrary to the idea behind the error theory. The results are interpreted as negating the error theory.

D. Comment

There is little support for the idea of errors in proteins. The evidence is strong that no changes in error frequency occur with increasing passage number. On the contrary, there is growing evidence that altered proteins in tissue culture result from cytoplasmic factors. To the extent that this is the case, the mechanism involved appears to differ from that in intact animals although it is premature to draw firm conclusions. Could the observed changes in enzymes be due to a slowed protein turnover as proposed for intact animals? There has been little or no work done on the turnover of individual enzymes in early- versus late-passage cells. For proteins in general, the turnover picture is contradictory, and no conclusions can be drawn at this time (Chapter 8, Section IV).

X. ENZYMES IN RED BLOOD CELLS

If enzymes in old organisms become altered postsynthetically, then the aged red blood cell might provide a good model in which to study the process, because once the cell matures, no new enzyme is synthesized. Thus, there can be no change of sequence, and any modification must be posttranslational. This model differs from study of red cells from old donors (Chapter 2, Section VII).

A number of investigators have examined enzymes in relation to erythrocyte age. It was observed early that the electrophoretic mobility of certain enzymes changes with the age of red cells. Fornaini *et al.* (1969) reported that G6PD in old rabbit and human red cells showed changes of properties: lowered specific activity, increased thermolability, and increased K_m. Kahn *et al.* (1974) examined G6PD in a number of blood tissues and found that based upon immunotitration, the specific activity of the enzyme decreased in old red cells. They also found changes in the electrophoretic mobility of G6PD upon electrofocusing. The "old" erythrocytes showed bands of enzyme with a lowered isoelectric point. The changes were shown to be a result of factors in the tissues. Small proteolytic changes in the C-terminal end of the molecule were reported to be responsible for the differences in the enzyme isolated from leukocytes versus red cells (Kahn *et al.*, 1977c). Still later, Kahn *et al.* (1978h) attributed the changes in G6PD in old red cells to an enzymatically formed degradation product of NADP which binds to the enzyme.

Goldstein and Moerman (1978) observed that fresh progeric erythrocytes contain substantially more heat labile G6PD and 6-phosphogluconate dehydrogenase than control samples. The activity of G6PD in progeric red cells declines with the age of the cells, much as does the activity in normal erythrocytes. Recently, Racz *et al.* (1979) attributed the differences observed to technical problems, stating that the differences in heat stability are due to different endogenous concentrations of NADP. However, most, if not all investigators added external cofactor. Moreover, the enzyme was purified in a number of instances.

Not all enzymes show an altered structure in old red cells. In some cases, enzyme is simply lost, presumably either by denaturation beyond recognition by the antiserum, or by proteolysis. Thus, Kahn *et al.* (1977d) found that the decreased activity of glucosephosphate isomerase and phosphoglycerate kinase in old cells was a result of less enzyme being present as determined by immunotitration. On the other hand, the enzymes, pyruvate kinase, G6PD, and 6-phosphogluconate dehydrogenase had a lowered specific activity because of the presence of cross-reacting material.

There are many other examples of enzymes changing properties with the age of the cells. These papers include nucleoside monophosphate kinases (Jamil *et al.*, 1978); aldolase (Mennecier *et al.*, 1979); hexokinase (Fornaini *et al.*, 1978); galactokinase (Magnani *et al.*, 1980).

The activity of a number of membrane-bound enzymes in aging red cells also has been investigated. Adenylcyclase has been reported to decline (Pfeffer and Swislocki, 1976). However, Na^+, K^+-ATPase, glyceraldehyde-3-phosphate dehydrogenase and NADH-ferricyanide reductase showed no change with cell age (Kadlubowski and Agutter, 1977).

The results from studies of enzyme changes in aging red cells generally agree with the conclusions drawn from the nature of altered enzymes, namely, that depending on their structure, some enzymes are lost beyond detection (proteolysis, denaturation), and others form stable or detectable intermediates, which may be active, partly active, or inactive. To this extent, the results from studies with red blood cells can be said to parallel the concept of posttranslational changes in somatic tissues of old animals. The parallel, however, may not hold up to more profound scrutiny, as protein turnover (i.e., replacement) is involved in the latter. Moreover, the tissue milieux differ as well as the possibility that proteases my exist in red cells, whose specific function is to degrade certain enzymes. An analogous situation may not exist in other cells. In short, one should be cautious about stretching the red cell model too far.

XI. OTHER CHANGES IN PROTEINS

Several other mechanisms for changes in proteins with age have been proposed, including deamidation and racemization. Although these changes undoubtedly occur, their role in aging appears to be relatively restricted in scope.

A. Deamidation

Deamidation of proteins was suggested as playing a role in the aging process by Robinson *et al.* (1970). A well known case of this process is the hydrolysis of the asparaginyl residue in aldolase, converting a subunit to the β form, which is more prevalent in older rabbits.

The proteins of the lens of the eye are thought to turn over extremely slowly if at all, during life. The oldest tissue is found in the center (nucleus) of the lens with progressively younger material toward the periphery. The reason for this situation is that new fiber cells are formed from epithlial cells after which protein synthesis stops. New cells are formed layer by layer, each being covered in turn, by younger cells. Thus, an age-related gradient of lens material can be obtained for experiments by taking successive layers from a single lens or by using lenses of different ages. Any changes in protein must, therefore, be posttranslational. Van Kleef *et al.* (1976) demonstrated that with increasing age, there was increasing deamidation and peptide chain degradation with age in bovine α-crystallin. Subsequently, Kramps *et al.* (1978) showed similar changes in human α-crystallin. Zigler (1978) studied bovine β-crystallin, a heterogeneous group of lens proteins. The results suggest that post-synthetic modifications of this material follow the same pattern as for α-

crystallin. It is interesting that Skala-Rubinson et al. (1976) observed electrophoretic changes in G6PD, triosephosphate isomerase, and nucleoside phosphorylase in old bovine and human lens material that were reminiscent of those observed in old red blood cells. The greater degree of change occurred in the older parts of the lens. Deamidation is suggested as the cause of the increased anodic mobility. McKerrow (1979) goes so far as to suggest that the altered enzymes, which have been described, could be due to this process. Curiously, the author states that the methods employed in the studies would not detect deamidation of a "few" amide residues. He goes on to say that without careful isoelectric focusing, one cannot exclude deamidation as a factor in aging proteins. Goren et al., (1977), Sharma et al. (1976), Sharma et al. (1980), and Reiss and Rothstein (1975) all performed "careful isoelectric focusing." Moreover, the loss of a single amide group in aldolase (Gürtler and Leuthardt, 1970) and in phosphoglycerate kinase (Fujii et al., 1980) is detectable by this process. In fact, no evidence whatsoever has been found that in altered enzymes obtained from old tissues, the electrophoretic pattern is altered. This is not to say that deamidation is not involved in protein metabolism or even in the aging process. However, on the basis of the evidence available, it is either a relatively slow occurrence that takes place in very long-lived proteins or it has a limited involvement in enzymes. In the latter case deamidated molecules would not accumulate to a damaging degree, but would be routinely replaced by turnover. There is no evidence that soluble protein becomes generally more acidic with age as would be expected if deamidation were occurring.

B. Racemization of Aspartic Acid

D-Aspartate was shown to accumulate with age at about 0.1% per year in tooth enamel (Helfman and Bada, 1975) and in dentine (Helfman and Bada, 1976) and in fact, the authors proposed that the amount of D-aspartate in teeth could be used to estimate the age of long-lived mammals. Masters et al. (1977) suggested that racemization in the lens would be more suitable for this purpose since they found that there was a steady increase in the D/L-aspartic acid ratio with age in human lenses. The ratio decreased in peripheral samples, as would be expected since these areas are the most recently synthesized (Section XI,A). Cataract formation in various diseased states had no effect, except for brunnescent cataracts, which showed a much higher than expected D/L-aspartate ratio (58–97% above normal). The increased ratio may be related to development of brown pigment during cataract formation. The rate of accumulation of D-aspartate in lens nuclear proteins is 0.14% per year. In a subsequent and more extensive report (Masters *et al.*, 1978), the authors observed an enhanced racemization of water-insoluble proteins of the lens, whereas purified α-, β-, and γ-crystallins have low D/L-aspartate ratios.

Helfman *et al.* (1977) discussed the possibility that racemized amino acids may cause conformational changes in long-lived proteins and that this phenomenon may be related to aging. Collagen, particularly, would be susceptible. Although the authors specify that this idea may be a factor in long-lived mammals, it begs the question as to what happens in short-lived animals. The rate of racemization is too slow to be a serious factor in aging, particularly when one considers that most biologically active proteins are replaced by turnover.

C. Shift in Isozyme Composition

It is well recognized that during development, many proteins change from fetal to adult form. One is, therefore, justified in asking if there are any isozymic forms of enzymes that are specific for aged animals. Although there is adequate evidence that the relative amount of certain isozymes changes with age, this phenomenon appears to be relatively minor in a quantitative sense. Clearly, regulation of synthesis (or breakdown) is altered, but not in a dramatic way. Early work by Singh and Kanungo (1968) on isozymes of lactic dehydrogenase in several tissues unfortunately dealt with rats only up to 96 weeks of age. The authors reported a decrease in muscle type of lactic dehydrogenase subunits with age, but no increase in heart type. Later work on malate dehydrogenase and tyrosine aminotransferase considered animals only to about 70 weeks of age and should not be considered "aging." Patnaik (1979) found only B-type cytoplasmic alanine aminotransferase in the liver of 100-week-old rats, whereas at 52 weeks, both the immature A and B types are found. Reiss and Rothstein (1975) showed a considerable alteration in the pattern of isozymes of isocitrate lyase in aged versus young *T. aceti*. Wang and Mays (1978) noted quantitative differences in the monomeric, dimeric, and tetrameric forms of G6PD in young and old rats. They also noted sex and strain differences (Wang and Mays, 1977).

D. Changed Susceptibility of Proteins to *in Vitro* Proteolysis

Soluble liver proteins from young and old rats (4–6 versus 18–27 months) were treated *in vitro* with a number of proteolytic enzymes (papain, trypsin, pepsin, pronase and cathepsin D, and lysosomal extract) from rat liver. The lysosomal extract acted about 20% faster on the young preparations (Wiederanders *et al.*, 1978). Individual proteases showed little difference. The results could be misleading. For example, some protein particularly subject to lysosomal enzyme attack could be present in the young preparations. A somewhat similar experiment was conducted with early- versus late-passage WI-38 cells. Bradley *et al.* (1975) found an increased susceptibility of cell protein extracts (to pronase and trypsin) only at the last population doubling. Thus, proteins in early phase III cells were not different, in this respect, from phase II cells. The meaning of experiments with such crude preparations is obscure.

E. Comment

At this time, it is clear that altered enzymes may be formed in old animals because of postsynthetic changes. The finding that the changes are conformational in nematode enolase has yet to be extended to altered enzymes from higher animals. It also remains to be proved that slowed protein turnover is the underlying mechanism. No firm evidence has been provided that supports the idea of errors, but much strong evidence has been reported against it. The time has come when without new concrete support for the idea, errors should no longer be a major focus in considering the effect of aging on proteins.

In tissue culture, altered proteins, at least in part and perhaps in total, appear to be caused by cytoplasmic factors. A reasonable amount of evidence, when taken together, points in this direction although individual reports tend to be fragmentary. Thus, the situation may be different *in vitro* and *in vivo*. It is too early to tell.

Beyond obvious experiments dealing with changes in structure, it is becoming more and more important that investigations of altered proteins move to studies of regulatory mechanisms that control protein synthesis and degradation. It hardly needs to be noted that these are areas that ultimately involve gene expression.

REFERENCES

Alekseev, S. B., Mamaev, V. B., Stepanova, L. G., Kalinina, L. I., and Andzhaparidze, O. G. (1979). *Biokhimiya* **44**, 1251–1255.
Anderson, P. J. (1976). *Can. J. Biochem.* **54**, 194–196.
Barrows, C. H., and Kokkonen, G. C. (1980). *Age* **3**, 53–58.
Barrows, C. H., Roeder, L. M., and Olewine, D. A. (1962). *J. Gerontol.* **17**, 148–150.
Beauchene, R. W., Roeder, L. M., and Barrows, C. H. (1967). *J. Gerontol.* **22**, 318–324.
Bradley, M. O., Dice, J. F., Hayflick, H. L., and Schimke, R. I. (1975). *Exp. Cell Res.* **96**, 103–112.

Buchanan, J. H., and Stevens, A. (1978). *Mech. Ageing Dev.* **7,** 321–334.
Buchanan, J. H., Bunn, C. L., Lappin, R. I., and Stevens, A. (1980). *Mech. Ageing Dev.* **12,** 339–353.
Burrows, R. B., and Davison, P. F. (1980). *Mech. Ageing Dev.* **13,** 307–317.
Butzow, J. J., McCool, M. G., and Eichorn, G. L. (1981). *Mech. Ageing Dev.* **15,** 203–216.
Capasso, J. M., and Zimmerman, J. A. (1980). *Exp. Gerontol.* **15,** 161–165.
Chappel, A., Hoogenraad, N. J., and Holmes, R. S. (1978). *Biochem. J.* **175,** 377–382.
Chetsanga, C. J., and Liskiwskyi, M. (1977). *Int. J. Biochem.* **8,** 757–756.
Coniglio, J. J., Lium D. S. H., and Richardson, A. (1979). *Mech. Ageing Dev.* **11,** 77–90.
Danner, D. B., Schneider, E. L., and Pitha, J. (1978). *Exp. Cell Res.* **114,** 63–67.
Danot, G. M., Gershon, H., and Gershon, D. (1975). *Mech. Ageing Dev.* **4,** 289–299.
Dreyfus, J. C., Rubinson, H., Schapira, F., Weber, A., Marie, J., and Kahn, A. (1977). *Gerontology* **23,** 211–218.
Duncan, M. R., Dell'Orco, R. T., and Guthrie, P. L. (1977). *J. Cell Physiol.* **93,** 49–56.
Edelmann, P., and Gallant, J. (1977). *Proc. Natl. Acad. Sci. U.S.A.* **74,** 3396–3398.
Ekstrom, R. E., Liu, D. S. H., and Richardson, A. (1980). *Gerontology* **26,** 121–128.
Evans, C. H. (1976). *Differentiation* **5,** 101–105.
Evans, C. H. (1977). *Exp. Gerontol.* **12,** 169–171.
Finch, C. E. (1972). *Exp. Gerontol.* **7,** 53–67.
Foote, R. S., and Stulberg, M. P. (1980). *Mech. Ageing Dev.* **13,** 93–104.
Fornaini, G., Leoncini, G., Segni, P., Calabria, G. A., and Dacka, M. (1969). *Eur. J. Biochem.* **7,** 214–222.
Fornaini, G., Magnani, M., Dacka, M., Bossu, M., and Stocchi, V. (1978). *Mech. Ageing Dev.* **8,** 249–256.
Fry, M., and Weisman-Shomer, P. (1976). *Biochemistry* **15,** 4319–4329.
Fujii, H., Krietsch, W. K. G., and Yoshida, A. (1980). *J. Biol. Chem.* **255,** 6421–6423.
Gallant, J., and Palmer, L. (1979). *Mech. Ageing Dev.* **10,** 27–38.
Gallant, J. A., and Prothero, J. (1980). *J. Theor. Biol.* **83,** 561–578.
Gershon, D. (1970). *Exp. Gerontol.* **5,** 7–12.
Gershon, H., and Gershon, D. (1970). *Nature (London)* **227,** 1214–1217.
Gershon, H., and Gershon, D. (1973a). *Proc. Natl. Acad. Sci. U.S.A.* **70,** 909–913.
Gershon, H., and Gershon, D. (1973b). *Mech. Ageing Dev.* **2,** 33–41.
Gershon, D., and Gershon, H. (1976). *Gerontology* **22,** 212–219.
Ghai, R. D., and Lane, R. S. (1980). *Fed. Proc. Fed. Am. Soc. Exp. Biol.* **39,** 1683.
Gockel, S. F., and Lebherz, H. G. (1981). *J. Biol. Chem.* **256,** 3877–3883.
Gold, G., and Widnell, C. C. (1974). *Biochim. Biophys. Acta* **334,** 75–85.
Goldstein, S., and Moerman, E. J. (1975). *Nature (London)* **255,** 159.
Goldstein, S., and Moerman, E. J. (1976). *Interdiscip. Top. Gerontol.* **10,** 24–43.
Goldstein, S., and Moerman, E. J. (1978). *Prog. Clin. Biol. Res.* **21,** 217–228.
Goren, P., Reznick, A. Z., Reiss, U., and Gershon, D. (1977). *FEBS Lett.* **84,** 83–86.
Grinna, L. S., and Barber, A. A. (1972). *Biochim. Biophys. Acta* **288,** 347–353.
Grinna, L. S., and Barber, A. A. (1975). *Exp. Gerontol.* **10,** 319–323.
Gupta, S. K., and Rothstein, M. (1976a). *Arch. Biochem. Biophys.* **174,** 333–338.
Gupta, S. K., and Rothstein, M. (1976b). *Biochim. Biophys. Acta* **445,** 632–644.
Gurtler, B., and Leuthardt, F. (1970). *Helv. Chim. Acta* **53,** 654–658.
Hardwick, J., Hsieh, W. M., Liu, D. S. H., and Richardson, A. (1981). *Biochim. Biophys. Acta* **652,** 204–217.
Hargrove, J. L., Diesterhaft, M., Nogucki, T., and Granner, D. K. (1980). *J. Biol. Chem.* **255,** 71–78.

Harley, C. B., Pollard, J. W., Chamberlain, J. W., Stanners, C. P., and Goldstein, S. (1980). *Proc. Natl. Acad. Sci. U.S.A.* **77,** 1885–1889.

Helfman, P. M., and Bada, J. L. (1975). *Proc. Natl. Acad. Sci. U.S.A.* **72,** 2891–2894.

Helfman, P. M., and Bada, J. C. (1976). *Nature (London)* **262,** 279–281.

Helfman, P. M., and Price, P. A. (1974). *Exp. Gerontol.* **9,** 209–214.

Helfman, P. M., Bada, J. L., and Shov, M. Y. (1977). *Gerontology* **23,** 419–425.

Holland, J. J., Kohne, D., and Doyle, M. V. (1973). *Nature (London)* **245,** 316–319.

Holliday, R., and Tarrant, G. M. (1972). *Nature (London)* **238,** 26–30.

Holliday, R., Porterfield, J. S., and Gibbs, D. D. (1974). *Nature (London)* **248,** 762–763.

Hosoi, K., Kobayashi, S., and Ucha, T. (1978). *Biochim. Biophys. Acta* **543,** 283–292.

Houben, A., and Remacle, J. (1978). *Nature (London)* **275,** 59–60.

Houben, A., Houbion, A., and Remacle, J. (1980). *Exp. Gerontol.* **15,** 629–631.

Hunter, L. (1980). *Anal. Biochem.* **101,** 78–87.

Jacobus, S., and Gershon, D. (1980). *Mech. Ageing Dev.* **13,** 311–322.

Jamil, T. P., Swallow, D. M., and Povey, S. (1978). *Biochem. Genet.* **16,** 1219–1232.

Joenje, H., Frants, R. R., Arwert, Fré., and Eriksson, A. W. (1978). *Mech. Ageing Dev.* **8,** 265–267.

Kadlubowski, M., and Agutter, P. S. (1977). *Brit. J. Haematol.* **37,** 111–125.

Kahn, A., Boivin, P., Vibert, M., Cottreau, D., and Dreyfus, J. C. (1974). *Biochimie* **56,** 1395–1407.

Kahn, A., Bertrand, O., Cottreau, D., Boivin, P., and Dreyfus, J. C. (1976). *Biochim. Biophys. Acta* **445,** 537–548.

Kahn, A., Gillouzo, A., Leibovitch, M. P., Cottreau, D., Bourel, M., and Dreyfus, J. C. (1977a). *Biochem. Biophys. Res. Commun.* **77,** 760–766.

Kahn, A., Guillouzo, A., Cottreau, D., Marie, J., Bourel, M., Boivin, P., and Dreyfus, J. C. (1977b). *Gerontology* **23,** 174–184.

Kahn, A., Bertrand, O., Cottreau, D., Boivin, P., and Dreyfus, J. C. (1977c). *Biochem. Biophys. Res. Commun.* **77,** 65–72.

Kahn, A., Boyer, C., Cottreau, D., Marie, J., and Boivin, P. (1977d). *Pediatr. Res.* **11,** 271–276.

Kahn, A., North, M. L., Cottreau, D., Giron, G., and Lang, M. J. (1978a). *Hum. Genet.* **43,** 85–89.

Kahn, A., Vibert, M., Cottreau, D., Skala, H., and Dreyfus, J. C. (1978b) *Biochim. Biophys. Acta* **526,** 318–327.

Kerr, J. S., and Frankel, H. M. (1976). *Int. J. Biochem.* **7,** 455–460.

Kirkwood, T. B. L. (1977). *Nature (London)* **270,** 301–304.

Kirkwood, T. B. L. (1980). *J. Theor. Biol.* **82,** 363–382.

Kramps, J. A., DeJong, W. W., Wollensak, J., and Hoenders, H. J. (1978). *Biochim. Biophys. Acta* **533,** 487–495.

Kurtz, D. I. (1975). *Biochim. Biophys. Acta* **407,** 479–484.

Lavie, L., Resnick, A. Z., and Gershon, D. (1982). *Biochem. J.* **202,** 47–51.

Lindena, J., Friedel, R., Rapp, K., Sommerfeld, U., Trautschold, I., and Deerberg, F. (1980). *Mech. Ageing Dev.* **14,** 379–407.

Linn, S., Kairis, M., and Holliday, R. (1976). *Proc. Natl. Acad. Sci. U.S.A.* **8,** 2812–2822.

Magnani, M., Cucchiarini, L., Dacha, M., and Gornaini, G. (1980). *Age* **3,** 39–41.

Mariotti, D., and Ruscitto, R. (1977). *Biochim. Biophys. Acta* **475,** 96–102.

Masters, P. M., Bada, J. L., and Zigler, J. S. Jr. (1977). *Nature (London)* **268,** 71–73.

Masters, P. M., Bada, J. L., and Zigler, J. S. Jr. (1978). *Proc. Natl. Acad. Sci. U.S.A.* **75,** 1204–1208.

McKerrow, J. H. (1979). *Mech. Ageing Dev.* **10,** 371–377.

Medvedev, Z., and Medvedeva, M. M. (1978). *Biochem. Soc. Trans.* **6,** 610.
Mennecier, F., Weber, A., Tudbary, C., and Dreyfus, J. C. (1979). *Biochimie* **61,** 79–85.
Menninger, J. R. (1977). *Mech. Age. Develop.* **6,** 131–142.
Mori, N., Mizuno, D., and Goto, S. (1979). *Mech. Age. Develop.* **10,** 379–398.
Obenrader, M. F., and Prouty, N. F. (1977). *J. Biol. Chem.* **252,** 2866–2872.
O'Bryan, D., and Lowenstein, L. M. (1974). *Biochim. Biophys. Acta* **339,** 1–9.
Orgel, L. E. (1963). *Proc. Nat. Acad. Sci. (U.S.A)* **49,** 517–521.
Orgel, L. E. (1970). *Proc. Nat. Acad. Sci. (U.S.A.)* **67,** 1476.
Orgel, L. E. (1973). *Nature* **243,** 441–445.
Ogrodnik, J. P., Wulf, J. H., and Cutler, R. G. (1975). *Exp. Geront.* **10,** 119–136.
Ohrloff, C., Bensch, J., Jaeger, M., and Hockwin, O. (1980). *Exp. Eye Res.* **31,** 573–579.
Parker, J., Pollard, J. W., Friesen, J. D., and Stanners, C. P. (1978). *Proc. Nat. Acad. Sci. (U.S.A.)* **75,** 1091–1095.
Paterniti, J. R., Jr., Lin, C. I. P., and Beattie, D. S. (1978). *Arch. Biochem. Biophys.* **191,** 792–797.   ✦
Paterniti, J. R., Jr., Lin, C. I. P., and Beattie, D. S. (1980). *Mech. Age. Develop.* **12,** 81–91.
Patnaik, S. K. (1979). *Cell Biol. Int. Reports* **3,** 607–614.
Pendergrass, W. R., Martin, G. M., and Bornstein, P. (1976). *J. Cell Physiol.* **87,** 3–14.
Petell, J. K., and Lebherz, H. G. (1979). *J. Biol. Chem.* **254,** 8179–8184.
Pfeffer, S. R., and Swislocki, N. I. (1976). *Arch. Biochem. Biophys.* **177,** 117–122.
Pitha, J., Adams, R., and Pitha, P. M. (1974). *J. Cell Physiol.* **83,** 211–218.
Pitha, J., Stork, E., and Wimmer, E. (1975). *Exp. Cell Res.* **94,** 310–314.
Pontremoli, S., Melloni, E., Salamino, F., Sparatore, B., Michetti, J., and Horecker, B. L. (1979). *Proc. Nat. Acad. Sci. (U.S.A.)* **76,** 6323–6325.
Pontremoli, S., Melloni, E., Salamino, F., Sparatore, B., Michetti, M., and Horecker, B. L. (1980). *Arch. Biochem. Biophys.* **203,** 390–394.
Racz, O., Biszku, E., and Straub, F. B. (1979). *Eur. J. Biochem.* **96,** 503–507.
Reiss, U., and Gershon, D. (1976a). *Eur. J. Biochem.* **63,** 617–623.
Reiss, U., and Gershon, D. (1976b). *Biochem. Biophys. Res. Commun.* **73,** 255–262.
Reiss, U., and Rothstein, M. (1974). *Biochem. Biophys. Res. Commun.* **61,** 1012–1016.
Reiss, U., and Rothstein, M. (1975). *J. Biol. Chem.* **250,** 826–830.
Reznick, A. Z., and Gershon, D. (1977). *Mech. Ageing Dev.* **6,** 345–353.
Reznick, A. Z., Lavie, L., Gershon, H. E., and Gershon, D. (1981). *FEBS Lett.* **128,** 221–224.
Robinson, A. B., McKerrow, J. H., and Cary, P. (1970). *Proc. Natl. Acad. Sci. U.S.A.* **66,** 753–757.
Rosenberger, R. F., Goskett, G., and Holliday, R. (1980). *Mech. Ageing Dev.* **13,** 247–252.
Ross, M. H. (1969). *J. Nutr.* **97** (Supplement 1), 563–601.
Ross, M. H., and Ely, J. O. (1954). *J. Franklin Inst.* **258,** 63–66.
Rothstein, M. (1975). *Mech. Ageing Dev.* **4,** 325–338.
Rothstein, M. (1977). *Mech. Ageing Dev.* **6,** 241–257.
Rothstein, M. (1979). *Mech. Ageing Dev.* **9,** 197–202.
Rothstein, M., Coppens, M., and Sharma, H. K. (1980). *Biochim. Biophys. Acta* **614,** 591–600.
Rubinson, H., Kahn, A., Boivin, P., Schapira, F., Gregori, C., and Dreyfus, J. C. (1976). *Gerontology* **22,** 438–448.
Savageau, M. A., and Freter, R. R. (1979). *Biochemistry* **18,** 3486–3493.
Schapira, F., Weber, A., Guillouzo, C., and Dreyfus, J. C. (1978). *In* "Liver and Ageing - 1978" (K. Kitani, ed.), pp. 47–54. Elsevier/North Holland, New York.
Schlenska, G. K., and Kleine, T. O. (1980). *Mech. Ageing Dev.* **13,** 143–154.
Schmucker, D. L., and Wang, R. K. (1980). *Exp. Gerontol.* **15,** 321–329.
Schofield, J. D. (1980). *Exp. Gerontol.* **15,** 533–538.

Schofield, J. D., and Hadfield, J. M. (1978). *Exp. Gerontol.* **13**, 147–157.

Shakespeare, V., and Buchanan, J. H. (1978). *J. Cell Physiol.* **94**, 105–116.

Sharma, H. K., and Rothstein, M. (1978a). *Biochemistry* **17**, 2869–2876.

Sharma, H. K., and Rothstein, M. (1978b). *Mech. Ageing Dev.* **8**, 341–354.

Sharma, H. K., and Rothstein, M. (1980). *Proc. Natl. Acad. Sci. U.S.A.* **77**, 5865–5868.

Sharma, H. K., Gupta, S. K., and Rothstein, M. (1976). *Arch. Biochem. Biophys.* **174**, 324–332.

Sharma, H. K., Prasanna, H. R., and Rothstein, M. (1980). *J. Biol. Chem.* **255**, 5043–5050.

Singh, S. N., and Kanungo, M. S. (1968). *J. Biol. Chem.* **243**, 4526–4529.

Skala-Rubinson, H., Vibert, M., and Dreyfus, J. C. (1976). *Biochim. Biophys. Acta* **70**, 385–390.

Somville, M., and Remacle, J. (1980). *Arch. Int. Physiol. Biochem.* **88**, B99–B100.

Steinhagen-Thiessen, E., and Hilz, H. (1976). *Mech. Ageing Dev.* **5**, 447–457.

Steinhagen-Thiessen, E., and Hilz, H. (1979). *Gerontology* **25**, 132–135.

Tomkins, G. A., Stanbridge, E. J., and Hayflick, L. (1974). *Proc. Soc. Exp. Biol. Med.* **146**, 385–390.

Van Kleef, F. S. M., Willems-Thijssen, W., and Hoenders, H. J. (1976). *Eur. J. Biochem.* **66**, 477–483.

Viceps-Madore, D., and Cristofalo, V. J. (1978). *Mech. Ageing Dev.* **8**, 43–50.

Wang, R. K., and Mays, L. L. (1977). *Exp. Gerontol.* **12**, 117–124.

Wang, R. K. J., and Mays, L. L. (1978). *Age* **1**, 2–7.

Weber, A., Gregori, C., and Schapira, F. (1976). *Biochim. Biophys. Acta* **444**, 810–815.

Weber, A., Guguen-Guillouzo, C., Szajnert, M. F., Beck, G., and Schapira, F. (1980a). *Gerontology* **26**, 9–15.

Weber, A., Szajnert, M. F., and Beck, G. (1980b). *Biochim. Biophys. Acta* **631**, 412–419.

Wiederanders, B., Ansorge, S., Bohley, P., Kirschke, H., Langner, J., and Hanson, H. (1978). *Mech. Ageing Dev.* **8**, 355–362.

Wilson, P. D. (1972). *Gerontologia* **18**, 36–54.

Wilson, P. D. (1973). *Gerontologia* **19**, 79–125.

Wilson, P. D. (1978). *Gerontology* **24**, 348–357.

Wilson, P. D., and Franks, L. M. (1971). *Gerontologia* **17**, 16–32.

Wojtyk, R. I., and Goldstein, S. (1980). *J. Cell Physiol.* **103**, 299–303.

Wright, H. T. (1980). *FEBS Lett.* **118**, 165–171.

Wulf, J. H., and Cutler, R. G. (1975). *Exp. Gerontol.* **10**, 101–117.

Yoshida, A. (1973). *Science* **179**, 532–537.

Yutani, K., Ogasahara, K. Sugino, Y., and Matsushiro, A. (1977). *Nature (London)* **267**, 274–275.

Zeelon, P., Gershon, H., and Gershon, D. (1973). *Biochemistry* **12**, 1743–1750.

Zigler, J. S. Jr. (1978). *Exp. Eye Res.* **26**, 537–546.

Zorzoli, A. (1955). *J. Gerontol.* **10**, 156–164.

Chapter **10**

Hormones

I. OVERVIEW

The process of neuorendocrine hormone action can be viewed as originating from the hypothalamus in response to some stimulus, perhaps neural, chemical, or physical. Specific releasing factors or inhibitors from this tissue invoke a response in the pituitary gland which, in turn, may release adrenocorticotropic hormone (ACTH), thyrotropic hormone (THS), follicle-stimulating hormone (FSH), luteinizing hormone (LH), prolactin (PRL), or growth hormone (GH). These substances stimulate hormone production in target glands or tissues such as adrenal cortex, thyroid, testes, and ovary, or react directly with a target tissue to elicit metabolic responses. Any age-related change in the response of the hypothalamus, the pituitary, the target organs, or the tissue which is the final recipient of a given signal, could have a profound effect on metabolism. Obviously, changes at the top of this sequence would have far-reaching effects even if the glands and tissues at the lower levels retain their ability for normal response. As far as we know, there does not seem to be a generalized decline during aging, in the response of the hypothalamus or pituitary, although changes related to specific stimuli may occur.

In addition to sequences of events generated by neuroendocrine stimuli, there are other reactions mediated by hormones with nonneural function, which are independent of or under less direct pituitary control. Of these, epinephrine, glucagon, and insulin have been studied with respect to aging.

It is obvious that study of age-related changes in hormone metabolism can be directed toward several loci. One common approach is the investigation of changes with age of receptors in target tissues. Binding decreases in about 70% of the tissues studied, but it is rare that receptor affinity is altered. In short, in these cases receptor numbers decrease. Other approaches involve the determination of basal levels of various hormones and the magnitude of the response of target tissues to various stimuli during aging. Although these determinations provide needed background information, they tell us little about the relationship between changes or lack of changes and aging.

There is little information available that deals with the metabolism either of hormones or of their receptors. Outside of a few studies of hormone-related enzymes, hormone turnover, and a single measurement of receptor turnover, these aspects of aging metabolism remain unexplored. This situation is easily understood because hormones and their actions represent a most complex and difficult area for study. Not only are the compounds active in miniscule amounts, but the response of different tissues to the same hormone may differ. Moreover, there may be hormone interaction resulting in inhibition or stimulation of production of other hormones. Since many of these functions are speculative in the first place, the additional factor of age adds a large burden for the investigator to overcome. It is thus not surprising that investigations to date have been mostly in the nature of baseline studies. Even these carry formidable difficulties. For example, study of sex hormones has the problem of semi-independent "aging" in the tissues involved in reproduction, particularly in females. Moreover, perusal of the data so far available shows that there is often a substantial variation of hormone levels among aging individuals even in the same age-group. Thus, the reported basal or stimulated levels of hormones often represent an average of wide-ranging normal values. Another problem is that diurnal variations of a substantial nature may exist, and these may change with age. Such is the case for TSH and corticosterone in rats. Other complicating features include the proposed ability of some of the hormones to function as feedback inhibitors, the effect of one hormone on the action of another, and even the effect of basal levels of certain hormones (e.g., androgens in rats) on the number of receptors. An overall, but often ignored factor may be the changed metabolic needs of old versus young animals, which may quite naturally modulate the hormone pattern. Such changes could readily be taken as an age-incurred defect, although they are not really a pathological consequence of the aging process. Even the handling of animals during experimental procedures can affect hormone levels and may do so differently with respect to age. These considerations may or may not explain some of the contradictory results obtained by different investigators seeking to delineate changes in hormone levels in aging organisms, but they certainly complicate interpretation of the results.

From the point of view of researchers, one of the more unfortunate findings is that the effect of aging on the levels of certain hormones varies from species to species. Thus, there is no change in the basal LH level of aged mice, but there is a decrease in aging male (but not female) rats. The degree of stimulation of LH levels by LHRH shows no change with age in mice, but a decrement in rats. Basal levels of testosterone (T) are not changed in aged mice, but they decrease in aged rats. Even different parts of some organs may behave differently. For example, various areas of the brain give different results with respect to β-adrenergic receptors and response to noradrenaline. Similarly, dorsolateral and ventral prostate tissues have quite different distributions of testosterone receptors. In certain cases, sex and the strain of animal may lead to differences in response: old Sprague-Dawley rats show a considerably delayed recovery of glucokinase levels after fasting, but Wistar rats do not. All of these factors suggest that there is no specific aging lesion in hormone metabolism that is applicable to all animals, unless it is well hidden and manifests itself in different ways in different animals. There is a strong implication that the relationship of hormone changes to aging cannot be studied in single model systems (e.g., rats or mice) and then broadly applied.

As mentioned, in a majority of cases, receptor numbers decline with age, thus reducing the response of the target tissue. However, there are also many examples in which receptor numbers are unchanged and even a few in which they increase. Thus, glucocorticoid receptors decrease with age in rat muscle, brain, and adipocytes, but not in rat liver. In a similar vein, β-adrenergic receptors decrease in adipocytes but not in cardiac tissue. The nature of receptor changes with age in brain seems to depend on the area being examined.

In considering the reason for a lowered response in aging, when it occurs, there are a number of indications that, particularly in the case of hormones which stimulate production of cAMP, the problem lies not in the loss of receptors but in the coupling mechanism through which hormone binding elicits the response. Reduced stimulation of lipolysis in adipocytes by catecholamines appears to be another example in which the control mechanism lies outside of the number of receptors.

Although it cannot be said that a clear picture of the effect of aging on hormone metabolism has taken shape, nonetheless, a good start has been made in laying down reasonably consistent baseline information. Progress of a more sophisticated nature must await advances in our basic understanding of hormone action which can in turn, be applied to studies of aging.

II. HORMONAL INTERACTION

The effect of aging on hormonal interaction surely must be of considerable importance. However, the overall relationships of hormones, one with another, are extremely complex and not well defined. To add the complexities of aging to this unsettled situation leaves us more with speculation than with fact. A number of proposals have been put forth to explain various observations, although few attempts have been made to develop experimentally derived correlations. For example, deKloet and McEwen (1976) propose a relationship between glucocorticoids and hippocampal and other limbic neurons, which would provide a connection between neural and endocrine changes. Although there are no direct experiments showing that brain and endocrine alterations during aging are quantitatively related, there have been proposals that brain–endocrine interactions may be pacemakers for aging (Finch, 1976). According to this idea, age-related endocrine or neural changes could cause changes in function that would lead to a series of increasing physiological imbalances and ultimately to physical deterioration. One such relationship is pointed up by Landfield *et al.* (1978). They assessed measurable, age-related changes in hippocampal structure in Fischer rats against plasma concentrations of adrenocorticoids and found a quantitative relationship between hippocampal pathology and adrenal activity. Somewhat in this vein, Tang and Phillips (1978) proposed that the increase they observed in the basal level of ACTH in aging rats ($2\frac{1}{2}$–26 months) might be correlated with degenerative changes in the adrenal cortex.

III. GLUCOCORTICOIDS

Glucocorticoids, which are produced in the adrenal cortex, have been implicated in several metabolic functions, including the induction of gluconeogenic and other enzymes in liver, lipolysis in adipocytes, and estrogen action on uterine tissues. The hormone is bound to a cytoplasmic receptor, which is then translocated to the nucleus where there is an appropriate synthesis of mRNA. One would thus expect rather substantial changes in the response of tissues to corticosteroids during development and growth. Indeed, such changes have been noted. However, from the point of view of aging, we are concerned with events subsequent to maturity. In this period, changes tend to be less substantial. In fact, basal peripheral plasma corticosterone levels do not change in the Sprague-Dawley rat ($2\frac{1}{2}$–26 months), and stress levels of the hormone are unaffected by age (Tang and Phillips, 1978). However, the usual adaptive increase in the hormone brought about by starvation is reduced with age (Britton *et al.* 1975; Adelman, 1975; Sartin *et al.*, 1980). Adrenal response to insulin, however, is reported to be unchanged in elderly men (Muggeo *et al.*, 1975).

A decreased response to glucocorticoids, where it occurs, seems to result from a reduction in the number of receptor sites rather than a change in their affinity, although the situation may be different in liver. Whether or not age-related changes occur appears to depend on the type of tissue. Thus, in the rat, some tissues show a decline in receptor number with age, whereas others do not. Differences between early mature, and senescent periods are exemplified in the study of Roth (1974), who showed that the numbers of cytosolic high-affinity binding sites for cortisol and dexamethasone decrease substantially in skeletal muscle, brain, and adipose tissue in adrenalectomized Sprague-Dawley male rats between 2 and 12 months of age, whereas in liver and prostate, the receptor numbers increase. Results with male Wistar rats between 13 and 25 months were similar, showing a decline in receptors (dexamethasone binding) in skeletal muscle, brain cerebral hemispheres, and adipose tissue. Unlike the earlier ages, in this later period, binding in the prostate declined, and remained unchanged in liver.

Other tissues also showed substantial reductions in glucocorticoid binding sites. Roth (1975) observed that splenic leukocytes from old male Wistar rats (24–26 months) showed a 57% reduction in cytosolic binding sites for cortisol compared to adult animals (12–14 months). Subsequently, Roth (1976) found similar reductions (55–65%) in the cytoplasmic receptor concentration of glucocorticoid binding sites in cortical neuronal perikarya using similar age-groups of Sprague-Dawley rats. Adrenalectomy did not affect the results, suggesting that endogenous steroids did not reduce the number of available receptors and, therefore, were present at best, only at very low levels. [In this regard, adrenalectomy has been reported to double the number of cytoplasmic sites available in liver (Liu and Webb, 1979). On the other hand, glucocorticoid administration decreased cytoplasmic receptors, which apparently translocated to the nucleus]. There were no obvious age-related differences in the physical or chemical properties of the neuronal preparations, and with age, the binding sites did not appear to be distributed differently between cytoplasm and nucleus.

More detailed studies of the effect of age on glucocorticoids were carried out on rat adipocytes (Wistar and Sprague-Dawley) by Roth and Livingston (1976). In these cells, glucocorticoid hormones stimulate lipolysis and inhibit glucose transport. The epididymal fat adipocytes represent a nonproliferating cell population in mature animals, so that the age of the animal is reflected in the age of the cells. Moreover, the cells represent a homogeneous population—an advantage which may not always be reflected in sampling other tissues. However, one must be careful to recognize that cell size influences metabolic rate in adipocytes from rats and humans. Cell size does not increase after maturity in Wistar rats but increases continually in Sprague-Dawleys.

Roth and Livingston (1976) observed that in Wistar rats, the concentration of receptors in adipocytes decreases with age. The glucocorticoid inhibition of glucose oxidation also decreased in both strains of rats. For example, the inhibition caused by dexamethasone is reduced from 22–25% in mature adipocytes (12–13 months) to 5–8% in senescent (24–26 months) cells. It should be noted that the animals were not adrenalectomized. In a subsequent paper, Roth and Livingston (1979), using only Wistar rats so that cell size would not be a factor, found that glucose transport (measured directly and with analogs) is no longer inhibited by dexamethasone after 12 months of age. Nonetheless, the hormone continues to inhibit glucose oxidation at least until 26 months of age, although the magnitude of the inhibition is progressively reduced. Therefore, during aging, the hormone acts to inhibit glucose oxidation in some way other than by blocking glucose transport. Inhibition of glucose phosphorylation also decreases with age, in parallel with inhibition of glucose oxidation suggesting that the former process may be a controlling factor.

The situation with respect to liver seems to be different from that in other tissues so far investigated. As noted above, Roth (1974) reported that there was no change with age in receptor numbers in aging rat liver. Latham and Finch (1976) also found no age-related change in glucocorticoid receptor numbers in liver from C57BL/6J mice (8–12 months versus 28–32 months). Curiously, the latter authors found that binding was optimal at pH 6.9. Little binding was obtained at the higher pH (7.5–8.0) usually used with rat preparations. In contrast to the above reports, Singer *et al.* (1973) observed a decrease in cortisone receptors with age in postmortem human livers (30–40 versus 66–80 years).

With regard to receptor properties, Bolla (1980) recently reported that although there is indeed no change in the concentration of receptor proteins in rat liver cytosol, their affinity for both corticosterone and dexamethasone decreases with age. Petrovic and Markovic (1975) had earlier provided indirect evidence for a changed affinity in rat liver during early development. Roth (1974) also noted a significant decrease in the binding affinity of cortisol in the liver of rats at 2 versus 12 months (older animals were not considered). However, Latham and Finch (1976) found no such change in mouse liver. Perhaps problems arise from the fact that, at least in rats, there are three hepatic receptors (one major and two minor) for dexamethasone detectable after labeling *in vivo* (Liu and Webb, 1979), of which, two are translocated to the nucleus. There appear to be five corticosterone binding proteins (Cake and Litwack, 1976), only one of which is translocated to the nucleus. Bolla (1980) observed substantial changes with age in the amount of nonspecific binding proteins and pointed out that besides a possible alteration in the structure of the specific receptor, a change in the concentration of these nonspecific binders could be responsible for apparent age-related differences in affinity constants.

The loss of receptors, where it occurs, must lie either in decreased synthesis or increased degradation. Chang and Roth (1980) incorporated labeled amino acids into adipocyte proteins which bind to an affinity column (dexamethasone- and deoxycorticosterone-Sepharose). Evidence that these proteins are indeed receptors was found in the fact that use of cycloheximide or pretreatment of the adipocyte cytosol with unlabeled steroids resulted in a loss of the proteins that bound to the column. Using the dexamethasone-Sepharose affinity column to isolate the labeled receptors, Chang *et al.* (1981) found that the rate of synthesis in 23- to 26-month-old Wistar rats was about 50% of the value for 9- to 12-month-old animals. There was considerable scatter and even overlap in some of the experiments, but the results were shown to be statistically valid. Since the incorporation of labeled amino acids into total proteins was unchanged in the two age-groups, the results indicate a specific reduction in the synthesis of the receptor. It should be mentioned that the results do not exclude the possibility that inactive (altered) receptors are being synthesized. Such receptors would not be seen by the assay. However, there is no evidence that senescence results in synthesis of "faulty" proteins (Chapter 9), and, therefore, this seems an unlikely event unless the change is postsynthetic.

The effect of age on glucocorticoid metabolism has been reviewed by Roth (1979a; 1979b; 1980) and Chang and Roth (1979).

IV. ADRENOCORTICOTROPIC HORMONE (ACTH)

There has been little direct work on ACTH with respect to aging. Tang and Phillips (1978) reported that basal ACTH levels increase with age ($2\frac{1}{2}$–26 months) in rats. No change was found in response (increase of ACTH) to a short ether stress ($2\frac{1}{2}$ min), but a 15-min stress gave a lessened response in old animals. The authors suggest that the increased basal levels of ACTH in old animals may be necessary to assure sufficient corticosterone production from a degenerating adrenal. Riegle (1973) reported that young rats had an increased corticosterone concentration after ACTH injection compared to old animals. There was also a reduced response to stress (restraint) in aged Long Evans rats (22–28 months), which is suggested to be due to a decreased sensitivity of the adrenocortical control system. The results of Tang and Phillips (1978) supported the idea that the pituitary–adrenal relationship becomes less sensitive with age.

Cooper and Gregerman (1976) observed a decreased stimulation of adenylate cyclase by ACTH in rat fat cell ghosts from 1 to 24 months of age.

V. CATECHOLAMINES IN NONNEURAL TISSUES

The catecholamines epinephrine, noradrenaline, and dopamine all derive from tryptophan via dihydroxyphenylalanine, and all function by interaction with membrane receptors coupled to activation of adenylate cyclase with subsequent production of cAMP. Protein kinases are then stimulated to phosphorylate certain proteins with consequent metabolic effects. Of these three hormones, epinephrine and norepinephrine are associated with the adrenal medulla and their receptors are called β-adrenergic receptors. Norepinephrine is additionally associated with brain as a neurotransmitter along with dopamine and other compounds such as γ-aminobutyric acid (GABA) and 5-hydroxytryptamine (serotonin).

In general, research dealing with the effect of aging on catecholamines revolves around three approaches; levels of hormones, properties of receptors, and responsiveness of the target tissue to hormone treatment. For nonneural tissues, there has been relatively little work on enzymes involved in the synthesis or metabolism of catecholamines with respect to age. In any case, fluctuations in total enzyme levels per se are probably not of great import. It should be noted, however, that Reis *et al.* (1977) reported large increases in several adrenal enzymes involved in hormone metabolism in old rats and mice (Section VI,A). Most work has concentrated upon receptors and their response to stimulation in the form of enhanced production of cAMP. Studies typically involve binding of labeled agonists and antagonists and is subject to the difficulties of using crude preparations, proper dosage of binding substrate, level of nonspecific binding, and problems with the animals themselves—diurnal variation and reaction to the stress of being prepared for experiments.

As with corticosteroids, β-adrenergic receptor concentrations have been reported to decrease in a number of tissues during senescence. Schocken and Roth (1977), using $(-)[^3H]$dihydroalprenolol for the binding studies, measured the concentration of these receptors in crude membrane fractions of human male lymphocytes. This agent has been shown to be bound to β-adrenergic receptors, which are coupled to adenylate cyclase activation, and its use avoids the more generalized binding shown by labeled catecholamines (Williams *et al.*, 1976). Shocken and Roth (1977) found that the affinity of the receptors was unchanged with age, but the receptor concentration (per milligram of protein) declined by nearly one-half from 24 to 81 years of age. Receptor number per lymphocyte also correlated inversely with age. Cell yield, viability, and protein content did not change.

β-Adrenergic receptors have also been reported to decrease with age in rat adipocytes. Giudicelli and Pecquery (1978) studied the effect of age on adipocyte β-adrenergic receptors in Wistar rats. They observed a high-affinity binding of $-[^3H]$dihydroalprenolol to membranes of the cells. The number of receptors increased with the increase in cell size up to 8 months of age. Subsequent to maturity in the rats, cell size remained constant but the number of receptors was calculated (assuming the cells are round) to fall from 40,800 at 8 months to 7200 per cell by 30 months of age. The stimulation of adenylate cyclase by norepinephrine and isoproterenol followed a parallel pattern, falling with age. As noted with human lymphocytes (Schocken and Roth, 1977), the dissociation constant of the receptors was not changed with age. Although the above experiments suggest a direct relationship between loss of receptor number and loss of response, there is much evidence that suggests that other factors are involved. Lipolytic response to catecholamine also declined with age, disappearing by 30 months, but losing 40% of the response by 8 months when receptor number is at a maximum. Yu *et al.* (1980) also noted a decline in lipolysis after 6 months in adipocytes from Fischer rats. Since the lipolytic response of the cells to catecholamine declined much earlier than the loss of receptors, it appears that in aging rats, lipolysis is regulated at some stage beyond the receptor–adenylate cyclase response. In this respect, basal and fluoride-stimualted activity did not change with age, showing that the reduction in response was not due to the amount of adenylate cyclase which could potentially function to produce cAMP. Furthermore, Krall *et al.* (1981) found evidence from reconstitution experiments that the age-related decrease in the response of adenylate cyclase to guanyl nucleotide in human lymphocytes may be due to alterations in guanyl nucleotide coupling factors. In addition, O'Connor *et al.* (1981) recently provided evidence that the diminished contractile response of cardiac muscle to isoproterenol is a result of alterations in the catalytic subunit or coupling protein of the adenyl cyclase complex. A further indication that the loss of lipolytic stimulation is not related to changes in receptors was provided by Cooper and Gregerman (1976), who studied the effect of age on epinephrine-stimulated adenyalte cyclase in adipocytes from Wistar rats between 1 and 24 months of age. An age-related decline in response starting after 6 months, which parallels a fall in basal levels of the enzyme was observed. Before that time, the cells were still growing in size and showed an enhanced stimulation of cyclase activity by glucagon, epinephrine and fluoride. The authors found that the total available enzyme was reduced in old cells. They, therefore, suggested that a component of the enzyme complex other than receptors is lost with age.

Very recently, Dax *et al.* (1981) provided evidence that the problem of epinephrine-stimulated lipolysis lies beyond the receptor–enzyme complex altogether. They studied lipolysis in adipocytes from male Wistar rats and found no change in β-adrenergic receptor numbers from 2 to 24 months of age (50,000–90,000 sites/cell). Binding affinity ([^{3}H]dihydroalprenolol) and maximal stimulation of adenylate cyclase were also unchanged. Yet, there was an age-related decrease of maximal epinephrine-sensitive lipolysis (50%), even though the old adipocytes showed a greater sensitivity than mature cells to the hormone. The authors noted that because of the greater sensitivity, a submaximal dose of hormone will give an erroneous measure of the true ability of the old cells to respond. To explain the loss of receptors reported by Giudicelli and Pecquery (1978) (see above), Dax *et al.* (1981) point out several concerns: that old adipocytes are fragile and thus may have provided biased samples; that although in practical terms, there is considerable variation from preparation to preparation, no statistics were provided; that for the Scatchard plots, computations of best fit are needed along with more readings at concentrations under 10 nM or the affinities will give low values. Indeed, the values found by Giudicelli and Pecquery were 2–5 times less than those found by Dax *et al*. Although the arguments remain relevant, one should note that the old animals of Giudecelli and Pecquery were 30 months of age versus 24 months for those of Dax and co-workers. The latter group concluded that the age-related decrease in lipolysis was due to alteration in the lipolytic pathway distal to the receptor–adenylate cyclase complex.

As to the relationship of adipocyte size to lipolytic response, Gonzalez and DeMartinis (1978) also reported a decrease lipolytic response of adipocytes to epinephrine with age. However, although cell sizes do not vary greatly in Fischer rats between 12–28 months of age, when the authors carefully selected cells of the same average size from Fischer rats of 3, 12, and 28 months of age, they obtained a slightly *increased* stimulation of glycerol production at 28 versus 12 months. It is difficult to resolve this result with respect to the work of Cooper and Gregerman (1976), Yu *et al*. (1980), and Giudicelli and Pecquery (1978). These investigators took into account the mean cell sizes of their preparations but nonetheless, observed a decline in lipolysis with age.

It is of interest to note that Yu *et al*. (1980) found that the decline in stimulation of lipolysis by epinephrine, which began at 6 months in normally fed rats (Fischer), was considerably delayed in epididymal adipocytes and slowed in perirenal adipocytes in rats fed a restricted diet.

It is well documented that in cardiac tissue, responses to β-adrenergic stimuli are reduced in old animals. As one example, old rats show a diminished response to catecholamines (LaKatta *et al*., 1975). Recently, Guarnieri *et al*. (1980) compared the contractile response of myocardium to catecholamine stimulation in young (7–9 months) and old (22–25 months) Wistar rats. There were no baseline differences in β-adrenergic receptors, cAMP levels, or cAMP-dependent protein kinase activity. Under maximal stimulation by isoproterenol, the latter two parameters increased equally (about twofold) in both young and old animals. However, the force of contraction was 40% less in the latter. This effect appeared to be due to factors occurring after protein kinase activation of the system. Calcium influx may be involved, as a higher concentration of the metal increased the contractibility of the old preparations to the "young" value. Zitnik and Roth (1981) again observed that there was no significant change in the concentration or affinity of rat heart β-adrenergic receptors. The authors noted that the ratio of sialic acid to protein content was 33% higher in senescent preparations, suggesting that for these membranes, protein concentration may not be the best standard of measurement for receptor concentration.

β-Adrenergic receptors in liver provide a different picture. Kalish *et al.* (1977) studied epinephrine and glucagon-sensitive adenylate cyclases in livers of Fischer rats at 3, 12, and 24 months of age. Crude homogenates were utilized as it was found that washed particulates had a reduced cyclase activity and showed a reduced stimulation of activity with both hormones. With respect to age, there was a 1.4-fold increase in the basal level of the cyclase, and epinephrine caused a substantially increased activation of the enzyme at 24 versus 12 months in both sexes. The increased response to epinephrine in old animals does not result from an altered affinity of the receptors, but could be explained by an increase in the number of epinephrine-sensitive adenylate cyclase complexes or by an increase in the velocity of the enzyme. The authors noted that, on storage, both basal and stimulated cyclase activities showed an age-related decrease in stability, the 24-month-old preparations being the least stable. Addition of common protease inhibitors offered no protection from loss of activity. Thus, there is at least inferential evidence for membrane changes in the old animals that could alter the enzyme properties. From the information available, it appears that the epinephrine-stimulated increase of adenylate cyclase activity with age in liver differs from the situation in other issues, including certain brain tissues described in Section VI.

VI. CATECHOLAMINES IN NEURAL TISSUES

Clearly, age-related changes in neurotransmitter metabolism, whatever the cause, would have far-reaching consequences in the regulation of hypothalmic signals for production of pituitary hormones. In fact, it has been proposed that changes in the regulation of neurotransmitters are deeply involved in aging. For example, age-related changes in neurons or loss of relatively few cells in the hypothalamus could unbalance normal endocrine responses and result in a cascade effect on metabolism and physiology (Finch, 1976). Of course, we are still left with the question as to how the original changes which would affect neurotransmitter production or response, came about.

As to the substances involved, acetylcholine is the neurotransmitter for all cell sites except postganglionic sympathetic terminals. The latter utilize norepinephrine. Other transmitters include dopamine and serontonin and in the hypothalamus, perhaps histamine and GABA. Samorajski (1977) and Finch (1977) have provided useful reviews of neurotransmitters and aging.

A.   Enzyme Levels

Changes in neurotransmitter levels could be brought about by alterations in the amounts of the enzyme responsible for their synthesis. As an example, in humans, monoamine oxidase seems to increase with age, but enzyme levels of dihydroxyphenylalanine (DOPA) decarboxylase, tyrosine hydroxylase (rate-limiting for synthesis of dopamine and noradrenaline), and choline acetyltransferase were found to decline in different degrees, or not to change at all, depending on the part of the brain being considered. McGeer and McGeer (1975) interpreted these results as indicating that some areas of the brain age more than others. However, there are obviously problems posed by the use of postmortem material so that reported changes in enzymes and particularly in levels of small molecules (dopamine, norepinephrine) in human tissues are subject to some uncertainty.

Using Fischer rats (4 versus 26 months) and CB6F$_1$ mice (4 versus 28 months), Reis et al. (1977) examined several areas of young and old brain for age-related changes in enzyme levels. Tyrosine hydroxylase, aromatic L-amino acid decarboxylase, dopamine β-hydroxylase, phenylethanolamine-N-methyltransferase, and choline acetyltransferase were assayed in different brain regions and peripherally in adrenal medulla and sympathetic ganglia. In rats, there were small regional changes of activity in brain, both increases and decreases, but these amounted to less than 20%. Mice showed no change. McGeer et al. (1971a) had earlier reported a 35% decline of tyrosine hydroxylase in neostriatum of aged Wistar rats, but found no change in brain minus this tissue and cerebellum. In contrast to the results they obtained with brain tissue, Reis et al. (1977) found that peripheral catecholamine systems in adrenals of the aged rats and mice showed large increases (1.5- to 2.5-fold) in choline acetyltransferase, tyrosine hydroxylase, and aromatic L-amino acid decarboxylase activities, but not in dopamine β-hydroxylase. The increases in tyrosine hydroxylase and the decarboxylase were shown by immunotitration to be due to the presence of increased amounts of enzyme. Somewhat similar increases in enzymes were obtained with rat superior cervical ganglion. Thus, the relatively constant levels of the enzymes involved in synthesis of catecholamines and acetylcholine observed in aging rat brain were not reflected in the peripheral tissues. The increased capacity for biosynthesis in the latter regions does not, of course, necessarily mean that higher levels of neurotransmitters are normally produced in vivo.

Finch (1973) also reported that there was no change in hypothalamic or striatal levels of DOPA decarboxylase activity in aged C57BL/6J mice (12 versus 28 months). However, the author observed sharply reduced levels (30–40%) of dopamine in the striatum and a reduced conversion, *in vivo*, of labeled tyrosine and DOPA to catecholamines in the striatum, brain stem, hypothalamus, and cerebellum of the old mice. Norepinephrine levels were unchanged. Thus, there appears to be a selective effect of age, which varies both with cell population and with neurotransmitter. Turnover of norepinephrine in the hypothalamus and dopamine in the striatum were slowed sharply with age. The author cautions that changes in pool sizes were not taken into account in the turnover experiments. Simkins *et al.* (1977) also estimated the turnover of dopamine and norepinephrine, but in this case, they used male Wistar rats (3–4 versus 21 months). The authors followed depletion rates of the compounds after blocking synthesis by inhibition of tyrosine hydroxylase. Serotonin turnover was estimated by determining the rate of increase after blocking degradation by monoamine oxidase inhibition. Both the steady state concentration and the depletion rate for dopamine in old animals were lower in medial basal hypothalamus. There was no change in either measurement in the olfactory tubercle. For norepinephrine, both parameters were lower in hypothalamus.

Masuoka *et al.* (1979) reported that aging mice developed numerous large, fluorescent spots related to catecholamines rather than to lipofuscin (Chapter 3, Section III). They suggest that the spots may result from lesions in the handling of catecholamines by old tissue.

In summary, it appears that levels of enzymes involved in the metabolism of brain neurotransmitter are not dramatically altered with age. Although there are indications that the metabolism of these substances may be slowed with age, the data so far available are too sparse to enable one to draw any general conclusions, particularly given the possibilities for regional differences and problems deriving from external events— handling of the animals, possible changes in cell population, technical differences in handling tissues, and perhaps animal strains used.

B. cAMP Production

cAMP is produced by the activation of adenylate cyclase when an appropriate hormone is bound to its receptor. Thus, effects of age could involve not only hormone and receptor levels, but also the basal level of adenylate cyclase and the signal or coupling mechanism between receptor and enzyme. Fluoride is typically utilized to stimulate adenylate cyclase in a noncoupled manner, i.e., to determine the maximal amount of enzyme.

In general, the nonhormonal status of the system in brain tissue appears to be unaffected by age. That is, levels of cAMP and by and large, adenylate cyclase activity and fluoride-stimulated levels of the enzyme are unchanged. Response to hormones varies with the part of the brain being tested. For convenience, the results are summarized in Table 10.I. In more detail, the response of adenylate cyclase to norepinephrine and dopamine in rat cerebellum, hippocampus, cortex, and caudate tissue at 3 versus 24 months in Sprague-Dawley rats was examined by Walker and Walker (1973). Basal activity of adenylate cyclase was significantly higher in crude homogenates of caudate and cerebellum from old compared to young animals and essentially unchanged in cortex and hippocampus. Each tissue showed a different pattern of dose response, particularly to norepinephrine. In general, all of the old tissues showed generally small differences in adenylate cyclase activity with increasing amounts of hormone, whereas young tissues, except for the cortex, showed large changes. Since no age-related deficiency in adenylate cyclase was observed after stimulation by fluoride (except in caudate), the lack of response to noradrenaline in old tissue would seem to lie in a loss of receptors or in receptor–enzyme coupling.

Zimmerman and Berg (1974) reported a 75% decline in cAMP in the cerebral cortex of Fischer rats between 3 and 6 months of age. Subsequently Zimmerman and Berg (1975) observed no age-dependent changes in adenylate cyclase activity with or without fluoride up to 24 months of age. Therefore, insufficiency of the enzyme does not explain the 75% drop in cAMP content at early ages, which the authors had earlier reported. Moreover, phosphodiesterase, which perhaps might differentially hydrolyze cAMP in young and old animals, did not change with age. Other investigators have not observed a similar large drop in cAMP. Thus, Schmidt and Thornberry (1978) found no change in cAMP level with age. These authors pointed out a number of technical procedures that might account for the original findings. Puri and Volicer (1977) also examined the levels of cAMP, adenylate cyclase, and phosphodiesterase in striatum from CD-F rats of 4–30 months of age. They too, found no age-related differences in cAMP levels or in adenylate cyclase activity with or without fluoride stimulation. Phosphodiesterase decreased slightly at older ages. However, dopamine-stimulated production of cAMP was reduced at 12 and 24 months and disappeared by 30 months. The lowered values observed for cAMP were not caused by increased action of phosphodiesterase. Since, from the fluoride data, adequate amounts of adenylate cyclase are present, the problem must lie in the stimulatory mechanism. Perhaps there is a loss of receptor activity as suggested by the results of Jonec and Finch (1975), who found that synaptosomes prepared from hypothalamus and striatum of old mice (28–30 months versus 8

Table 10.I Adenylate Cyclase Activity in Various Regions of Brain[a,b]

Animal[c]	Age (months)	Brain area	Basal AC	Response to			Reference[d]
				F	NE	DA	
Rat (SD)	3 versus 24	Caudate	+	–	–	–	1
Rat (SD)	3 versus 24	Cerebellum	+	NC	–	–	1
Rat (W)	12 versus 24	Cerebellum	NC		–[e]		4
Rat (SD)	3 versus 24	Hippocampus	NC	NC	NC[f]	NC[g]	1
Rat (W)	12 versus 24	Hippocampus	NC		NC[h]		4
Rabbit	5 months versus 5 years	Anterior limbic cortex	NC		–	–	6
Rabbit	5 months versus 5 years	Frontal cortex	NC		–	–	6
Rat (SD)	3 versus 24	Cortex	NC	+	–		1
Rat (W)	12 versus 24	Cortex	NC		NC		4
Rat (F)	1–24	Cortex	NC	NC			5
Rat (SD)	2–3 versus 20–24	Striatum	NC			–	2
Rat (F)	4–30	Striatum	NC	NC		–	3
Rat (W)	12 versus 24	Striatum	NC			NC	4
Rabbit	5 months versus 5 years	Brain stem	NC			–	6
Rat (W)	12 versus 24	Hypothalamus	NC		NC		4
Rat (W)	12 versus 24	Hypothalamus	NC		NC		4
Rabbit	5 months versus 5 years	Hypothalamus	NC		–	–	6
Rat (SD)	2–3 versus 20–24	Nucleus accumbens, substantia nigra, tuberculur olfactorium	NC			–	2

[a] +, Increase with age; –, decrease with age; NC, no change with age.

[b] Abbreviations: AC, adenylate cyclase; F, fluoride; NE, norepinephrine; DA, dopamine.

[c] F, Fischer; SD, Sprague-Dawley; W, Wistar.

[d] Key to references: (1) Walker and Walker, 1973; (2) Govoni et al. 1977; (3) Puri and Volicer, 1977; (4) Schmidt and Thornberry, 1978; (5) Zimmerman and Berg, 1975; (6) Makman et al., 1979.

[e] Slight decrease, but over 50% lower than three-month-old animals.

[f] Old starts higher but was not stimulated by increased concentrations of NE.

[g] Old was slightly lower.

[h] Old was slightly increased.

months), took up reduced amounts of dopamine at low concentrations. In these experiments, norepinephrine uptake was unaffected by age in the hypothalamus, but was reduced significantly in the striatum, once again demonstrating that aging effects can vary both for tissue and type of receptor.

Schmidt and Thornberry (1978), using chopped tissue slices from the brain of male Wistar rats of 3, 12, and 24 months of age, also measured dopamine and norepinephrine-stimulated cAMP accumulation. At 12 versus 24 months, neither basal nor norepinephrine-stimulated levels showed much age-related difference in brain stem, cerebral cortex, hippocampus, hypothalamus, and cerebellum. Basal cGMP levels dropped slightly in cerebellum, but stimulation of this nucleotide by kainic acid was considerably diminished between these ages. Stimulation of cAMP by this agent was slightly reduced. Dopamine caused no age-related difference in stimulation of adenylate cyclase activity in striatum. The authors point out that Puri and Volicer (1977) obtained comparatively low values for dopamine stimulation in their "young" preparations and, therefore, the loss of stimulation they observed in old tissue might have been a result of nonoptimal experimental conditions. Schmidt and Thornberry (1978) suggested in particular that the Mg^{2+} concentration was too high. They also point out that the lowered response to increased levels of norepinephrine and dopamine shown by old tissue reported by Walker and Walker (1973) (see above) could have been a result of technical problems. However, Govoni et al. (1977) also observed a decreased response to dopamine-stimulated cAMP production in four brain areas of aged rats. They reported that stimulation of adenyalte cyclase activity by dopamine was reduced sharply in old Sprague-Dawley rats (2–3 versus 20–24 months) in homogenates from striatum, nucleus accumbens, substantia nigra, and tuberculum olfactorium. Thus, the question of whether or not there is a substantially diminished response to dopamine in brain tissue of old animals has not been resolved.

Makman and co-workers (Makman et al.,1979;Thal et al.,1980)obtained results generally similar to those above using mature and old rabbits (mean life-span, 7 years). Basal adenylate cyclase activity was unchanged with age in the brain regions studied (hypothalamus, frontal cortex, and anterior limbic cortex). However, stimulation of adenylate cyclase by dopamine, norepinephrine, or histamine was decreased sharply (generally up to 65%) with increased age (5 months versus 5 years) in these areas. Similar results were obtained with dopamine stimulation in striatum. However, Gpp(NH), a GTP analog, showed no age-related alteration in its degree of stimulation, suggesting that the above-noted losses were specific for certain receptors. Moreover, since the retina showed no loss of adenylate cyclase stimulation with aging, the age-related loss of response in brain was tissue-specific. In this regard, Govoni *et al.* (1977) reported that in rat retinal homogenates, dopamine stimulation of adenylate cyclase in old samples was almost 2 times that noted for young animals. As indicated above, such a stimulation was not observed in the retina from old rabbits (Makman *et al.*, 1979).

The above work, summarized in Table 10.I, is indicative that in certain tissues there is a reduced catecholamine function in old animals. Presumably, the decline is due to reduced numbers of receptors, to a changed affinity, or to a loss of coupling of receptor to adenylate cyclase. Each of these alternatives has been considered as a contributing factor to reduced adenylate cyclase response in aging brain. With regard to receptor number, the evidence is quite clear that there is a reduction without change of affinity, as shown by the work described below.

C. Receptors

As with other hormones, many of the recent efforts to study aging effects on catecholamine metabolism have been concerned with receptors. These are typically studied by measuring the binding of radioactive ligands. Several synthetic derivatives are available which for various reasons, usually specificity, are used in these studies. Loss of synaptic receptors, a decrease in their affinity, or change in response could be the cause of neurobiological losses perhaps originating in aging brain. In general, results indicate a loss in receptor number.

Greenberg and Weiss (1978) found that, in Fischer rat corpus striatum and cerebellum, β-adrenergic receptors ([³H]dihydroalprenolol binding) decreased steadily after an initial increase up to 6 months of age. Nonspecific binding sites increased with age, but this result was not obtained in mice by Severson and Finch (1980). In pineal gland, after an early drop, there was no change in the number of binding sites between 3 and 24 months. There was no change in affinity with age. The authors observed that the increased density of pineal β-adrenergic receptors which resulted from exposure to light in young animals was absent in 24-month-old animals. To explain these results, Weiss *et al.* (1979) proposed that in old age, certain tissues have a reduced ability to increase the number of receptors in response to decreased adrenergic activity. Thus, the reduced number of receptors would be responsible for the reduced stimulation of adenylate cyclase. Maggi *et al.* (1979) also observed a reduced β-adrenergic receptor level (dihydroalprenolol binding) between 12 and 24 months in Fischer rat cerebellum and brain stem, but not in cerebral cortex. Receptor affinities were unchanged. Human tissue gave similar results, cerebellum but not cortex showing a reduced amount of binding. The authors pointed out that norepinephrine stimulation of adenylate cyclase declines by 12 months in cerebellum (according to Schmidt and Thornberry, 1978), and thus it precedes the decline in receptor binding. Recently, Severson and Finch (1980) made an extensive study of dopaminergic binding in rodent brain. They used both spiroperidol and 2-amino-6,7-dihydroxy-1,2,3,4-tetrahydonaphthalene (ADTN), a strong dopamine agonist in young and old mice (C57BL/6J and C3HeB/FeJ). The experiments with spiroperidol showed a substantial decline with age in receptor number in striatum with no change in binding affinity, results they also obtained using Sprague-Dawley rats. Similar findings were obtained in the hypothalamus of C57BL/6J mice (8 versus 28 months), but in olfactory bulbs the results are not clear because of serotonin binding sites. Age-related differences in membrane sedimentation during the isolation procedure were not responsible for the observed changes. Binding of ADTN in striatal membranes of mice gave results similar to those obtained with spiroperidol, but the age-related loss of receptors was even greater (30–50%). Again, affinity was unchanged. The use of apomorphine (agonist) and (+)-butaclamol (antagonist) for binding studies affirmed the lack of change in binding affinity. One should bear in mind that loss of receptors as determined in this type of experiment may stem from an age-related decrease in striatal neurons that contain the dopaminergic binding sites. The authors point out that spiroperidol binding in sites other than striatum may involve serontonin receptors. For example, serontonin competes better than dopamine for spiroperidol binding sites in hippocampus but not in striatum.

In the same vein, Misra *et al.* (1980) reported that binding of labeled spiperone (dopamine binding) in the striatum of young (5 months) and old (25 months) Fischer rats decreased by 39% in the latter without change in affinity. Dihydroalprenolol binding (β-adrenergic) in cerebral cortex decreased (43%), again without a change in affinity. Joseph *et al.* (1978) also reported that there was a 33% reduction in dopamine receptor levels (haloperidol binding) in the substantia nigra of old Wistar rats (6 versus 25 months). Recently, Levin *et al.* (1981) again observed a progressive decrease in dopamine receptor concentration in striatal membrane preparations from Wistar rats (3–6, 12, and 24 months) based upon ADTN binding. Affinity of the receptors was unchanged. Interestingly, as for glucorticoid receptors in adipocytes, loss of dopaminergic receptors was ameliorated under conditions of dietary restriction (Levin *et al.*, 1981). Table 10.II provides a summary of the effects of age on catecholamine binding in rat brain.

Makman *et al.* (1979) and Thal *et al.* (1980), working with rabbits,

Table 10.II Effects of Age on the Amount of Receptor Binding in Rat Brain[a,b]

	DA	Reference[c]	NE	Reference[c]	NS	Reference[c]
Striatum	−[d]	4,6,7	−	1,3	+, NC	1,3
Cerebellum			−[d]	1,2	+	1
Pineal			NC	1		
Stem			NC	2		
Cortex			−	4		
Cortex			NC[d]	2		
Hypothalamus	−	3				
Olfactory	NC	3				
Substantia nigra	−	5				

[a] DA, Dopamine; NE, norepinephrine; NS, nonspecific binding.

[b] +, Increases with age; −, decreases with age; NC, no change with age.

[c] References: (1) Greenberg and Weiss, 1978; (2) Maggi et al., 1979; (3) Severson and Finch, 1980; (4) Misra et al., 1980; (5) Joseph et al., 1978; (6) Levin et al., 1981; (7) Memo et al., 1980.

[d] Also in rabbit (Makman et al., 1979).

observed that the age-related reduction in response to dopamine corresponds to a decreased number of receptors as determined by spiroperidol binding. Thus, resuspended membranes from the three regions of brain (striatum, frontal cortex, and anterior limbic cortex) all showed a reduced number of binding sites, somewhat parallel to the degree of age-related loss of stimulation of adenylate cyclase for the respective regions. Since affinity of the spiroperidol binding was unchanged and the neurons seem able to make normal amounts of presynaptic components (there were no age-related changes in dopamine concentration or in choline acetylase activity), the reduced response in aged rabbit brain appears to be mediated by a decrease in receptor number.

From the above, it can be seen that there is unusually good agreement that in brain tissue, monoamine receptors decrease with age and that affinity is unchanged. Yet, in contrast to these seemingly clear cut results, Govoni et al. (1980) presented evidence that in Sprague-Dawley rats, receptor number (spiroperidol binding) in striatum was unchanged, but affinity was decreased fivefold with age (4 versus 24–30 months). Nonspecific binding was unchanged. Curiously, binding in the pituitary yielded opposite results, old animals showing an increase of 50% in specific binding. In tuberculum olfactorium, basal levels were unchanged, but stimulation of adenylate cyclase by dopamine was reduced in old animals. Members of the same research group (Memo et al., 1980) carried out another study using two different ligands, [³H]spiroperidol and (MI)[³H]sulpiride. In this report, spiroperidol binding sites showed a significant decrease with age and no change in binding affinity. Sulpiride binding is specific for dopaminergic sites not coupled to adenylate cyclase. This type of receptor showed no age-related changes.

One other claim of an age-related change in binding affinity was made by Marquis *et al.* (1981), who recently reported a significantly decreased binding of haloperidol in mouse striatum between 24 and 32 months of age. However, there was almost a doubling in receptor binding between 10 and 24 months (and an almost ninefold increase between 4 and 24 months) so that the 32-month level, even after the late decline, remained greater than that for mature animals. In rats, receptor binding in striatum was reported to increase substantially (threefold) and receptor affinity to decrease sevenfold. These rather dramatic changes are not in agreement with any other reports.

VII.   METABOLISM OF OTHER NEURORECEPTORS

Most of the neuroreceptor and neurotransmitter work has focused on catecholamines, with little effort expended on other systems.

A.   Acetylcholine

Freund (1980), using C57BL/6J mice from 5 to 30 months of age, found that crude, low-speed supernatants from whole brain without the cerebellum declined in synaptic cholinergic receptors between 18 and 30 months of age, based on binding of radioactive quinuclidinyl benzilate, a cholinergic antagonist. There was no change in receptor affinity. At earlier ages (between 5 and 18 months), there was no significant change in receptor number. There was no apparent loss of neurons, based upon RNA–DNA ratios, no diurnal effects, and no change with age of nonspecific binding.

There is no evidence for a generalized change in enzymes involved in acetylcholine metabolism. Even where losses in brain tissue are reported, they are of the order of 20% in crude preparations (Finch, 1977). Reis *et al.* (1977) found little change in choline acetyltransferase in several brain regions of old rats, although the enzyme increased considerably in the adrenals and sympathetic ganglia of both rats and mice. Choline acetylase activity did not change in aged rabbit striatum (Makman *et al.*, 1979). Vijayan (1977) found no change of acetylcholinsterase and choline acetyltransferase in aging mouse cerebellum (3 versus 24 months), although there was a small reduction in hippocampus. Meek *et al.* (1977) using rat brain nuclei, observed a reduced choline acetyltransferase activity with age only in the caudate. Choline and acetylcholine levels were unchanged. In this study, the young animals were only 35–45 days old (versus 24 months) so that it is uncertain whether the results can be interpreted in terms of aging.

B. γ-Aminobutyric Acid

In general, there appear to be few age-related changes in GABA or in serontonin. Early studies show little change in levels with age in brain (Finch, 1977). More recently, Simkins *et al.* (1977) found steady state concentrations of serotonin to be unchanged with age in rat brain. Meek *et al.* (1977) found a reduction in serotonin in rat brain nuclei but as noted above, the "young" animals were 35–45 days of age.

As to receptors, work thus far performed agrees that GABA binding is unchanged with age in most tissues studied. Maggi *et al.* (1979) found no changes in cerebellum, brain stem, or cortex of Fischer rats. Govoni *et al* (1980) found no changes in cerebral cortex, cerebellum, striatum, and nucleus accumbens. They did, however, note a sharp decrease in substantia nigra and hypothalamus (55 and 39%, respectively). Thal *et al.* (1980) noted that the high-affinity binding of spiroperidol in rabbit cortex (unlike striatum) appears to be serotonergic rather than dopaminergic, based on displacement studies. Severson and Finch (1980) suggest a similar binding in olfactory bulbs and hypothalamus.

VIII. GLUCAGON

Relatively few investigations have been performed on the effect of age on glucagon metabolism. Particularly lacking are binding studies. Moreover, work has been limited to a few tissues. Therefore, few conclusions have been drawn with respect to the relationship of glucagon to aging. Dudl and Ensinck (1977) reported that arginine infusion did not alter the release of glucagon in elderly humans, or were basal concentrations changed. In rats, the picture may be different, as Klug *et al.* (1979) observed that fasting for 3 days followed by administration of glucose causes an increase in glucagon serum levels of 24-month animals (Sprague-Dawley) which appears sooner (2 versus 3 h) and is greater (net increase, 0.95 versus 0.40 ng/ml) than that for 12-month-old animals. In 2-month-old animals, glucagon levels were lowered by the treatment.

As with other hormones, adipocytes have been utilized for aging studies on glucagon. There has been considerable concern regarding the effect of cell size (DeSantis *et al.*, 1974; Holm *et al.*, 1975) and, for that matter, the size of the donor animals (Manganiello and Vaughan, 1972). However, from the point of view of aging studies, these are not serious considerations as in Fischer and Wistar rats, adipocytes show little size change after 6 months of age, although several of the investigations include animals younger than this.

Cooper and Gregerman (1976) observed that in adipocytes from Wistar rats, glucagon stimualtion of adenylate cyclase had disappeared by 12 months of age. Cell size was constant after 6 months of age so that this factor was not responsible for the reduction in hormone response. Holm *et al.* (1975) reported that the glucagon-mediated stimulation of lipolysis was lost by 15 weeks of age and was not related to the size of the adipocytes. Livingston *et al.* (1974) investigating the mechanism of the loss of lipolytic response found that when stimulated maximally with other agents (e.g., cAMP, epinephrine) both small and large cells responded equally well. Thus, the reason for the loss of stimulation of lipolysis does not lie in the lipolytic enzyme system. Rather, it appears to lie partly in a reduced number of glucagon receptors in the large cells (DeSantis *et al.*, 1974). It may also lie in part in an increased level of phosphodiesterase, which hydrolyzes cAMP or in a loss of a coupling component of the receptor–cyclase complex.

Bertrand *et al.* (1980) observed very little lipolytic response to glucagon in epididymal and perirenal adipocytes from normally fed Fischer rats after 6 and up to 30 months of age. When the diet of the animals was restricted, the cells from 6-month-old animals showed a greatly enhanced response which then dropped steadily until 24 months with a subsequent rise at 30 to 36 months. These results make it appear that the "aging" loss of lipolytic response is delayed by the restricted diet. Since the restricted rats, on average, had smaller adipocytes and this situation might have an effect on the metabolism of the cells, the authors compared groups of cells of similar size from 6-month-old dietarily "unrestricted" and 12-month-old "restricted" animals. The cells from the restricted animals, although they were older, nonetheless showed a markedly greater lipolytic response (over 30-fold) to glucagon.

Liver particulate preparations from old Wistar rats (12 versus 24 months) showed no loss with age of glucagon-stimulated adenylate cyclase activity (Kalish *et al.*, 1977).

IX. INSULIN

There is little mechanistic information available on the effect of aging on insulin metabolism. In old subjects, administration of glucose results in blood levels that remain elevated for extended periods of time. The reason for this situation might include a deficiency in the signal for control of insulin production, a deficiency of insulin receptors, inadequate postreceptor activity, or perhaps the production of defective insulin. Increased production of proinsulin would be particularly confusing as it would give positive results in the immune assay, but would not provide a biological response.

The bulk of studies conclude that in aged humans, there is no impairment of the ability to produce insulin. According to a thorough review of the effect of aging on carbohydrate metabolism (Davidson, 1979), it is doubtful that abnormalities in insulin secretion could alone account for the impairment of glucose tolerance seen in many aged subjects. A similar conclusion was reached by McGuire *et al.* (1979). Further, Dudl and Ensinck (1977) concluded that basal levels of insulin were unchanged with aging in human subjects, and there were no differences in fasting or in glucose-induced release of insulin. Andres and Tobin (1975), using a "glucose clamp" technique, concluded that in humans, the sensitivity of β cells to glucose diminishes with age but not the ability to produce insulin. In this procedure, plasma glucose concentration is fixed and the insulin response measured.

With respect to insulin receptors, Pagano *et al.* (1981) reported that compared to young subjects (32 years) the number of insulin receptors in fat cells from aged human subjects (70 years) was reduced 57 and 40% for high- and low-affinity receptors, respectively. There was no change of receptor affinity with age. The subjects (both sexes) were of normal weight, and the adipocytes were of similar size. This is an important consideration as there is much speculation about changes in adipocyte metabolism with respect to weight of the donor and size of the cells. For example, Smith (1971) observed that larger adipose cells (human) were less sensitive than small ones to insulin stimulation of lipid synthesis. In contrast to the results with adipocytes, Helderman (1980) examined insulin receptors in T lymphocytes from young (mean age, 28 years) versus old men (mean age, 72 years). No age-related change in either affinity or receptor number was found. The estimated receptor affinity (1.7 nM) was similar to that reported for other human tissues (fat, muscle, and liver).

Using another approach, Hollenberg and Schneider (1979) found that there was no significant change in insulin receptors in early-passage skin fibroblasts (7–13 doublings) from young (22–31 years) versus old (65–80 years) human donors. Quite to the contrary, Rosenbloom *et al.* (1976) had earlier concluded that in human fibroblasts (mid-passage), specific insulin binding increased with the age of the donor (3 months to 70 years).

There appeared to be an age-related increase in binding affinity as greater concentrations of insulin were required for 50% competition with 1 nM labeled insulin. There seems no obvious reason for the disparate results. Confluent cells were used which should show the maximal number of receptors.

Freeman *et al.* (1973) reported that in fasted or fed Sprague-Dawley rats, insulin concentrations in portal blood fall 50–60% between 12 and 24 months of age. The levels are much higher than those for peripheral blood as the portal location represents a direct path from pancreas to liver. On the other hand, serum insulin was reported by Lewis and Wexler (1974) to increase markedly between 6–8 and 15–18 months in male Sprague-Dawley rats. The pattern of insulin response to glucose stimulation also differs with age (Gold *et al.*, 1976). Reaven *et al.* (1979) examined the glucose-stimulated response of isolated islets of Langerhans in parallel with stereological examination of intact pancreas from Sprague-Dawley rats. They found that between 2 and 18 months, the number of β cells more than doubled and their insulin content doubled. However, insulin response to increasing glucose concentrations was lower at 12 and 18 months than at 2 and 6 months in the isolated islets. The difference in insulin proliferation between 12- and 18-month-old islets was very small even though the latter were larger. Maximal output of the older islets was decreased, indicating that perhaps older tissue cannot respond adequately to a glucose stimulus. The authors suggest that the increase in β cell mass they observed in the pancreas of old animals is a compensatory mechanism. Unfortunately 18 months was the oldest age used in these experiments. Remacle *et al.* (1975) also had observed an age-related increase in β cell number and size in rats up to 24 months of age. Kitahara and Adelman (1979) examined isolated pancreatic islets of Langerhans from Sprague-Dawley rats and discovered that small and large islets show quite different secretory responses. Upon glucose stimulation, the large islets secrete up to fivefold more insulin in both young (2 month) and old (24 months) animals. Secretion in both types of cells is inhibited with age, but more severely in the small islets. The authors speculate that the increase in number of large islets in old animals represents a compensatory mechanism for the sharp reduction in insulin secretion by small islets.

The few studies available suggest that in rodents, there is little change in insulin receptors with age. Freeman *et al.* (1973) reported that, although the number of binding sites decreased sharply (50%) in rat hepatic plasma membrane between 2 and 12 months, the difference between 12 and 24 months was small (14%) (statistics not provided). Binding constants were unchanged. Similarly, Sorrentino and Florini (1976) found no change in binding capacity or affinity for insulin in liver or heart membranes of C57BL/6J mice from 2 to 31 months of age. The authors noted that there was a very large individual variation. It is interesting to note that although serum insulin is elevated during pregnancy and that rat adipocytes (Flint *et al.*, 1979) and rat hepatocytes (Flint, 1980) have increased numbers of receptors, the animals exhibit insulin resistance. Thus, receptor number is not necessarily a controlling element in the insulin equation.

As to metabolic effects, Gommers *et al.* (1977) reported that there is no insulin effect on CO_2 production from glucose in isolated rat diaphragm from young (3 months) versus old (24 months) rats. Similarly, lipid synthesis is not differently affected by insulin in isolated adipose tissue (Jeanjean *et al.*, 1977). These results are interpreted as showing that insulin sensitivity is unchanged after development.

X. PITUITARY HORMONES

As is the case with other hormones, there is little research on the relationship between pituitary hormones and aging that can be applied in a direct biochemical context. Most of the work that has been done involves measurement of basal hormone levels and response to various stimuli in young versus old animals. A summary of selected results is provided in Table 10.III. In general, there is agreement that, in the mouse, neither pituitary hormone levels nor the response to LHRL or TRH change with age (Finch *et al.*, 1977). The fact that the stimulation of LH by LHRH in isolated pituitary cells (experiment 7, Table 10.III) is the same for mature (12 months) and old (28 months) tissue confirms that the lack of change *in vivo* was not due to circulatory or other problems in old animals.

As can be seen from Table 10.III, the situation in rats, which one would expect to be analogous to that in mice, seems to be quite different. Furthermore, there are sex-related differences. Plasma levels of LH are reported to be unchanged in senescent female rats (experiment 12, Table 10.III) but reduced in old males (experiment 13) and isolated pituitaries (experiment 11). For both sexes, there is a substantial decrease in the response of LHRL *in vivo*. Basal PRL levels also differ with age in male and female rats (experiments 15 and 16). The results are complicated by the fact that the young female rats could have been in different stages of the estrous cycle.

In humans, reports on the effect of age on pituitary hormones are somewhat conflicting. Several, although not all, authors report an increase with age in basal levels of FSH and LH (experiments 21–26).

Table 10.III Effect of Age on Pituitary Hormone Levels[a,b]

Experiment	Animal or tissue[c]	Sex	Age (months)	Hormone	Effect of age on basal level	Treatment	Response	Reference[d]
1	Mouse (C57BL/6J)	M	12 versus 28	LH	NC	LHRH	NC	1
2	Mouse (C57BL/6J)	M	12 versus 28	FSH	NC	Castration	NC	1
3	Mouse (C57BL/6J)	M	12 versus 28	TSH	NC	TRH	NC	1
4	Mouse (C57BL/6J)	M	12 versus 28	PRL	NC			1
5	Mouse (C57BL/6J)	M	12 versus 28	GH	NC			1
6	Mouse (C57BL/6J)	M	12 versus 28	LH	NC	Castration	–	1
7	Cultured mouse pituitary cells	M	12 versus 28	LH	+	LHRH	NC	1
8	Cultured mouse pituitary cells	M	12 versus 28	FSH	NC	LHRH	NC	1
9	Rat (W)	M	12 versus 28	FSH	NC			1
10	Rat (SD)	M	3 → 24	LH	–[e]			2
11	Rat (LE) pituitaries	M	4 versus 26	LH	–			3
12	Rat (LE)	F	4–6 versus 23–30	LH	NC	LHRH	–	4
13	Rat (LE)	M	4–6 versus 23–30	LH	–	LHRH	–	5,6
14	Rat (LE)	M	4–6 versus 23–30	LH	–	Stress[f]	–	5,7
15	Rat (LE)	F	4–6 versus 23–30	PRL	+			7
16	Rat (LE)	M	4–6 versus 23–30	PRL	NC			5
17	Rat (LE)	M	4–6 versus 22–30	PRL	+	L-DOPA	–[g]	5
18	Rat (LE)	M	8 → 21	LH	–			8

19	Rat (SD)	M	2 versus 24	TSH	NC	TRH	NC	9
20	Rat (F)		5,12,29	FSH	–			10
				LH	–			
				PRL	+			
21	Women		60–100 years	FSH, LH	NC	LHRH	NC	11
22	Men		20–79 years	FSH	NC	GnRH[h]	–	12
23	Men		20–79 years	LH	+	GnRH[h]	–[i]	12
24	Men		20–89 years	FSH, LH	NC	GnRH[h]	–	13
25	Men		18–70 years	FSH, LH	+	LHRH[j]	NC	14
26	Men		50 versus 65 years	FSH, LH	+	LHRH	+	15

[a] LH and LHRH, Luteinizing and releasing hormone; FSH, follicle-stimulating hormone; TRH, thyroid-releasing hormone; TSH, thyrotropin; PRL, prolactin; GH, growth hormone; L-DOPA, L-dihydroxyphenylalanine.

[b] +, Increase with age; –, decrease with age; NC, no change.

[c] F, Fischer; LE, Long Evans; SD, Sprague-Dawley; W, Wistar.

[d] Key to references: (1) Finch et al., 1977; (2) Pirke et al., 1978; (3) Kaler and Neaves, 1981; (4) Riegle et al., 1977; (5) Shaar et al., 1975; (6) Watkins et al., 1975; (7) Riegle and Meites, 1976; (8) Gray, 1978; (9) Klug and Adelman, 1979; (10) Bethea and Walker, 1979; (11) Scaglia et al., 1976; (12) Snyder et al., 1975; (13) Haug et al., 1974; (14) Hasimoto et al., 1973; (15) Rubers et al., 1974.

[e] Slight.

[f] Handling, anesthesia, and blood sampling.

[g] The lowering of the PRL level by L-DOPA was slower in the old animals.

[h] Synthetic gonadotropin-releasing hormone.

[i] There was a large degree of individual variation.

[j] The time course was slowed.

Reports of responsiveness to LHRH are also not consistent: some investigators find a decline, others no change or an increase. Snyder *et al.* (1975) make the point that when one considers the higher starting level in old subjects, the *degree* of change is reduced with age. Kovacs *et al.* (1977), from morphological studies, concluded that there were no changes in properties or number of prolactin cells in humans, both male and female over 80 years of age. Contradictions in measurements of hormonal levels and response to stimulation in aging humans are not unique to the pituitary. Similar variations have been reported in the case of thyroid hormones. A decline, a lack of change, and increase have been reported (Section XI).

The effect of one hormone on the response of another is not clearly understood. With the added uncertainty of aging, survey experiments cannot be expected to clarify the nature of interdependent events. It does appear that, on balance, there is a lessening with age in the degree of response of the pituitary. Whether this decrease is due to receptor changes (number or affinity) or to postreceptor problems is simply not known at this time. The ups and downs reported for basal hormone levels may reflect altered metabolic states rather than a lessened ability to respond to a stimulus.

XI. THYROID HORMONES

Circulation of the thyroid hormones thyroxine (T_4) and triiodothyronine (T_3) depends in part on the thyroid stimulating hormone thyrotropin (TSH), which is regulated by the thyrotropin-releasing hormone (TRH). Studies of the effect of age on the thyroid gland have been mostly concerned with changes in levels of these components.

In humans, the effect of age on response to TRH (release of TSH) is not clear. Snyder and Utiger (1972a) reported a decline in old males but not in females (1972b). Sakoda *et al.* (1973) found no age-related effects in cohorts of mixed sex. O'Hara *et al.* (1974) found an increased response in the old age-group, also of mixed sex. As Klug and Adelman (1979) pointed out, interpretation is difficult in part because the dose-response curve may vary with the different age-groups and variables which regulate pituitary sensitivity to TRH were not considered.

As to the levels of thyroid-related hormones, TSH was found to be in the normal range in elderly subjects (Hesch *et al.*, 1976). Several investigators reported decreases with age in the level of total T_3 and T_4 in humans based on specific immunoassays (Rubenstein *et al.*, 1973; Hermann *et al.*, 1974; Hesch *et al.*, 1976). These decreases would be expected because basal metabolic rate declines with age, and this function is presumably mediated by thyroxine. The amount of thyroid-binding globulin in serum was reported both to increase (Hesch *et al.*, 1976) and remain unchanged (Hermann *et al.*, 1974). There are also conflicting reports on the levels of free T_3. Degradation of thyroxine is also reported to slow in aged humans (Gregerman *et al.*, 1962).

In 2-month-old male Sprague-Dawley rats, serum TSH followed a circadian rythm, which essentially had disappeared by 12 months of age (Klug and Adelman, 1979). (Interestingly, the pattern of circadian periodicity of serum corticosterone was little changed with age, although the lowest levels observed were somewhat elevated in old rats.) Average TSH levels were unchanged up to 24 months. There was also no age-related change in response to TRH administration. Similar conclusions were reached by Finch *et al.* (1977) in mice (12 versus 28 months). On the other hand, Eleftheriou (1975) reported a decreased thyroid sensitivity with age to TSH, and Frolkis *et al.* (1973) also reported a decreased functional activity of thyroid in aged rats.

As Klug and Adelman (1979) pointed out, immunoreacting TSH does not necessarily reflect the biological level of the hormone. The authors had previously observed an age-dependent decrease in the ratio of these two measurements in rats (see below).

In the case of thyroid hormones, Gregerman and Crowder (1963) reported a decrease with age in protein-bound iodine in the rat. Frolkis *et al.* (1973) also reported a decrease with age in serum T_4. Subsequently, Eleftheriou (1975) reported on protein-bound iodine in two strains of mice. The results of this work indicate that one should be careful of strain differences: although protein-bound iodine decreased consistently with age between 2 and 30 months, the effect was much greater in DBA/2J than in C57BL/6J mice.

With the advent of specific antigens, differences in T_3 and T_4 could be readily determined. It was found that serum levels of total and free T_3 and T_4 decrease in aging, but the decrease precedes senescence, occurring almost entirely between 2 and 12 months in male Sprague-Dawley rats (Klug and Adelman, 1979). Zitnik and Roth (1981) also observed no differences between mature (9–13 months) and old (22–24 months) Wistar rats.

There have been a few studies dealing with the effects of age on the biochemical or metabolic function of thyroid-related hormones. Williams *et al.* (1977) provided evidence to show that thyroid hormones increase the number of β receptors in rat hearts. However, Zitnik and Roth (1981) found no significant age-related differences in this phenomenon, nor did they find a difference in the time course or extent of cardiac hypertophy brought about by pharmacologic doses of thyroxine. Florini *et al.* (1973) had earlier reported a lag period for this phenomenon in old versus young mice. Gregerman and Crowder (1963) found no change with age in fractional turnover rate of thyroxine in 3-, 12-, and 24-month-old Wistar rats, but total degradation is increased in the old rats by about 50%. Klug and Adelman (1977) provided evidence for the presence of large molecular forms of TSH in the pituitary and serum of Sprague-Dawley male rats. This large molecular weight material could be separated from the monomeric hormone by elution from a column of Sephadex G-200. The material is immunoreactive, but not biolgoically active and is present in increased amount in 24- versus 2-month-old rats. In fact, the product inhibited the release of thyroid hormones in the intact mouse bioassay system. Therefore, although the amounts are relatively small, the fact that increased amounts of this high-molecular-weight product is present in old animals suggests that it may be involved in age-related changes in response, especially given its inhibitory effect on production of thyroid hormone.

It would be premature to attempt to draw more than inferences from the data available. However, additional measurements of thyroid hormone levels or responses to stimuli would not appear to be a particularly rewarding approach to the study of aging: even if one could resolve the conflicting reports, there would be little added to our intrinsic understanding. More biochemical studies of a basic nature are needed but would be difficult to carry out at this time since the mechanism of action of thyroid hormones is not understood. Moreover, the effect of age-related changes on other hormones may play an important role, and little is known about such relationships. There remains much to be done but where to get a handle on the problem seems to be an overriding difficulty.

XII.   TESTOSTERONE

The effect of aging on sex hormone production and response is complicated in the female by the specialized development and eventual semi-demise of a reproductive system, which generates its own time pattern almost to the exclusion of the host organism. Thus, it is especially difficult, if not impractical, to distinguish the effects of aging from the planned obsolescence of the female reproductive system. Therefore, this section will deal only with the effect of age on sex hormones in males where the sexual system appears to age more in concert with the whole organism and where decrements in sex function can be arguably viewed as a direct concomitant of aging rather than as a programmed phasing out of a specific capacity. In considering aging effects, at least in the case of rodents, one must be careful to take into account such elements as diseased conditions and changes in the quantity of testicular tissue. For example, Nelson *et al.* (1975) found no change in plasma levels of testosterone in healthy old C57BL/6J mice, but a marked decrease in diseased senescent animals (5–11 versus 29–31 months). Bethea and Walker (1979) observed a drastic loss of Leydig cells in 29-month-old Fischer rats. However, this situation appears to be specific for this strain: Kaler and Neaves (1981) found no significant loss of cells in aged Sprague-Dawley rats (24 months) and Pirke *et al.* (1978) observed an increase in Leydig cell volume in aged (26–28 months) Wistar rats. However, three of five diseased old rats excluded from the study showed atrophy of the testes. There is not a good correlation between changes in body weight with age and testes size (Eleftheriou and Lucas, 1974).

A. Hormone Levels

As with other hormones, studies of testosterone deal mainly with changes in level or response to stimuli during aging as outlined in Table 10.IV, although some work on receptors and metabolism of various intermediates has been reported.

From Table 10.IV, it can be seen that age has little effect on basal levels of testosterone in mice or on its response to LH administration. *In vitro* incubation of intact testes with LH also showed no age-related change in the degree of increased proliferation of testosterone. In rats, the results are quite different. There is a consistent drop of plasma testosterone (about 50%) with age. Reasons for reduced testosterone levels may include decreased gonadotropin secretion, decreased testicular response to gonadotropin, or to increased testosterone metabolism and excretion. None of these possibilities has been clearly demonstrated to be the causative factor for the phenomenon. In this regard, Kaler and Neaves (1981) pointed out that although plasma testosterone levels declined 50% in Sprague-Dawley rats (experiment 8, Table 10.IV), the plasma volume was approximately doubled so that total circulating levels of the hormone were relatively unchanged. Taking volume into account, response to gonadotropin was also undiminished. Chan *et al.* (1977) noted a circadian rhythm in testosterone levels, which was still retained at 18 months of age, although generally reduced in level. There were no tumors present in the testes of the old Long-Evans rats used in these experiments.

In beagle dogs, mean plasma levels of testosterone were unchanged, although the values for individuals in the age-groups (2.5, 4.5, and 11 years) were quite variable.

Table 10.IV Examples of the Effect of Age on Testosterone Levels[a]

Animal or tissue[b]	Age (months)	Effect of age on T levels	Treatment	Age effect on response	Reference
1. Mouse (C57BL/6J)	8–11 versus 29–31	NC[c]			1
2. Mouse (DBA/2J)	2 → 28	NC			2
3. Mouse (C57BL/6J)	12 versus 28	NC[d]	LH	–(NS)	3
4. Mouse testes	12 versus 28	+	LH	NC	3
5. Rat (W)	3–6 versus 25	–			4
6. Rat (LE)	3–6 versus 20–30	–	hCG	–[e]	5
7. Rat (LE)	4 versus 18	–	hCG	–	6
8. Rat (SD)	3 → 24	–[f]	hCG	–	7
9. Testes	3 → 24	NC[g]	hCG	NC	7
10. Rat (W)	3 versus 26–28	–	hCG		8
11. Testes	3 versus 26–28		hCG	NC	8
12. Rat (LE)	6 → 36	–	hCG		9
13. Rabbit testes	6 → 36	–	[h]	–	10
14. Dog (Beagle)	4–5 versus 11.0	NC[i]			11
15. Human male	50 versus 65 years	–	hCG		12
16. Human male	20–79 years	NC[k]	hCG	–(NS)[j]	13

[a] +, Increases; –, decreases; NC, no change; NS, not significant. [b] W, Wistar; LE, Long Evans; SD, Sprague-Dawley. [c] Diseased mice had a marked decrease. [d] Testicular T was one-third less (NS). [e] Over a 7-day period of hCG administration, the level of T in old animals rose to the level of young animals. [f] Increased weight of old animals suggests *total* amount of T is unchanged. [g] Expressed per cell. [h] Gonadotropin mixture. [i] On a per cell basis. Old prostates are hyperplastic. [j] Absolute values lower. Relative increase, considering lower basal level was similar to that of the controls. [k] 80- to 89-year-old group showed a reduction. [l] Key to references: (1) Nelson *et al.*, 1975; (2) Eleftheriou and Lucas, 1974; (3) Finch *et al.*, 1977; (4) Ghanadian *et al.*, 1975; (5) Miller and Riegle, 1978; (6) Chan *et al.*, 1977; (7) Kaler and Neaves, 1981; (8) Pirke *et al.*, 1978; (9) Gray, 1978; (10) Ewing *et al.*, 1972; (11) Shain and Nitchuck, 1979a; (12) Rubens *et al.*, 1974; (13) Haug *et al.*, 1974.

In humans, it would seem that plasma levels of testosterone and response to gonadotropic stimulus are not greatly reduced, at least not until very old age. Even where a reduction is found, many old men in a given sampling retain a normal range of the hormone. Rubens *et al.* (1974) concluded that the reduction they observed in humans is testicular in origin (experiment 16) and may in part be related to impaired blood supply.

B. Receptors

Testosterone receptors in rat prostate tissue have been studied by Shain and co-workers (Shain *et al.*, 1975; Shain and Boesel, 1977; Boesel *et al.*, 1980). These investigators found a 50% decrease in total androgen receptor content in ventral and dorsolateral prostate in old (about 25 months) versus mature AX (Wistar-derived) and Sprague-Dawley rats. In ventral prostate, both cytoplasmic and nuclear receptors decreased with age. In dorsolateral prostate, cytoplasmic receptors are at too low a level for detection, but nuclear receptors decrease. In general, considerations such as increased cell number, cell hypertophy, and changes in kinetic or physical properties were shown not to be responsible for the reduction in binding. In the most recent report (Boesel, *et al.*, 1980), it was shown that administration of exogenous testosterone to old animals raised the number of prostate receptors (ventral and dorsolateral) to "young" levels. Thus, the decreased level of androgen receptors in old animals may be a result of the decreased androgen level circulating in aged rats (Table 10.IV). In contrast to the situation in rats, prostatic androgen receptor level was found to be unchanged in old beagles on a per cell basis (Shain and Boesel, 1978), even though the prostates of old animals were hyperplastic.

Pirke *et al.* (1978) found that binding of human chorionic gonadotropin (hCG) was significantly reduced in testicular tissue from old Wistar rats (3 versus 26–28 months). In spite of the reduction in hCG receptors, old testes tissue *in vitro* was identical with young tissue in its production of testosterone in response to hCG and dibutyryl cAMP. However, when a more potent stimulus (an NADPH generating system) was used, young tissue produced about 2 times as much testosterone. The results indicate that there is a reduced capacity in old tissue that is only seen at maximal demand levels.

C. Metabolism

The metabolism of testosterone involves its conversion to the reduced product, 5α-dihydrotestosterone, mediated by Δ^4-3-ketosteroid-5α-oxidoreductase. The level of this enzyme is dependent upon the plasma androgen titer. In senescent AXC rats (36 months) compared to mature animals (6 months), ventral prostate tissue produced significantly less 5α-dihydotestosterone from labeled testosterone, but produced more 4-androstenedione and 5α-androstane-3,17β-diol. In senescent dorsolateral prostate, these changes were not found. Administration of testosterone restored the metabolism of senescent ventral prostate tissue to "young" patterns (Shain and Nitchuk, 1979b).

In old beagles, prostate tissue showed an increased proportion of reduced products, principally 5-α-dihydotestosterone and less 4-androstenedione (Shain and Nitchuk, 1979a).

XIII.   ADAPTIVE RESPONSE TO HORMONES

The control of metabolic pathways by hormones sometimes involves increases in individual enzymes brought about by *de novo* synthesis, i.e., enzyme induction. This type of change is clearly different from post-translational changes in proteins mediated by hormones which stimulate production of cAMP. Such events bring about phosphorylation of a key protein, which then triggers a sequence of enzymatically modulated events.

One example of hormone-mediated enzyme induction is glucokinase in liver. Elevated blood glucose results in an increase in insulin production, which, in turn, brings about an increase in the amount of glucokinase. Adelman (1970a) observed that restoration of glucokinase activity, which drops precipitously during fasting, takes place more rapidly after glucose administration in young (2 months) Sprague-Dawley rats, than in old (24 months) animals. Direct injection of large doses of insulin resulted in stimulation of glucokinase activity in old animals to the "young" level. These experiments showed that there is a time lag in the production of the enzyme in old animals rather than an inability to obtain a full measure of response. In fact, the duration of the delay increases progressively in proportion to chronological age up to the 24 months measured (Adelman, 1972). Perhaps the delayed induction of glucokinase is related to the fact that glucose administration results in an age-related increase in serum glucagon levels (Klug *et al.*, 1979), because glucagon inhibits glucokinase induction.

Similar time lags in aged rats were observed in the response of tyrosine aminotransferase after administration of ACTH (Adelman, 1970b) and in microsomal NADPH-cytochrome *c* reductase as induced by phenobarbital (see Adelman, 1972, for a review of earlier work). Jacobus and Gershon

(1980) also reported a delay in the induction of tyrosine aminotransferase and ornithine decarboxylase in livers of old (20–26 months) as compared to young (3–6 months) and mature (12 months) C57BL/6J mice after administration of dexamethasone or growth hormone, respectively. The figures are 4 versus 6 hours for the peak of growth hormone-stimulated ornithine decarboxylase activity and 5 versus 7–8 hours for dexamethasone-stimulated tyrosine aminotransferase production. Curiously, Weber *et al.* (1980) observed a greater tyrosine aminotransferase response to hydrocortisone in old Wistar rats (3–6 versus 27–31 months) at the single time point of 5 hours after injection. Isolated hepatocytes showed no significant age-related change in the amount of the enzyme induced by dexamethasone.

Frolkis *et al.* (1979) observed changes in the induction of glucose 6-phosphatase and fructose 1,6-diphosphatase by hydrocortisone in male albino rats (2–3, 8–12, and 24–28 months of age). Liver, kidney, and spleen each yielded individual patterns, induction in old animals being lower, higher, or the same. With ACTH, liver showed a delayed response in glucose 6-phosphatase production, but fructose 1,6-diphosphatase was unaffected by age.

As can be seen in Chapter 5, Section III, the effect of age on induction of microsomal enzymes, when considered on a broad basis with respect to the kinds of test animals, stimuli and enzymes involved, varies from decreased response, to no change, to full but delayed response. Davis and Pfeifer (1973) observed major differences in the behavior of glucokinase in Wistar rats, compared to the Sprague-Dawley rats used by Adelman and co-workers. There were also sex-related differences. Initial glucokinase activities were much lower in the Wistar rats and varied greatly with season. After fasting, enzyme levels were increased 31–45% over fed values for both young (4 months) and mature (14 months) female animals by 24 hours after refeeding, compared to 48 hours for the Sprague-Dawleys rats. The differences may result from altered metabolic patterns evolved to cope with the larger and fatter Sprague-Dawley rats. In short, the internal milieu as well as the external environment is always an important factor in attempting to analyze hormonal effects and responses in intact animals. Such considerations as the prevention of glucokinase recovery by glucagon and epinephrine after fasting, or the ACTH-stimulated induction of tyrosine aminotransferases being subject to adrenal function are illustrative of the complexity of the system. In fact, there are many indications that changed response is not a problem of the target organ. For example, insulin receptor levels in liver change little if at all with age in mice or after maturity in rats (Section IX). Nonetheless, direct injection of cortisol, insulin, or glucagon into old rats abolished the delayed lag period for induction of tyrosine aminotransferase and gave degrees of enzyme induction similar to those for young animals (Adelman and Freeman, 1972); similar results were obtained in young and old mice after injection of corticosterone or insulin (Finch *et al.*, 1969); isolated hepatocytes from rats (2 versus 24 months) show no diminution of tyrosine aminotransferase stimulation with age after treatment with hydrocortisone (Britton *et al.*, 1976) or dexamethasone (Weber *et al.*, 1980); newly synthesized liver cells in partially hepatectomized old rats retain the "old" pattern of glucokinase induction (Adelman, 1970c), presumably because of external influences.

There are few other reports of the effect of age on hormone-induced induction of enzymes. Wu (1977) observed that glutamine synthetase activity measured in crude homogenates was depressed 32% in livers of young (3 months) but not old (28 months) Fischer rats after glucagon administration.

Isoproterenol is a β-adrenergic agonist, which, about 20 hours after administration to young rats, stimulates DNA synthesis and cell division in salivary glands. The peak of stimulation (measured as increased specific activity of DNA after injection of labeled thymidine) was delayed in time in both submandibular and parotid glands of old Sprague-Dawley rats (2 to 24 months). In fact, the delay was proportional to the age of the animals (Adelman, 1972), as was the case for glucose-mediated glucokinase stimulation (see above). By 12 months of age, a decreased magnitude of incorporation was also observed, although this effect may have been a result of changes in thymidine pool size. Older animals were not studied in this respect. Curiously, at low dosage, isoproterenol showed a greater response in the 12-month-old than in the young (2 months) animals. The magnitude of the stimulation of DNA synthesis is dramatically affected by the time of day at which the isoproterenol is injected in both young and mature animals, presumably reflecting their nutritional status. Fasting practically abolishes the response and glucocorticoids can delay and reduce its magnitude (Roth *et al.*, 1974). These results indicate that the isoproterenol-stimulated DNA synthesis is regulated by endogenous glucocorticoids.

Piantanelli *et al.* (1978) examined the effect of age on isoproterenol-induced DNA synthesis in mice, again only to maturity (13 months), and found a decreasing level of response with increasing age. In this case, there was no difference in time of response as reported for rats. Different patterns of stimulated DNA synthesis were obtained for males and females. In liver and spleen, response to isoproterenol was variable, no specific peak time of DNA synthesis being apparent. Thus, if conclusions may be based on a single report, the pattern of response appears to be species specific, tissue specific, and perhaps sex related. Unfortunately, there are no data available for senescent mice. Adelman (1979) has recently provided an overview of age-related changes in adaptive response.

XIV. TISSUE CULTURE

There have been relatively few studies of hormone metabolism in early-versus late-passage cells. Hydrocortisone, because of its ability to extend the doubling capacity of some cell types (Chapter 2, Section III), is of particular interest. The life-extension effect appears to be specific for the hormone or close structural relatives. Two recent investigations of receptors have been made (Kalimi and Seifter, 1979; Rosner and Cristofalo, 1981). Both agree that there is a substantial decrease (40–50%) in specific glucorcorticoid receptor sites in late- versus middle-passage WI-38 cells, that translocation of receptors to the nucleus is reduced, and that affinity is unaltered. Kalimi and Seifter utilized quiescent cells. Rosner and Cristofalo carried out their experiments with proliferating cells to maximize binding which is cell cycle dependent.

Polgar *et al.* (1978) found that basal cAMP levels are greater in late-passage human embryo lung fibroblasts. The degree of stimulation of cAMP production by epinephrine increases with passage number. Haslam and Goldstein (1974) had earlier obtained similar results using human skin fibroblasts.

Recently, Goldstein and Harley (1979) reported that to effect 50% and 95% of maximal stimulation of DNA synthesis, late-passage cells require more nonsuppressible insulin-like hormone than cells at early-passage. Uptake of 2-deoxyglucose showed less hormone stimulation in "old" cells, although basal uptake was higher.

XV. COMMENT

For most of the hormones studied, it appears that there is a reduction in receptor number with age. It does not seem that this decrease, when it occurs, is entirely responsible for lowered response. In particular, changes in the coupling mechanism between β-adrenergic receptors and production of cAMP may also be involved. Moreover, there are indications that age-related decreases in receptor number (e.g., catecholamine binding in adipocytes) do not always correspond to the time of metabolic changes. Thus, the mechanism of altered hormonal function with age may be multifarious.

It is clear that there is considerable variation with age among different tissues, or even different areas of the same tissue, both as to the amount of hormone binding and to the response to stimuli. Unfortunately, for some hormones, there are also species differences, which may preclude general application of results obtained from "model systems." This situation represents a serious handicap to rapid advance of our understanding of the role of hormones in aging. Finch (1976) stated that there was not enough information to evaluate conjectures that many phenomena of aging may ultimately be traced to neuroendocrine and autonomic loci. Although much more information is now available, its nature is still exploratory, and the statement thus remains true.

REFERENCES

Adelman, R. C. (1970a). *J. Biol. Chem.* **245**, 1032–1035.
Adelman, R. C. (1970b). *Nature (London)* **228**, 1095–1096.
Adelman, R. C. (1970c). *Biochem. Biophys. Res. Commun.* **38**, 1149–1153.
Adelman, R. C. (1972). *Adv. Gerontol. Res.* **4**, 1–23.
Adelman, R. C. (1975). *Fed. Proc.* **34**, 179–182.
Adelman, R. C. (1979). *Fed. Proc. Fed. Am. Soc. Exp. Biol.* **38**, 1968–1971.
Adelman, R. C., and Freeman (1972). *Endocrinology* **90**, 1551–1560.
Adelman, R. C., Stein, G., Roth, G. S., and Englander, D. (1972). *Mech. Ageing Dev.* **1**, 49–59.
Andres, R., and Tobin, J. D. (1975). *Adv. Exp. Biol. Med.* **61**, 239–249.
Bertrand, H. A., Masoro, E. J., and Yu, B. P. (1980). *Endocrinology* **107**, 591–595.
Bethea, C. L., and Walker, R. F. (1979). *J. Gerontol.* **34**, 21–27.
Boesel, R. W., Klipper, R. W., and Shain, S. A. (1980). *Steroids* **35**, 157–177.
Bolla, R. (1980). *Mech. Ageing Dev.* **12**, 249–259.
Britton, G. W., Rotenberg, S., and Adelman, R. C. (1975). *Biochem. Biophys. Res. Commun.* **64**, 184–188.
Britton, G. W., Britton, V. J., Gold, G., and Adelman, R. C. (1976). *Exp. Gerontol.* **11**, 1–4.
Cake, M. H., and Litwack, G. (1976). *In* "Biochemical Actions of Hormones" (G. Litwack, ed.), Vol. III, pp. 317–390. Academic Press, New York.
Chan, S. W. C., Leathem, J. H., and Esashi, T. (1977). *Endocrinology* **101**, 128–133.
Chang, W. C., and Roth, G. S. (1979). *J. Steroid Biochem.* **11**, 889–892.
Chang, W. C., and Roth, G. S. (1980). *Biochim. Biophys. Acta* **632**, 58–72.
Chang, W. C., Hoopes, M. T., and Roth, G. S. (1981). *J. Gerontol.* **36**, 386–390.
Cooper, B., and Gregerman, R. I. (1976). *J. Clin. Invest.* **57**, 161–168.
Davidson, M. B. (1979). *Metabolism* **28**, 688–705.
Davis, L. C., and Pfeifer, W. D. (1973). *Biochem. Biophys. Res. Commun.* **54**, 726–731.
Dax, E. M., Partilla, J. S., and Gregerman, R. I. (1981). *J. Lipid Res.* **22**, 934–943.
deKloet, R., and McEwen, B. S. (1976). *In* "Molecular and Functional Neurobiology" (W. H. Gispen, ed.), pp. 257–295. Elsevier, Amsterdam.
DeSantis, R. A., Gorenstein, T., Livingston, J. N., and Lockwood, D. H. (1974). *J. Lipid Res.* **15**, 33–38.
Dudl, R. J., and Ensinck, J. W. (1977). *Metabolism* **26**, 33–41.
Eleftheriou, B. D. (1975). *J. Gerontol.* **30**, 417–421.

Eleftheriou, B. E., and Lucas, L. A. (1974). *Gerontologia* **20,** 231–238.

Ewing, L. L., Johnson, B. H., Desjardius, C., and Clegy, R. F. (1972). *Proc. Soc. Exp. Biol. Med.* **140,** 907–911.

Finch, C. E. (1973). *Brain Res.* **52,** 261–276.

Finch, C. E. (1976). *Q. Rev. Biol.* **51,** 49–83.

Finch, C. E. (1977). *In* "Handbook of the Biology of Aging" (Hayflick and Finch, eds.), pp. 262–280. Van Nostrand-Rheinhold. Princeton, New Jersey.

Finch, C. E., Foster, J. R., and Mirsky, A. E. (1969). *J. Gen. Physiol.* **54,** 690–712.

Finch, C. E., Jones, V., Wisner, J. R. Jr., Sinha, Y. N., DeVellis, J. S., and Swerdloff, R. S. (1977). *Endocrinology* **101,** 1310–1317.

Flint, D. J. (1980). *Biochim. Biophys. Acta* **628,** 322–327.

Flint, D. J., Sinnett-Smith, P. A., Clegg, R. A., and Vernon, R. G. (1979). *Biochem. J.* **182,** 421–427.

Florini, J. R., Saito, Y., and Manowitz, E. J. (1973). *J. Gerontol.* **28,** 293–297.

Freeman, C., Karoly, K., and Adelman, R. C. (1973). *Biochem. Biophys. Res. Commun.* **54,** 1573–1580.

Freund, G. (1980). *Life Sci.* **26,** 371–375.

Frolkis, V. V., Verzhovskaya, N. V., and Valueva, G. V. (1973). *Exp. Gerontol.* **8,** 285–296.

Frolkis, V. V., Bezrukov, V. V., and Muradian, K. K. (1979). *Exp. Gerontol.* **14,** 65–76.

Ghandian, R., Lewis, J. G., and Chisholm, C. D. (1975). *Steroids* **25,** 753–762.

Giudicelli, Y., and Pecquery, R. (1978). *Eur. J. Biochem.* **90,** 413–419.

Gold, G., Karoly, K., Freeman, C., and Adelman, R. C. (1976). *Biochem. Biophys. Res. Commun.* **73,** 1003–1010.

Goldstein, S., and Harley, C. B. (1979). *Fed. Proc. Fed. Am. Soc. Exp. Biol.* **38,** 1862–1867.

Gommers, A., Dehez-Delhage, M., and Jeanjean, M. (1977). *Gerontology* **23,** 127–133.

Gonzalez, J., and DeMartinis, F. D. (1978). *Exp. Aging Res.* **4,** 455–477.

Govoni, S., Luddo, P., Spano, P. F., and Trabucchi, M. (1977). *Brain Res.* **138,** 565–570.

Govoni, S., Memo, M., Saiani, L., Spano, P. F., and Trabucchi, M. (1980). *Mech. Ageing Dev.* **12,** 39–46.

Gray, G. D. (1978). *J. Endocrinol.* **76,** 551–552.

Greenberg, L. H., and Weiss, B. (1978). *Science* **201,** 61–63.

Gregerman, R. I., and Crowder, S. E. (1963). *Endocrinology* **72,** 382–392.

Gregerman, R. I., Gaffney, G. W., and Shock, N. W. (1962). *J. Clin. Invest.* **41,** 2065–2069.

Guarnieri, T., Filburn, C. R., Zitnik, G., Roth, G. S., and Lakatta, E. G. (1980). *Am. J. Physiol.* **239,** H501–H508.

Hashimoto, T., Miyai, K., Izumi, K., and Kumahara, Y. (1973). *J. Clin. Endocrinol. Metab.* **37,** 910–916.

Haslam, R. J., and Goldstein, S. (1974). *Biochem. J.* **144,** 253–263.

Haug, E. A., Aakvaag, A., Sand, T., and Torjesen, P. A. (1974). *Acta Endocrinol.* **77,** 625–635.

Helderman, J. H. (1980). *J. Gerontol.* **35,** 329–334.

Hermann, J., Rusche, H. J., Kroll, H. J., Hilger, P., and Druskemper, H. L. (1974). *Horm. Metab. Res.* **6,** 239–240.

Hesch, R. D., Gatz, J., Pape, J., Schmidt, E., and von zar Muhlen, A. (1976). *Eur. J. Clin. Invest.* **6,** 139–145.

Hollenberg, M. D., and Schneider, E. L. (1979). *Mech. Ageing Dev.* **11,** 37–43.

Holm, G., Jacobsson, B., Bjorntorp, P., and Smith, U. (1975). *J. Lipid Res.* **16,** 461–467.

Jacobus, S., and Gershon, D. (1980). *Mech. Ageing Dev.* **12,** 331–312.

Jeanjean, M., Dehez-Delhage, M., and Gommers, A. (1977). *Gerontology* **23,** 134–141.

Jonec, V., and Finch, C. E. (1975). *Brain Res.* **91,** 197–215.

Joseph, J. A., Berger, R. E., Engel, B. T., and Roth, G. S. (1978). *J. Gerontol.* **33,** 643–649.

Kaler, L. W., and Neaves, W. B. (1981). *Endorcinology* **108,** 712–719.

Kalish, M. I., Katz, M. S., Pineyro, M. A., and Gregerman, R. I. (1977). *Biochim. Biophys. Acta* **483,** 452–466.

Kalimi, M., and Seifter, S. (1979). *Biochim. Biophys. Acta* **583,** 352–359.

Kitahara, A., and Adelman, R. C. (1979). *Biochem. Biophys. Res. Commun.* **87,** 1207–1213.

Klug, T. L., and Adelman, R. C. (1977). *Biochem. Biophys. Res. Commun.* **77,** 1431–1437.

Klug, T. L., and Adelman, R. C. (1979). *Endocrinology* **104,** 1136–1142.

Klug, T. L., Freeman, C., Karoly, K., and Adelman, R. C. (1979). *Biochem. Biophys. Res. Commun.* **89,** 907–912.

Kovacs, K., Ryan, N., Horvath, E., Penz, G., and Ezrin, C. (1977). *J. Gerontol.* **32,** 534–540.

Krall, J. F., Connelly, M., and Tuck, M. L. (1981). *Biochem. Biophys. Res. Commun.* **99,** 1028–1034.

Lakatta, E. G., Gerstenblith, G., Angell, C. S., Shock, N. W., and Weisfeldt, M. L. (1975). *Circ. Res.* **36,** 262–269.

Landfield, P. W., Waymire, J. C., and Lynch, G. (1978). *Science* **202,** 1098–1102.

Latham, K. R., and Finch, C. E. (1976). *Endocrinology* **98,** 1480–1489.

Levin, P., Janda, J. K., Joseph, J. A., Ingram, D. K., and Roth, G. S. (1981). *Science,* **214,** 561–562.

Lewis, B. K., and Wexler, B. C. (1974). *J. Gerontol.* **29,** 139–144.

Liu, S. L., and Webb, T. E. (1979). *Biochem. J.* **180,** 187–193.

Livingston, J. N., Cuatrecasas, P., and Lockwood, D. H. (1974). *J. Lipid Res.* **15,** 26–32.

Maggi, A., Schmidt, M. J., Ghetti, B., and Enna, S. J. (1979). *Life Sci.* **24,** 367–374.

Makman, M. H., Ahn, H. S., Thal, L. J., Sharpless, N. S., Dvorkin, B., Horowitz, S. G., and Rosenfeld, M. (1979). *Fed. Proc. Fed. Am. Soc. Exp. Biol.* **38,** 1922–1926.

Manganiello, V., and Vaughan, M. (1972). *J. Lipid Res.* **13,** 12–16.

Marquis, J. K., Lippa, A. S., and Pelham, R. W. (1981). *Biochem. Pharmacol.* **13,** 1876–1878.

Masuoka, D. T., Jonsson, G., and Finch, C. E. (1979). *Brain Res.* **169,** 335–341.

McGeer, E. G., and McGeer, P. L. (1975). *In* "Neurobiology of Aging" (J. M. Ordy and K. R. Brizzee, eds.), pp. 287–305. Plenum, New York.

McGeer, E. G., Fibiger, H. C., McGeer, P. L., and Wickson, V. (1971a). *Exp. Gerontol.* **6,** 391–396.

McGeer, E. G., McGeer, P. L., and Wada, J. A. (1971b). *J. Neurochem.* **18,** 1647–1658.

McGuire, E. A., Tobin, J. D., Berman, M., and Andres, R. (1979). *Diabetes* **28,** 110–120.

Meek, J. L., Bertilsson, L., Cheney, D. L., Zsilla, G., and Costa, E. (1977). *J. Gerontol.* **32,** 129–131.

Memo, M., Lucchi, L., Spano, P. F., and Trabucchi, M. (1980). *Brain Res.* **202,** 488–492.

Miller, A. E., and Riegle, G. D. (1978). *J. Gerontol.* **33,** 197–203.

Misra, C. H., Shelat, H. S., and Smith, R. C. (1980). *Life Sci.* **27,** 521–526.

Muggeo, M., Fedele, D., Tiengo, A., Molinari, M., and Crepaldi, G. (1975). *J. Gerontol.* **30,** 546–551.

Nelson, J. F., Latham, K. R., and Finch, C. E. (1975). *Acta Endocrinol.* **80,** 744–752.

O'Connor, S. W., Scarpace, P. J., and Abrass, I. B. (1981). *Mech. Ageing Dev.* **16,** 91–95.

O'Hara, H., Kobayashi, T., Shiraisi, M., and Wada, T. (1974). *Endocrinol. Jap.* **21,** 377–386.

Pagano, G., Cassader, M., Diana, A., Pisu, E., Bozzo, C., Ferrero, F., and Lenti, G. (1981). *Metabolism* **30,** 46–49.

Petrovic, J. S., and Markovic, R..Z. (1975). *Dev. Biol.* **45,** 176–182.

Piantanelli, L., Brogli, R., Bevilacqua, P., and Fabris, N. (1978). *Mech. Ageing Dev.* **7,** 163–169.

Pirke, K. M., Vogt, H. J., and Geiss, M. (1978). *Acta Endocrinol.* **89,** 393–403.

Polgar, P., Taylor, L., and Brown, L. (1978). *Mech. Ageing Dev.* **7,** 151–160.

Puri, S. K., and Volicer, L. (1977). *Mech. Ageing Dev.* **6**, 53–58.
Reaven, E. P., Gold, G., and Reaven, G. M. (1979). *J. Clin. Invest.* **64**, 591–599.
Reis, D. J., Ross, R. A., and John, T. H. (1977). *Brain Res.* **136**, 465–474.
Remacle, C., Hauser, N., Jeanjean, M., and Gommers, A. (1975). *Exp. Gerontol.* **12**, 207–214.
Riegle, G. D. (1973). *Gerontologia* **11**, 1–10.
Riegle, G. D., and Meites, J. (1976). *Proc. Soc. Exp. Biol. Med.* **151**, 507–511.
Riegle, G. D., Meites, J., Miller, A. E., and Wood, S. M. (1977). *J. Gerontol.* **32**, 13–18.
Rosenbloom, A. L., Goldstein, S., and Yip, C. C. (1976). *Science* **193**, 412–415.
Rosner, B. A., and Cristofalo, V. J. (1981). *Endocrinology* **108**, 1965–1971.
Roth, G. S. (1974). *Endocrinology* **94**, 82–90.
Roth, G. S. (1975). *Biochim. Biophys. Acta* **399**, 145–156.
Roth, G. S. (1976). *Brain Res.* **107**, 345–354.
Roth, G. S. (1979a). *Fed. Proc. Fed. Am. Soc. Exp. Biol.* **38**, 1910–1914.
Roth, G. S. (1979b). *Mech. Ageing Dev.* **9**, 497–514.
Roth, G. S. (1980). *Proc. Soc. Exp. Biol. Med.* **165**, 188–192.
Roth, G. S., and Livingston, J. N. (1976). *Endocrinology* **99**, 831–839.
Roth, G. S., and Livingston, J. N. (1979). *Endocrinology* **104**, 423–428.
Roth, G. S., Karoly, K., Adelman, A., and Adelman, R. C. (1974). *Exp. Gerontol.* **9**, 13–26.
Rubens, R., Dhont, M., and Vermeulen, A. (1974). *J. Clin. Endocrinol. Metab.* **39**, 40–45.
Rubenstein, H. A., Butler, V. P., Jr., and Werner, S. C. (1973) *J. Clin. Endocrinol. Metab.* **37**, 247–253.
Sakoda, M., Otsuki, M., Kusaka, T., and Baba, S. (1973). *Folia Endocrinol. Jap.* **49**, 1177–1185.
Samorajski, T. (1977). *J. Am. Geriat. Soc.* **25**, 337–348.
Sartin, J., Chaudhuri, M., Obenrader, M., and Adelman, R. C. (1980). *Fed. Proc. Fed. Am. Soc. Exp. Biol.* **39**, 3163–3167.
Scaglia, H., Medina, M., Pinto-Ferreira, A. L., Vazques, G., Gual, C., and Perez-Palacios, G. (1976). *Acta Endocrinol.* **81**, 673–679.
Schmidt, M. J., and Thronberry, J. F. (1978). *Brain Res.* **139**, 169–177.
Schocken, D. D., and Roth, G. S. (1977). *Nature (London)* **267**, 856–858.
Severson, J. A , and Finch, C. E. (1980). *Brain Res.* **192**, 147–162.
Shaar, C. J., Euker, J. S., Riegle, G. D., and Meites, J. (1975). *J. Endocrinol.* **66**, 45–51.
Shain, S. A., and Boesel, R. W. (1977). *Mech. Ageing Dev.* **6**, 219–232.
Shain, S. A., and Boesel, R. W. (1978). *J. Clin. Invest.* **61**, 654–660.
Shain, S. A., and Nitchuk, W. M. (1979a). *Mech. Ageing Dev.* **11**, 23–25.
Shain, S. A., and Nitchuk, W. M. (1979b). *Mech. Ageing Dev.* **11**, 9–22.
Shain, S. A., Boesel, R. W., and Axelrod, L. R. (1975). *Arch. Biochem. Biophys.* **167**, 247–263.
Simkins, J. W., Mueller, G. P., Huang, H. H., and Meites, J. (1977). *Endocrinology* **100**, 1672–1678.
Singer, S., Ito, H., and Litwack, G. (1973). *Int. J. Biochem.* **4**, 569–573.
Smith, U. (1971). *J. Lipid Res.* **12**, 65–70.
Snyder, P. J., and Utiger, R. D. (1972a). *J. Clin. Endocrinol. Metab.* **34**, 380–385.
Snyder, P. J., and Utiger, R. D. (1972b). *J. Clin. Endocrinol. Metab.* **34**, 1096–1098.
Snyder, P. J., Reitano, J. E., and Utiger, R. D. (1975). *J. Clin. Endocrinol. Metab.* **41**, 938–945.
Sorrentino, R. N., and Florini, J. R. (1976). *Exp. Aging Res.* **2**, 191–205.
Tang, F., and Phillips, J. G. (1978). *J. Gerontol.* **33**, 377–382.
Thal, L. J., Horowitz, S. G., Dvorkin, B., and Makman, M. H. (1980). *Brain Res.* **192**, 185–194.
Vijayan, V. K. (1977). *Exp. Gerontol.* **12**, 7–11.
Walker, J. B., and Walker, J. P. (1973). *Brain Res.* **54**, 391–396.
Watkins, B. E., Meites, J., and Riegle, G. D. (1975). *Endocrinology* **97**, 543–548.

Weber, A., Guguen-Guillouzo, C., Szajert, M. F., Beck, G., and Schapira, F. (1980). *Gerontology* **26,** 9–15.

Weiss, B., Greenberg, L., and Cantor, E. (1979). *Fed. Proc. Fed. Am. Soc. Exp. Biol.* **38,** 1915–1921.

Williams, L. T., Jarett, L., and Lefkowitz, R. J. (1976). *J. Biol. Chem.* **251,** 3096–3104.

Williams, L. T., Lefkowitz, R. J., Watanabe, A. M., Hathaway, D. R., and Besch, H. R. Jr. (1977). *J. Biol. Chem.* **252,** 2787–2789.

Wu, C. (1977). *Biochem. Biophys. Res. Commun.* **75,** 879–885.

Yu, B. P., Bertrand, H. A., and Masoro, E. J. (1980). *Metabolism* **29,** 438–444.

Zimmerman, I., and Berg. A. (1974). *Mech. Ageing Dev.* **3,** 33–36.

Zimmerman, I. D., and Berg, A. P. (1975). *Mech. Ageing Dev.* **4,** 89–96.

Zitnik, G., and Roth, G. S. (1981). *Mech. Ageing Dev.* **15,** 19–28.

Chapter 11

Dietary Restriction

I. OVERVIEW

The only means so far discovered of substantially increasing the maximal life-span of animals is by restricting caloric intake or "hyponutrition." McCay (1935) and co-workers first demonstrated this effect over 45 years ago. Later workers described the effects of caloric restriction on aging and disease and in some cases, made efforts to investigate quantitative and qualitative aspects of the diet. Recently, more detailed studies have been undertaken by Masoro and co-workers with regard to the effects of restricted food intake not only on life-span and disease, but on certain aspects of lipid metabolism. Although hyponutrition experiments cannot be categorized as "biochemical," the relationship is obvious. Simply put, delayed mortality and a delay in specific biochemical changes known to occur during aging may well confirm their relationship.

There are interesting questions to be asked of experiments with animals on a restricted dietary regime. Are metabolic changes that occur in old rats delayed in animals with an extended life-span? Do altered enzymes occur, and if so, are they formed at later ages if life-span is increased? Does DNA repair capability change? Do deleterious peroxidative processes decrease? A difficulty with the results so far is that there are so many possible permutations and combinations of dietary conditions and relatively so few studies. Do the proportions of protein, carbohydrate, and fat in the diet have an effect? Should the food restriction be continuous or intermittent and for how long a period? Do dietary needs change with age and are these changes affected by hyponutrition? At what age should the dietary restriction be applied? What should the degree of restriction be and what is its relation to the age of the animal? That is, if 50% caloric restriction functions well at weaning, perhaps 30% restriction is better after 6-months of age. Even from this short list of obvious questions, it is obvious that the work performed to date is simply a beginning. Nonetheless, one paramount fact already stands out clearly—that caloric restriction extends the maximal life-span of animals, vertebrate, and invertebrate, to a substantial degree. Moreover, such dietary regimens clearly reduce the incidence and delay the time of onset of diseases and tumors. It also appears that dietary restriction delays biochemical changes associated with aging, but there is as yet, relatively little information in this regard. A sharper focus on how much restriction, when it must be applied, and the role of dietary composition is beginning to emerge, but slowly. It must be admitted that we have as yet, no understanding of the basic mechanism underlying the observed increase in life-span.

II. HYPONUTRITION AND LIFE-SPAN

The early report of McCay *et al.* (1935) that dietary restriction in rats results in greatly extended life-span has been amply confirmed by later investigators. Berg and Simms (1960) restricted food intake by 33 and 46% of the ad libitum amounts in Sprague-Dawley rats with little adverse effect on skeletal growth (90–95% of the values shown by unrestricted animals). Weight was reduced by roughly 25 and 40%, respectively, at about 27 months of age in males and females. Life-span was increased and the onset of disease was delayed. Ross (1961) altered casein and carbohydrate levels as well as restricting the food intake of Sprague-Dawley rats. All of the restricted diets resulted in substantially extended life-spans. Ross (1972) later observed that a short period of restriction (from 21 to 70 days of age) was beneficial in extended life-span even if followed by feeding ad libitum. Imposition of restriction was decreasingly effective when applied at increasing ages. Fernandes *et al.* (1976) observed an increased life-span for the short-lived mouse (NxB × NZW)F$_1$ after caloric restriction. The mice are prone to autoimmune renal disease. With DBA/2f mice, protein restriction rather than caloric restriction was more effective in increasing the life-span. Weindruch *et al.* (1979) pointed out that the mice in these experiments were housed four to a cage, which could result in unequal food distribution. Stuchlikova *et al.* (1975) reported that a 50% restriction in calories gave an increased longevity in Wistar rats. Even short-term restriction (from postweaning to 2–3 months of age) was effective. The authors also reported that food restriction extended the maximal life-span of mice and golden hamsters. In each case, early restriction for the first year, followed by a full diet yielded the longest survival.

Recently, Cheney *et al.* (1980) studied dietary restriction by a reduced frequency of feeding in weaned and preweaned C57BL/6J mice. Although there was considerable variation, a general decrease or delay in lymphoma was observed and life-span was extended, although not greatly in some groups. Masoro *et al.* (1979) observed that reduction of the food intake to 60% of the ad libitum amounts in barrier-reared Fischer 344 rats increased survival enormously. Weight loss was about 35% at 24 months. At 28 months of age, only 12% of the restricted rats had died versus 97% of the controls. Masoro *et al.* (1980) subsequently reported that the mean, median, and maximal life-spans of the control animals (fed ad libitum) were 701 ± 10, 714, and 963 days, respectively. In the restricted animals, the corresponding values were 986 ± 10, 1047, and 1435 days. This level of food restriction (60% of the ad libitum value) clearly had dramatic effects on longevity.

The means of restricting diet does not seem to be critical. Food can be offered for short periods (Leveille, 1972; Tucker *et al.*, 1976) or fed every other day (data from personal communication reported by Barrows and Kokkonen, 1978), or perhaps by reduction of protein intake. In fact, a number of investigators have considered the effect of protein restriction instead of or in addition to caloric restriction. As can be seen below, the amount of protein restriction that is beneficial is not in agreement. Miller and Payne (1968) had earlier shown that a high starch (and, therefore, low protein) diet provided in unlimited amounts to female rats, which had been fed a stock diet for 4 months, resulted in increased longevity. Reduction of protein intake from 23 to 12% after female Wistar rats reached 16 months of age increased average survival but further reduction (8% and 4%) gave only the same value as the control diet (Barrows and Kokkonen, 1975). Ross and Bras (1973) found that rats on a high protein diet (51 or 22%) did better than animals receiving a lower level (10%). Ross and Bras (1975) subsequently observed that rats eating a high protein diet early in life (self-selected) and then a low protein diet (self-selected) lived longer than rats that selected the lower protein level from the beginning. There are other reports that suggest lower levels of protein in the diet are advantageous. Miller and Payne (1968) found that 4% protein gave better longevity than 12% protein in the diet. Stoltzner (1977) reported improved average and slightly improved maximal survival in BALB/c male mice receiving 4 versus 24% casein; Leto *et al.*

(1976a) also reported improved average and maximal survival in C57BL/6J males fed 4 compared to 26% casein in the diet. Goodrick (1978) studied the effect of diets containing normal amounts of protein (26% casein) and low levels (4%) in both A/J, C57BL/6J, and F_1 hybrid mice. He observed that the slower the growth rate and the longer the duration of growth, the longer the life-span. Thus, the slower growth of the animals on the low protein diet was beneficial in this respect. Body weight did not correlate with life-span.

Ross and Bras (1971, 1973, 1974) found that the incidence of various types of tumors in Sprague-Dawley rats varied with the protein level of the diet. Leto *et al.* (1976a) observed reduced body temperature and higher oxygen consumption in mice fed a low protein diet (4% casein). Although it is clear that protein intake affects longevity, Ross (1976) points out that only carbohydrate intake gives a constant relationship to life-span.

Perhaps analogous to a protein deficiency is the use of a tryptophan-deficient diet. Segall and Timeras (1976) fed such a diet to Long-Evans rats for periods of 2 months to nearly 2 years. After return to a normal diet, the animals grew to normal size. They showed a much delayed onset of physiological function, producing litters throughout the period from 17 to 28 months of age, long after reproductive ability ceased for the control animals. There was also a delay in the appearance of tumors.

The effects of various levels of food restriction (caloric or protein) on body weight are provided by most authors. As one would expect, weights are generally reduced. Barrows and Kokkonen (1980) reported that in rats and particularly in mice, a large reduction occurs only when protein is less than 8% of the diet.

For many years, it has been assumed that providing a restricted diet to mature, rather than postweaned, animals did not improve life-span. In fact, Barrows and Roeder (1965) reported a slightly decreased survival if nutrient intake was cut by 50% in 13- or 19-month-old rats. Ross (1972), when he imposed a severe dietary restriction on animals of 70, 300, and 365 days of age (about one-third of the ad libitum level) obtained progressively decreased life-spans. However, it should be noted that less severe restriction (53% of the ad libitum amount) seemed to be advantageous. Ross and Bras (1975) concluded that the dietary habits established early in the life correlated with the life-span. On the other hand, Nolen (1972) observed increased longevity in Sprague-Dawley rats which were restricted to 60 and 80% of the ad libitum consumption starting after 3 months of age. Moreover, Stuchlikova *et al.* (1975) reported that after 1 year on restricted diets, ad libitum feeding yielded greatly improved longevity in rats, mice, and hamsters. The converse experiment, that is animals fed ad libitum for 1 year and then restricted, also showed extended life-spans.

The evidence is conclusive that dietary restriction applied directly after weaning will extend the life-span of rodents. It is also clear that the protein composition of the diet plays an important role in longevity, and evidence appears to be mounting that dietary changes applied later in life may be effective in extending life-span, although there are inconsistencies in the data. One of the difficulties may well be that there are changes in dietary requirements with age. Cohen (1979), although not discussing hyponutrition, makes a number of interesting points that would be well for the investigator to bear in mind. For example, he points out that commercial diets were designed to promote vigor, rapid growth, and early maturation. They were never designed for maximal longevity and might actually be deleterious in this respect. In fact, these diets promote obesity in some rat stocks. Thus, the composition of the diet adds an important factor to considerations of longevity. Even the housing arrangements can cause large differences in the characteristics of the animals. Fischer 344 rats, housed in groups, do not gain weight after 12 months of age. However, individually housed animals can become obese and weigh 900 g versus 400–500 g for group-housed animals.

Not only does hyponutrition extend the life-span of rodents, but of a number of invertebrates as well (Barrows and Kokkonen, 1978). In this regard, the life-span of the free-living nematode *Caenorhabditis elegans* was extended up to 52%. The earlier the restriction, the more effective the prolongation of life-span (Klass, 1977).

An important effect of hyponutrition in rats is a delayed onset of tumors and other pathological conditions. Berg and Simms (1960) found sharply reduced levels of glomerlular nephritis, periarteritis, and myocardial degeneration in male Sprague-Dawley rats restricted to 33 and 46% of the ad libitum diet. Tumors in these rats had a lower frequency of occurrence. Recently, Masoro *et al.* (1980) reported that in barrier-reared Fischer rats, nephrosis and testicular tumors, which are chronic in this strain, were delayed and reduced in severity under a restricted dietary regime containing 60% of the ad libitum amount. Tucker *et al.* (1976) found that by 24 months of age, male Wistar rats fed restricted diets did not show the typical increase in proteinurea noted in animals fed ad libitum. Transport of *p*-aminohippuric acid was enhanced in kidney slices of old restricted animals. The immune system also seems to be favorably affected (Weindruch *et al.*, 1979). As noted above, Ross and Bras (1973, 1974, 1975) found that protein composition of the diet is also a considerable factor in the prevalence and distribution of tumors.

III. BIOCHEMICAL EFFECTS

In addition to effects on disease and pathological conditions, hypo-nutrition can delay or prevent age-related changes in biochemical pro-cesses. However, it seems reasonable to consider that the need to cope with the unusual dietary regimen rather than the effects of aging might be the dominating factor as the tissues adjust their metabolic function. Clearly, conservation of nitrogen in cases of low protein diet, or for that matter, a need to oxidize more sugar in a high carbohydrate diet, will be the factors to which enzyme levels are attuned. Thus, to compare en-zyme levels or other adaptable or inducible biochemical parameters of animals fed quite differently (either by use of caloric restriction or changes in the composition of the diet) in an attempt to see if "re-stricted" animals retain "young" enzyme characteristics could be mis-leading. Leto *et al.* (1976b) found that in protein-restricted rats, levels of some liver, kidney, and heart enzymes were unchanged and some were lower. Where there was an initial difference in levels between the con-trol and restricted animals, the subsequent pattern of change with age was very similar, although the difference in magnitude remained.

One biochemical–physiological area that is being investigated in detail by Masoro and co-workers is the effect of hyponutrition on lipid metabo-lism in Fischer 344 rats. Lipolytic response of adipocytes (epididymal and perirenal, respectively) to glucagon in restricted animals is substan-tially enhanced in 6-month-old animals and drops thereafter, increasing again after 30 months (Bertrand *et al.*, 1980a). Animals of this age fed ad libitum show no response to the hormone, in agreement with reports that stimulation in rat adipocytes is lost by 4 months. Lipolytic response to epinephrine is also enhanced by dietary restriction (Yu *et al.*, 1980). Postabsorptive serum cholesterol and phospholipid concentrations in-crease and free fatty acid concentrations decrease with age in Fischer rats fed ad libitum. Animals on a restricted diet reflect these changes later in life and generally to a lesser degree. At all ages, serum triglycer-ide levels were lower in the restricted animals (Liepa *et al.*, 1980). Bertrand *et al.* (1980b) observed that both restricted and nonrestricted animals increased in total adipose mass until about 75% of the life-span was attained and then declined. Since the restricted animals live longer, the change from gain to loss occurred later. Moreover, gains were much smaller. In epididymal depots of rats fed ad libitum, the number of adipocytes increased significantly after about 18 months of age. For restricted animals, the pattern was somewhat similar but of lesser mag-nitude. In perirenal depots, numbers increased sharply with age in the animals fed ad libitum. In the restricted animals, the increase was modest, and there was a decline after 24 months. The relationship between percentage of life-span and fat content was similar for both groups, although the restricted rats lived much longer. That is, there was no significant correlation between length of life and percentage of body fat.

IV. COMMENT

Hyponutrition clearly plays a dramatic role in extending the life-span of animals. The onset of a number of metabolic changes that normally arise with increasing age are delayed or reduced. An example is the delayed increase in serum cholesterol in Fischer rats. Other age-related effects literally disappear—for example, the early loss of lipolytic response of adipocytes to glucagon. It is obvious that dietary restriction brings about major changes in intermediary metabolism, some of them apparently set irreversibly at an early age. The modification of disease processes is particularly fascinating.

Now that the broad outlines of the effects of food restriction on life-span are apparent, there remains a large amount of long-term work to be performed in order to bring about a basic understanding of the processes involved. It would be invaluable if some sort of standard of restriction could be created, its effects well documented, against which other experimental attempts at increasing longevity could be rated. Although a number of biochemical effects of food restriction have been reported, these are still too few to serve this purpose. Of course, an ongoing problem is that we have only a superficial understanding about aging changes in normally fed animals. Thus, we are forced to deal with changes brought about by dietary restriction as phenomenological events without much perception as to their meaning.

The use of dietary restriction provides an exciting but tantalizing probe for biochemists interested in aging. Already, speculative articles about the possible role of dietary restriction on peroxidative damage, hormone balance, and changes in the immune system have appeared. With the recent entry of more investigators into the field, one can expect to see substantial advances in uncovering the relationship between hyponutrition and aging with a consequent profit for biochemical approaches to the subject.

REFERENCES

Barrows, C. H., and Kokkonen, G. C. (1975). *Growth* **39,** 525–533.
Barrows, C. H., and Kokkonen, G. C. (1978). *Age* **1,** 131–143.

Barrows, C. H., and Kokkonen, G. C. (1980). *Age* **3**, 53–58.

Barrows, C. H. Jr., and Roeder, L. M. (1965). *J. Gerontol.* **20**, 69–71.

Berg, B. N., and Simms, H. S. (1960). *J. Nutrition* **71**, 255–261.

Bertrand, H. A., Masoro, E. J., and Yu, B. P. (1980a). *Endocrinology* **107**, 591–595.

Bertrand, H. A., Lynd, F. T., Masoro, E. J., and Yu, B. P. Y. (1980b). *J. Gerontol.* **35**, 837–835.

Cheney, K. E., Liu, R. K., Smith, G. S., Leung, R. E. Mickey, M. R., and Walford, R. L. (1980). *Exp. Gerontol.* **15**, 237–258.

Cohen, B. J. (1979). *J. Gerontol.* **34**, 803–807.

Fernandes, G., Yunis, E. J., and Good, R. A. (1976). *Proc. Natl. Acad. Sci. U.S.A.* **73**, 1279–1283.

Goodrick, C. L. (1978). *J. Gerontol.* **33**, 184–190.

Klass, M. R. (1977). *Mech. Ageing Dev.* **6**, 413–429.

Leto, S., Kokkonen, G. C., and Barrows, C. H. Jr. (1976a). *J. Gerontol.* **31**, 149–154.

Leto, S., Kokkonen, G. C., and Barrows, C. H. Jr. (1976b). *J. Gerontol.* **31**, 144–148.

Leveille, G. A. (1972). *J. Nutrition* **102**, 549–556.

Liepa, G. V., Masoro, E. J., Bertrand, H. A., and Yu, B. P. (1980). *Am. J. Physiol.* **238**, E253–257.

Masoro, E. J., Bertrand, H., Liepa, G., and Yu, B. P. (1979). *Fed. Proc. Fed. Am. Soc. Exp. Biol.* **38**, 1956–1961.

Masoro, E. J., Yu, B. P., Bertrand, H. A., and Lynd, F. T. (1980). *Fed. Proc. Fed. Am. Soc. Exp. Biol.* **39**, 3178–3182.

McCay, C. M., Crowell, M. F., and Maynard, L. A. (1935). *J. Nutrition* **10**, 63–79.

Miller, D. S., and Payne, P. R. (1968). *Exp. Gerontol.* **3**, 231–234.

Nolen, G. A. (1972). *J. Nutrition* **102**, 1477–1494.

Ross, M. H. (1961). *J. Nutrition* **75**, 197–210.

Ross, M. H. (1972). *Am. J. Clin. Nutr.* **25**, 834–838.

Ross, M. H. (1976). *In* "Nutrition and Aging" (M. Winick, ed.), pp. 43–57. Wiley, New York.

Ross, M. H., and Bras, G. (1971). *J. Nat. Cancer Inst.* **47**, 1095–1113.

Ross, M. H., and Bras, G. (1973). *J. Nutrition* **103**, 944–963.

Ross, M. H., and Bras, G. (1974). *Nature (London)* **250**, 263–265.

Ross, M. H., and Bras, G. (1975). *Science* **190**, 165–167.

Segall, P. E., and Timiras, P. S. (1976). *Mech. Ageing Dev.* **5**, 109–124.

Stoltzner, G. (1977). *Growth* **41**, 337–348.

Stuchlikova, E., Juricova-Horakova, M., and Deyl, Z. (1975). *Exp. Gerontol.* **10**, 141–144.

Tucker, S. M., Mason, R. L., and Beauchene, R. E. (1976). *J. Gerontol.* **31**, 264–270.

Weindruch, R. H., Kristie, J. A., Cheney, K. E., and Walford, R. L. (1979). *Fed. Proc. Fed. Am. Soc. Exp. Biol.* **38**, 2007–2016.

Yu, B. P., Bertrand, H. A., and Masoro, E. J. (1980). *Metabolism* **29**, 438–444.

Epilogue

In Chapter 1 of this volume, I quoted a description of aging by Shakespeare, which is as relevant today as when it was written some 380 years ago. It is fitting then, to close the book with Shakespeare's Twelfth Sonnet, which presents a more philosophical view of aging and one which also has not changed with time.

When I do count the clock that tells the time,
And see the brave day sunk in hideous night,
When I behold the violet past prime,
And sable curls all silver's o'er with white;
When lofty trees I see barren of leaves,
Which erst from heat did canopy the herd,
And summer's green all girded up in sheaves,
Borne on the bier with white and bristly beard,
Then of thy beauty do I question make,
That thou among the wastes of time must go,
Since sweets and beauties do themselves forsake
And die as fast as they see others grow;
And nothing 'gainst Time's scythe can make defense
Save breed,* to brave him when he takes thee hence.

If the difficult events sometimes associated with aging are to be avoided; if we are to age gracefully in modest health and achieve a dignified end, the answer lies not in treatment, but in the avoidance of needing treatment. It is the biochemist who must bear the responsibility for true progress, who must pursue with insight and dedication the biological mystery that is aging. From what is written in this volume, it can be seen that there is much, much to be done. Our science is "young"; we have only made a start. Maturation of aging research is yet to come. This is one case where we want to speed the aging process. Biochemists—to your laboratories!

*Child.

Index